THE
ROYAL SOCIETY
1660–1940

THE
ROYAL SOCIETY
1660–1940

*A History of its Administration
under its Charters*

BY

SIR HENRY LYONS, F.R.S.

CAMBRIDGE
AT THE UNIVERSITY PRESS
1944

To

THE PRESIDENT & FELLOWS
OF
THE ROYAL SOCIETY OF LONDON

this Account of Their Trust is
respectfully dedicated

CAMBRIDGE
UNIVERSITY PRESS

32 Avenue of the Americas, New York NY 10013-2473, USA

Cambridge University Press is part of the University of Cambridge.

It furthers the University's mission by disseminating knowledge in the pursuit of education, learning and research at the highest international levels of excellence.

www.cambridge.org
Information on this title: www.cambridge.org/9781107492813

© Cambridge University Press 1944

First published 1944
First paperback edition 2015

A catalogue record for this publication is available from the British Library

ISBN 978-1-107-49281-3 Paperback

CONTENTS

NOTE

SIR HENRY LYONS had long given to the Royal Society a high place in his regard, and had served it with a steadfast devotion and with a cheerful efficiency all his own. His experience of the Society as its treasurer had awakened in him a keen interest in the changes which its organization and the conduct of its affairs had undergone through the centuries, and it was fitting that the last offering of his affection should be this administrative history. His diary shows that he made the first rough draft in January 1940, and, with characteristic energy and enthusiasm, undimmed by a crippling disability and failing health, he gave to the work all that war-time disturbances then left to him of leisure in his retirement. It is sad, indeed, to think that he did not live to see the publication of the book, which he had finished in every detail and passed for the press. Apart from its intrinsic interest and historical value, there are many friends and colleagues of Henry Lyons, including those of the Royal Society to whom he has dedicated it, who will treasure this book as a memorial to one whose long life was so filled with the joy of good work done for science.

H. H. DALE

PRESIDENT R.S.

THE ROYAL SOCIETY
LONDON
SEPTEMBER 1944

INTRODUCTION

THREE CENTURIES AGO at the time of the civil wars a small group of learned men, who were interested in the Experimental, or New Philosophy as it was then called, made it their practice soon after 1640 to meet occasionally in London for talk and discussion at the lodgings of one of their number, or at a tavern conveniently near Gresham College where they often attended the professors' lectures. Regular meetings were not always practicable in those disturbed times, but Dr J. Wallis has recorded that as many of the members as could do so were in the habit of attending them weekly.

On the restoration of the monarchy in 1660 those who were in London resumed their meetings that had been discontinued in 1658, and others who had been at Oxford joined them; by the end of the year they and a number of their friends having similar interests resolved to constitute themselves a Society of Philosophers, which they succeeded in doing. In December their project received the approval of King Charles II and the promise of his support, which was followed a few months later by his permission to use the title of the 'Royal Society'. From such small beginnings did the Society arise.

The first attempt to record the Society's history was made at the request of the Council, who in 1664 appointed a committee to supervise it, and entrusted its preparation to Dr Thomas Sprat, one of the Fellows; later, he was appointed Bishop of Rochester. His *History of the Royal Society* was not so much an account of its doings as an explanation of the need for the Society's existence; he concerns himself for the most part with defending its Fellows from the attacks of those who supported the Aristotelian philosophy, and refuting their criticisms. To quote his own words: 'The objections and cavils of the detractors of so noble an institution did make it necessary for me to write of it, not in the way of a plain history, but as an apology.' Sprat's book was first published in 1667, only five years after the king had granted to the Society its First Charter.

The next attempt to provide a history was made in 1756 when Dr Thomas Birch, a trustee of the British Museum and a Secretary of the Society from 1752 to 1765, published four quarto volumes dealing with the period from 1660 to 1687 only. He quotes largely from the Council's minutes of these years, and reproduces some of the papers which were communicated to the early meetings of the Society; the more important items of the business which was transacted at these meetings are also recorded.

Dr W. Thomson's *History of the Society*, which was published in 1812, is as the author says 'an attempt to elucidate the *Philosophical Transactions*'.

It is devoted to sketches of the progress of science, and to analyses of papers which had been published in the *Transactions*.

Between 1844 and 1848, J. C. Weld, who was the Assistant Secretary of the Society from 1843 to 1861, compiled the only comprehensive history of it which exists. Soon after taking up his duties he realized the need for some account of past events in the Society's history, and of the decisions which the Councils had adopted from time to time. In the absence of any such record much time was often wasted in searching through the Minute-books of the Councils and the Journals of the Society for information relating to such decisions and the circumstances which had led up to them. He carried out his work carefully and efficiently, but he decided to close it at the year 1830 when H.R.H. the Duke of Sussex was elected President. By 1847 the statutes relating to the election of candidates had been revised, and it would have been difficult then to forecast with any certainty the effect which the new procedure would have on the future development of the Society.

The object of this work is not to present to the reader a history of three centuries of British science, and still less to describe the most important advances in it which the more eminent Fellows of the Society have contributed to the advancement of Natural Knowledge; its aim is to record in some detail how the Society's Councils have administered its affairs under its Charters for the past two and three-quarter centuries, during which it has gained the high repute that it now enjoys in the scientific world.

The minutes of the meetings of Council contain a record of most of the discussions that took place at them and the decisions which were reached; the reasons for bringing such subjects before the Council are less fully stated, indeed they are often omitted altogether; nevertheless these minutes together with the Journals of the Society's meetings constitute the most reliable sources of the Society's past history which we possess. The Council minutes also record for each year the names of the twenty-one Fellows who were elected members of Council; the dates on which the Councils met; and lastly the names of the councillors who were present at each meeting. During the first two centuries of the Society's existence some 2300 such Councils were elected at the successive Anniversary meetings, and in their recorded minutes we possess a mass of statistical material of great value. From the names of the councillors elected for each year we learn what proportion of scientific men were members of that Council, and by inference whether the Society's scientific aims were likely to have been well or but indifferently cared for; the number of meetings which took place during the year may be taken as a measure of the zeal with which the officers were dealing with the Society's affairs; and the average number of the councillors who attended the meetings is a good indication of the amount of interest which the councillors as a body were taking in the Society's business. The data have been collected from these Councils, and carefully analysed, with the result that a large amount of valuable information has been collected and utilized in the following chapters.

After 1820 conditions changed greatly, for then almost all the councillors were scientific men. Councils met regularly eleven or twelve times yearly, and the average attendance at a meeting increased from about ten or twelve to seventeen or eighteen or even more. A large amount of business was transacted at each meeting, and the minutes of these Councils are our principal source of information.

This account of the way in which the Society carried on its business at different periods will provide the groundwork for a fuller discussion of its influence on the advancement of science; it also records the conditions under which the more eminent of its Fellows carried out their researches and discoveries. The complete history of the Society has not yet been undertaken, and it may well require the united efforts of several workers to deal with so wide a field of activity.

Those who founded the Society soon realized that the number of scientific men in England was too small to maintain such an institution as they had in mind without some form of endowment, but neither the Crown nor the State was prepared to provide this; the only alternative was to admit into the Society men of substance and influence as well as those who had won distinction in other branches of knowledge. Thus it came about that from the first the Society consisted of two groups of members: the first included those who, carrying on the tradition of the founders of the Society, devoted themselves to the advancement of some branch of Natural Philosophy and strove to advance knowledge in this field; the other group was composed of men whose interests lay in history, literature, art, archaeology, or even in travel and exploration as well as of statesmen and diplomats. It was on the energy and learning of the former group that the growth of the Society's scientific reputation depended; it was to the wealth of the latter group that Sprat had looked as the source of the financial endowments which the Society would need for many years. Intellectually and socially the two groups had little in common. There was no intention of forming an institution which would include activities such as those of the French Academy as well as of the Académie des Sciences in Paris.

Of these two categories the non-scientific members increased more rapidly than the men of science, and before long they were twice as numerous as the latter in the Society. Until 1820 they provided about two-thirds of the members of the Councils, and thereby seriously hampered the scientific activity of the institution for a century and a half. It was the researches carried out on their own initiative by the genius and industry of the most eminent men of science in the Society which built up its reputation and prestige, and continued to promote them even while they were year after year in a minority on its Councils.

The history of an institution which has been in existence for three centuries must be based upon a large number of small details and occurrences, but viewed broadly that of the Royal Society falls into three distinct periods: the first includes the last forty years of the seventeenth century when those who had set themselves to advance the study of the New Philosophy planned and founded the Society, and after surmounting many difficulties they achieved a considerable measure of success; the second period includes the eighteenth century and the first twenty years of the nineteenth when the majority, consisting of the non-scientific Fellows, exercised a restraining influence on the Society's activity and hindered its steady growth as a scientific institution; after 1820 the control of the Society passed into the hands of men of science with the result that it quickly became what its founders had originally intended it to be, an institution devoted wholly to the advancement of scientific knowledge.

Until 31 December 1751 the Julian or Old Style Calendar was in use in this country, and dates before then are given in it; later dates are given in the Gregorian Calendar (see p. 180).

ACKNOWLEDGMENTS

I would gratefully acknowledge the privilege granted to me by the President and Council of the Royal Society of dedicating this work to the Fellows of the Society; and also for their permission to consult the Council minutes, the Journal-books, and the various publications of the Society for much information which is not otherwise obtainable.

I would express my indebtedness to Sir D'Arcy Thompson, F.R.S., for information on many historical points; also to Dr K. C. Baily of Trinity College, Dublin, for information relating to John Winthrop (junior), an early Fellow of the Society. Mr W. H. Robinson, the Librarian of the Royal Society, also has assisted me with much information, especially that relating to the administrative staff of the Society. Little about them has been published hitherto, though the help which this branch of the Society's staff rendered at all times to the officers of the Society was of great value to them. They were often men of good education who availed themselves of their opportunities to increase their knowledge while on the Society's staff, and several of them on retiring from their administrative duties were elected to the Fellowship of the Society.

H. L.

1943

CHAPTER I

THE INVISIBLE COLLEGE: BEFORE 1660

THE ORIGIN of the Royal Society is to be traced in the general change in intellectual outlook which was taking place in southern and western Europe during the fifteenth and sixteenth centuries and exercised a notable influence on thought and learning in the seventeenth. The old scholastic ideas, though still widely held, were already giving way to more liberal views, and many learned men were prepared to discard the mediaeval tradition of ecclesiastical authority.

By the latter part of the sixteenth century many had begun to feel that the explanations of natural phenomena offered by traditional authority no longer met their needs; they were seeking a surer guidance; they desired to measure, to weigh and to control the conditions among which they lived, and, as Bacon wrote, 'to extend more widely the limits of the power and the greatness of man'. This desire the 'new philosophy', with its scientific explanations based on reliable evidence, seemed to meet, for it removed to a great extent the fear of the unknown and of the supernatural, as well as giving explanations of such phenomena as eclipses, comets and earthquakes which had hitherto been accepted as signs of divine or demoniac wrath and vengeance. Even the lesser superstitions, witchcraft and so forth, were before long to lose much of the power they once had of disturbing the minds of educated men.[1] In the new world that Copernicus, Galileo and others had displayed reason could now guide men's minds to satisfactory explanations with greater certainty than had hitherto been attainable. Scholastic thought had been primarily metaphysical but Galileo and those who worked on similar lines had shown the wider fields of view which were open to those who concerned themselves with quantities rather than with qualities, and with facts which had been verified by repeated experiments. This new learning quickly spread to France, England and elsewhere, through the correspondence which was carried on between scientific men in those countries, as well as by the visits of men of learning, many of whom travelled extensively in order to meet their colleagues in other lands.

About 1629 it came to the knowledge of Cardinal de Richelieu that a group of men of letters was meeting in Paris, and he conceived a plan of organizing them so as to form an authoritative body to deal with questions relating to the literature and language of France. Out of this grew the Académie Française which dates from 1635 and has since then

[1] Cf. *A History of Science*, by Sir W. Dampier, F.R.S., Chapter III. Cambridge, 1942.

devoted itself to promoting literary work, its activities being limited to linguistic studies and especially to safeguarding the purity of the French language.

About the same time a Minorite Father, Marin Mersenne (1588–1648), founded and maintained until his death another series of conferences which aimed at bringing together such mathematicians and physicists as Gassendi, Descartes, the Pascals, father and son, and others. He had been educated by the Jesuits at La Flêche where he met Descartes, and was led by him to take up the study of mathematics and the natural sciences. From 1635 until his death in 1648 his conferences and his correspondence linked together all the more important men of the scientific world of his day. He boldly stated that 'for him the technique and method of the new sciences were of importance because they freed man's mind from error'. He had numerous friends in England, and his letters to Theodore Haak, Seth Ward, who afterwards became Bishop of Salisbury, Samuel Hartlib, Sir Charles Cavendish, Sir Kenelm Digby, John Pell the mathematician, Sir William Petty and others provided them with valuable information about the activities of scientific men of France and Italy at that time. They in return sent to him from England books, news and information for his friends in France.

Among those who doubted whether the Académie Francaise as it was constituted could effectively represent all branches of French learning was Henri-Louis Habert de Montmor, a wealthy patron of letters and learning. About 1654 informal gatherings were being held in his house and a year or two later these developed into weekly meetings of the Montmor Academy, some of which were attended by Henry Oldenburg and Lord Ranelagh when they were in Paris in 1659; they returned to London in 1660 and later became Fellows of the Royal Society. In 1664 these meetings were no longer held in de Montmor's house but were continued for a time in that of Thévenot. Soon after this Louis XIV's minister, Colbert, came to the conclusion that some academy was needed which should include men of distinction selected from various professions, and steps were taken in 1666 to form one. His decision, probably due in the first instance to a desire to emulate the success of his predecessor who had founded the Académie Française, was certainly hastened by the reports which were reaching him of the activities of the Royal Society in London; this had organized itself as a scientific society towards the end of 1660 and received its first charter of incorporation from King Charles II in 1662.

Early in the seventeenth century Francis Bacon (1561–1626), who had been educated at Trinity College, Cambridge (1573–5), began to put together his views on how the advancement of knowledge might be utilized to bring about an improvement in man's condition. The manuscript prepared by him in 1603 and entitled 'Of the Interpretation of Nature' is in the British Museum library, and contains the ideas which he developed in more complete form in his later writings. He was made Lord Chancellor in 1618 by James I, and was raised to the peerage as Lord Verulam. A few years later he was summoned before the House of Lords on a charge of bribery, to which he pleaded guilty, and was therefore removed from his post of Lord Chancellor.

Under the title of 'The Great Instauration' he had already begun to set out his views on the restitution of man from his fallen state to what he might have become. This was to consist of six parts: firstly, the Advancement of Learning; secondly, the Novum Organum, in which he recommends the method of induction, the employment of trained observers and verification by experiments; thirdly, the Sylva Sylvarum or a history of nature; the Scala Intellectus, an improved form of intellectual analysis, and an example of the new philosophy. The fifth and sixth parts were never written. In another work, the New Atlantis, written between 1614 and 1618, he described such an organization for advancing science as he thought would be most practicable and likely to succeed. In it there were to be a number of members or Fellows of whom twelve were to travel with the object of collecting books and plans from other lands; three others were to extract the notes and experiments recorded in the books; three others were to collect information relating to mechanical crafts and experimental science, as well as processes which have not yet been adopted in this country; three others were to analyse the information so collected and endeavour to deduce from it new generalizations; lastly, three Fellows would consider how these new laws and results might be best applied to 'use and practice for man's life and knowledge'. There was also to be a staff of apprentices and a number of assistants. It has been said that the term 'Fellows' for the members of the Royal Society was taken from Bacon's use of it in the New Atlantis and adopted in the English draft of the Charters; in the Latin text the word 'sodales' was used (see p. 40).

Bacon did not carry out experiments himself and did not fully realize that a skilled investigator carrying out his own experiments and learning from them as his work advanced was likely to obtain more valuable results than would be gained from material brought together by collectors and handed over to others for discussion. Nevertheless his learning and the eloquence of the appeal with which he put forward ideas that were much in advance of those of his time, aroused a world of thought already prepared for a change, and pointed out the road by which a fuller and a sounder knowledge of nature might be attained. He realized that the new knowledge which had for some time been spreading over western Europe provided the means for redressing the balance between the scholastic teaching of the past and the fuller appreciation of nature which need no longer be considered as being linked up with superstition and the satanic forces of evil. He rejected the view that scholastic tradition was all-sufficing, and he would transfer a large share of human interest from abstract speculation to the observation of nature, as when he says: 'Those, therefore, who determine not to conjecture and guess, but to find out and know; not to invent fables and romances of worlds, but to look into and to dissect the nature of this real world, must consult only things themselves'. He did not deny the existence of metaphysical truth, but would bring to the knowledge of the world of his day that fuller appreciation of experimental science which the age needed. This he did by precept and argument, leaving to other men like Galileo, Gilbert, Harvey, and their successors to advance the branches of science in which their own researches lay by employing the principles which he advocated.

He maintained that truth is twofold, there being truth in science as there is in religion, and held that science had been hampered by religion in an age in which religion had dominated opinion and thought. The mediaeval conception of science had regarded it as forbidden knowledge since the divine nature was considered to be wholly separate from, if not opposed to, natural manifestations; a sharp distinction was drawn between the natural and the supernatural. It was certainly with this in mind that those who drafted the charters of the Society laid special emphasis on its aims by defining them as being for promoting Natural Knowledge—the supernatural was not included. The new philosophy strongly supported this new confidence in nature and encouraged those who were promoting it.

By the middle of the seventeenth century there were a number of learned men to whom the new or experimental philosophy appealed strongly, and Bacon in his writings had set before them in an attractive and convincing form the possibilities which it offered of opening new fields of knowledge. Many of these were of Bacon's own University, Cambridge, which provided the majority of the small group of philosophers whose activities will be described. These men, and others like them, stirred by his writings and the scheme which he had designed for promoting the influence of science on man and his social development, were able by their teaching and example to introduce some of the younger men of their acquaintance to the study of this new experimental philosophy; a few years later some of these and others of like interests began to meet in London as occasion offered to exchange views and to discuss the many problems which were demanding solution or at least serious study.

During the troubled times before and after 1640 the government did not scruple to overrule the University authorities and to impose upon them its wishes. Commissions of Visitors armed with full powers could and did remove heads of Colleges and professors if their opinions were unacceptable to the ruling power, and replaced them by others who were more agreeable to it. Thus Dr J. Wilkins and Dr J. Goddard were appointed to the Wardenships of Wadham and Merton Colleges at Oxford in 1649 and 1651 respectively, while Dr J. Wallis and Dr Seth Ward were made Regius Professors of Geometry and Astronomy there; others moved to

London where they joined the philosophers who began to meet at Gresham College soon after 1640. The sympathies of many of those who supported the new learning were wholly with the Commonwealth, and this accounted for the immunity from interference which the philosophers usually enjoyed; their prudent resolve to allow neither political questions nor matters of religion to be debated at their meetings was also in their favour.

In the early part of the seventeenth century those who persisted in adhering to the rigid protestantism of the Swiss Church were harshly dealt with for refusing to use the State Prayer Book of Archbishop Laud. Many ministers resigned their livings, and considerable numbers emigrated to the new English colonies in America, including members of the Winthrop family who were greatly interested in the new philosophy and its possible developments as expounded by Francis Bacon, and had discussed it with several learned men of their acquaintance. The elder, John Winthrop (1588–1649), was the son of a clothier, who had built up a moderate fortune and then settled in Essex. He was educated at Trinity College, Cambridge, where he associated with those who held strong Calvinistic opinions, but he also made the acquaintance of others who held wider views and looked forward to improving their financial position by developing the resources of new lands in America. In 1629 at the age of forty-one he was living the life of a country landowner at Groton in Suffolk, and also held a small post under the government. In June he was deprived of his office under the Master of the Wards and Liveries, and later when he decided to sell his estate he only obtained for it £4200, which was considerably less than he expected. He wrote a full draft of the arguments for and against emigrating to New England as they appeared to him, and submitted copies of it to some of his friends and neighbours a week or two later.

On 28 July the General Court of the Massachusetts Plantation in the colony of New England had before it a proposal 'to transfer the government of the plantation to those that shall inhabit there and not continue in the same subordination to the Company here, as it now is'; on 26 August 1629 his name appears as one of the signatories to the agreement entered into at Cambridge by twelve of the leading friends of the Massachusetts Plantation to embark for New England 'to inhabit and continue there'. He was elected by the Company to be their Governor. Six months later John Winthrop, with a large number of his relatives, friends and dependents, emigrated to New England. They sailed on 22 March 1630 and landed in Massachusetts Bay on 17 August, where they set to work to found their settlement.

John Winthrop (junior), his eldest son, with whom the Society is more particularly concerned, was born in February 1606; he was educated at the Free Grammar School, Bury St Edmunds, and entered Trinity College, Dublin, in 1622, when he lived in Dublin with his uncle, Emmanuel

Downing. He did not graduate there but went to London a year or two later to study law, and was admitted to the Bar of the Inner Temple in 1625. In June 1628 he undertook a foreign tour and started by ship for Leghorn, afterwards extending his journey to Venice, Padua and to Constantinople. He made good use of his time and became acquainted with many merchants, learned men and others with whom he subsequently corresponded on the financial advantages to be gained from exploiting the products of other lands and in particular of the English colonies in America. After an absence of about fifteen months he returned to London in August 1629 and after staying for a while longer in London in order to settle his father's affairs and complete the sale of the Groton estate, he too sailed for New England where he arrived on 4 November 1631. Though he was fully occupied with work in the settlement, within a year he began to receive apparatus, chemicals and books which he had ordered from England, the first shipment having been made in March 1633. This was the result of what he had seen and learned during his foreign tour, where he had realized the importance of developing the natural resources of oversea possessions as soon and as fully as might be possible.

While acting as assistant to his father in Massachusetts between 1631 and 1649 he established a settlement of his own on the Connecticut river in 1632, and two years later, in the autumn of 1634, he sailed for England to carry out negotiations in the interests of the Massachusetts Bay colony. During this visit he was commissioned by Lord Saye and Sele and Lord Brooke in 1635 to found a colony in Connecticut for which they undertook to provide him with men and funds, and promised to obtain for him an official commission as Governor of it. During this visit he consulted with experts in England how industries in the colonies could best be established and encouraged; he may have met also a few of the leading natural philosophers, but on his next visit some years later he became acquainted with a number of others. He returned to the colony about the end of 1636 keenly interested in the production of copper, iron, glass, potash, tar, salt, alum and other products.

In 1641 he again sailed for England to obtain men and funds for the erection of iron works in Connecticut, and during this visit, which lasted for about two years, he had ample opportunity of meeting in London the philosophers who were already thinking over and discussing the possibility of co-operating with those who had emigrated to the colonies since at home political conditions were still unfavourable, the country being on the brink of civil war. With these men he had much in common, and must have greatly benefited by his conversations with them, thus adding to his information and increasing the definiteness of his plans for the future. The names of those whom he met have not been recorded, but Lord Brouncker, Dr W. Charlton, Sir Kenelm Digby, Dr J. Goddard and Dr G. Ent, and a few others, may well have been in London in 1642. Winthrop brought with him from his father an invitation to any of the philosophers, who might wish to emigrate in order to live and work free from persecution, to join him in Massachusetts where he offered them all the help and assistance that he as the Governor could command.

Dr Cromwell Mortimer, the Secretary of the Society, wrote to the grandson of John Winthrop (junior) on 15 August 1741, nearly a century later:

In concert with these (i.e. Boyle, Wilkins and Oldenburg) and other learned friends (as he often revisited England) he was one of those who first formed the Plan of the Royal Society, and had not the Civil Wars happily ended as they did Mr Boyle and Mr Wilkins with several other learned men would have left England and, out of esteem for the most excellent and valuable Governor, John Winthrop the younger, would have retired to his new-born colony and there established that Society for promoting Natural Knowledge which these gentlemen had formed, as it were, in embryo among themselves....

The feasibility of forming a scientific society as early as this may have been discussed, but it seems very unlikely that Robert Boyle and Dr J. Wilkins can have then agreed to emigrate, for Boyle was still travelling abroad for his education and was but fifteen years of age, and Dr J. Wilkins was but seventeen; Oldenburg did not come into contact with the philosophers until several years later.

The philosophers very wisely were content to follow the lines on which they had worked together for several years until such times as a more formal organization could be safely introduced. Although political conditions might be unfavourable for the formation of private societies, there was a considerable number of learned men whose interest had been aroused by the writings of Francis Bacon and others, as well as by what they heard from other countries; their common interests drew them together, and about 1645, or even earlier, they were meeting whenever suitable opportunities occurred. There was as yet no definite plan for the establishment of an academy or a society, and their ideas do not seem as yet to have advanced beyond the possible formation of a scientific society such as had been discussed with John Winthrop of New England; nevertheless a small number of intimate friends did arrange to meet and to exchange views on matters of common interest; regular meetings could not be attempted with safety, but it was due to the zeal and enthusiasm of a few eminent men with scientific interests that the movement was fostered which led, fifteen years later, to the foundation of the Royal Society.

When the Royal Society began a few years later, in a small way and with royal support and patronage, it was for many years bitterly attacked from many quarters for being irreligious, as aiming at infringing the prerogatives of the Universities, and as competing with or even seeking to supplant such technical foundations as the Royal College of Physicians.

Though such men were not as yet very numerous, it was natural that those who lived in or within reach of London, Oxford or Cambridge, should avail themselves of such opportunities as came in their way to meet together in order to discuss matters of a scientific or philosophical nature, if only to divert their minds from dwelling upon the disturbed political conditions which then prevailed throughout the country. As Bishop Sprat says in his *History*, 'their first purpose was no more than only the satisfaction of breathing a freer air, and of conversing in quiet with one another, without being ingag'd in the passions and madness of that dismal age'.... 'For such a candid and impassionate company as that was, and for such a gloomy season, what could have been a fitter subject than Natural Philosophy?'

As is often the case with institutions of long standing, the first steps which led to the founding of the Society are not very precisely known, but Dr John Wallis' 'Account of some passages of his own Life' written in January 1696/7 provides the most detailed account which has come down to us. Although about fifty years had passed since the events which it describes had taken place, Dr Wallis, who was one of the most notable of the men who took part in the movement to promote the new learning, was in a position to record what took place at London and Oxford during and after the years of the civil war. His account appears to be quite reliable and is supported by other information which has been preserved. He writes:[1]

About the year 1645, while I lived in London (at a time when, by our civil wars, academical studies were much interrupted in both our Universities), beside the conversation of divers eminent divines as to matters theological, I had the opportunity of being acquainted with divers worthy persons, inquisitive into natural philosophy, and other parts of human learning; and particularly of what hath been called the *New Philosophy* or *Experimental Philosophy*. We did by agreements, divers of us, meet weekly in London on a certain day and hour, under a certain penalty, and a weekly contribution for the charge of experiments, with certain rules agreed upon amongst us to treat and discourse of such affairs; of which number were Dr John Wilkins (afterwards Bishop of Chester), then chaplain to the Prince Elector Palatine in London, Dr Jonathan Goddard, Dr George Ent, Dr Glisson, Dr Merrett (Drs in Physick), Mr Samuel Foster, then Professor of Astronomy at Gresham College, or some place near adjoyning, Mr Theodore Haak (a German of the Palatinate, and then resident in London, who, I think, gave the first occasion and first suggested those meetings) and many others.

These meetings we held sometimes at Dr Goddard's lodgings in Wood Street (or some convenient place near), on occasion of his keeping an operator in his house for grinding glasses for telescopes and microscopes; sometimes at a convenient place (The Bull Head) in Cheapside, and (in term time) at Gresham

[1] Cf. Preface of T. Hearne's edition of *Peter Langtoft's Chronicle*, 1696–7.

College at Mr Foster's lecture (then Astronomer Professor there), and, after the lecture ended, repaired, sometimes to Mr Foster's lodgings, sometimes to some other place not far distant.

Our business was (precluding matters of theology and state affairs) to discourse and consider of *Philosophical Enquiries*, and such as related thereunto; as Physick, Anatomy, Geometry, Astronomy, Navigation, Staticks, Magnetics, Chymicks, Mechanicks, and Natural Experiments; with the state of these studies, as then cultivated at home and abroad. We then discoursed on the circulation of the blood, the valves in the Veins, the Venae Lacteae, the Lymphatick Vessels, the Copernican Hypothesis, the Nature of Comets and New Stars, the Satellites of Jupiter, the oval Shape (as it then appeared) of Saturn, the spots in the Sun, and its turning on its own Axis, the Inequalities and Selenography of the Moon, the several Phases of Venus and Mercury, the Improvement of Telescopes, and grinding of Glasses for that purpose, the Weight of Air, the Possibility or Impossibility of Vacuities and Nature's Abhorrence thereof, the Torricellian Experiment in Quicksilver, the Descent of heavy Bodies, and the degrees of Acceleration therein; and divers other things of like nature. Some of which were then but New Discoveries, and others not so generally known and embraced as now they are, with other things appertaining to what hath been called the New Philosophy which from the times of Galileo at Florence, and Sir Francis Bacon (Lord Verulam) in England, hath been much cultivated in Italy, France, Germany, and other parts abroad, as well as with us in England.

Those meetings in London continued, and (after the King's return in 1660) were increased with the accession of divers worthy and Honorable Persons; and were afterwards incorporated by the name of the Royal Society, etc., and so continue to this day.

About the year 1648/9 some of our company being removed to Oxford (first Dr Wilkins on his appointment by the Protector as Warden of Wadham College, then I, and soon after Dr Goddard) our company divided. Those in London continued to meet there as before (and we with them, when we had occasion to be there), and those of us at Oxford, with Dr Ward (since Bishop of Salisbury), Dr Ralph Bathurst (now President of Trinity College in Oxford), Dr Petty (since Sir William Petty), Dr Willis (then an eminent physician in Oxford), and divers others, continued such meetings in Oxford, and brought those Studies into fashion there; meeting first at Dr Petty's lodgings (in an apothecarie's house), because of the convenience of inspecting Drugs and the like, as there was occasion; and after his remove to Ireland (though not so constantly) at the lodgings of Dr Wilkins then Warden of Wadham College, and after his removal to Trinity College in Cambridge, at the lodgings of the Honourable Mr Robert Boyle, then resident for divers years in Oxford.

Wallis' account of the activities of himself and his colleagues is borne out by Robert Boyle, by whom they were called the Invisible College; he wrote on 22 October 1646, two years after he had returned to England from Geneva, to M. Marcombe in Paris:

The other humane studies I apply myself to are natural philosophy, the mechanics and husbandry, according to the principles of our new philosophical college that values no knowledge but as it has a tendency to use. And therefore I shall make it one of my suits to you that you would take the pains to inquire a little more thoroughly into the ways of husbandry, etc., practised in your parts: and when you intend for England, to bring along with you what good receipts or choice books of any of these subjects you can procure; which will make you extremely welcome to our Invisible College.

A few months later, when writing to Francis Tallents of Magdalene College, Cambridge, he says:

The best on't is that the corner-stones of the Invisible (or as they term themselves the Philosophical) college, do now and then honour me with their company, which makes me sorry for those pressing occasions that urge my departure. ...I will conclude their praises with the recital of their chiefest fault, which is very incident to almost all good things; and that is that there is not enough of them.

To all these men whom Wallis mentions the Society owes an undying debt of gratitude for the enthusiasm and the energy which they displayed in bringing about the Society's foundation and the definition of its aims.

The Visitations which the Commonwealth government sent to the Universities of Oxford and Cambridge to report upon the royalist sympathies which some of the college authorities and professors were said to hold were mainly responsible for several of the philosophers leaving London and Cambridge and establishing themselves at Oxford; these included Dr J. Wilkins, who was of a puritan family and was known to be a capable administrator; he replaced Dr John Pitt as Warden of Wadham College; Dr J. Wallis, the eminent mathematician, replaced Peter Turner, royalist, as Savilian Professor of Geometry; Dr J. Goddard accompanied the Protector as physician-in-chief to Ireland and Scotland, being nominated Warden of Merton College in 1651; Dr Seth Ward, a distinguished mathematician, was elected to succeed John Greaves, the Savilian Professor of Astronomy.

Dr John Wallis, to whom we owe this account, was a mathematician and classical scholar of great distinction and one of the most eminent men of his day; he energetically defended the Royal Society from the criticisms which were levelled against it in its early years. He was born at Ashford in Kent in 1616 and entered Emmanuel College, Cambridge, in 1632; he graduated in 1636–7 and was ordained in 1640. In 1647 he became minister of St Martin's Church, Ironmonger Lane, London. In November 1648 the royalist Savilian Professor of Geometry at Oxford, Peter Turner, was ejected by the Visitation which went to that University, and in June 1649 Wallis was appointed to succeed him. He continued to occupy the Savilian Chair of Geometry until his death. Despite his learning and the wide range of his scholarship, he was of a highly contentious temperament and was frequently involved in controversial exchanges of views with his contemporaries, especially with Pascal, Fermat and other French mathematicians.

Dr John Wilkins, who was the son of a puritan minister, graduated at Oxford in 1631 and having taken Holy Orders became chaplain to Lord Saye and Sele, and then in 1637 to the Elector Prince Palatine with whom he remained for some years. He was then in London where he met several of the philosophers and attended their meetings about 1644. On the outbreak of the civil war he took the covenant, and being recognized as a man of well-balanced judgment and a capable administrator he was appointed by Cromwell to be Warden of Wadham College, Oxford, on 13 April 1648 in place of Dr John Pitt who had been ejected on the recommendation of the Visitation of 1647; he therefore left London for Oxford, and was followed by others. John Evelyn, who visited Dr Wilkins at Wadham College in 1654, writes in his Diary on 13 July:

We all din'd at that most obliging and universally-curious Dr Wilkins's, at Wadham College. He was the first who shew'd me the transparent apiaries, which he had built like castles and palaces, and so order'd them one upon another as to take the honey without destroying the bees.... He had also contriv'd an hollow statue, which gave a voice and utter'd words by a long conceal'd pipe that went to its mouth, whilst one speaks through it at a good distance. He had above in his lodgings and gallery variety of shadows, dyals, perspectives, and many other artificial, mathematical, and magical curiosities, a way-wiser, a thermometer, a monstrous magnet, a conic and other sections, a balance on a demi-cycle, most of them of his owne and that prodigious young scholar, Mr Chr. Wren, who presented me with a piece of white marble, which he had stain'd a lively red, very deepe, as beautiful as if it had been natural.

He was appointed Master of Trinity College, Cambridge, in 1659; there, too, he introduced notable reforms, but after ten months he was ejected from his mastership at the Restoration. When the meetings of the philosophers in London were resumed in 1660, Dr Wilkins was one of those who were present at the historic meeting of 28 November 1660 at which it was decided to found a Society of Experimental Philosophy, and he was 'appointed to the Chaire'. Later he was named one of the two Secretaries of the Royal Society in both the First and Second Charters. He was appointed Bishop of Chester in 1668 on the recommendation of Dr Seth Ward, Bishop of Salisbury, who had obtained for him the living of St Lawrence Jewry, London, in 1662, and the precentorship of Exeter in 1667.

Drs Goddard, Ent, Glisson and Merret were all medical men of distinction who attended these early meetings at Gresham College. Jonathan Goddard was a skilful experimenter, and according to Aubrey, the diarist, was often called upon by his fellow-members to carry out the experiments and demonstrations which were required at their meetings. He was educated at Christ's College, Cambridge, where he took his M.B. degree in 1638; in 1646 he became a Fellow of the Royal College of Physicians in London. In 1649 he accompanied the Protector to Ireland as physician-in-chief, and in the following year he went with him to Scotland. In 1651 he was nominated Warden of Merton College, Oxford, a post which he held until 1660. He was one of the twelve who met on 28 November 1660 at Gresham College and decided to form a Society, a decision which a week later received the king's approval. He was Professor of Physic from 1655 to 1675 at Gresham College where he had a chemical laboratory, and is said to have been an admirable chemist. He was a member of Council in seven of the years between 1664 and 1673. Aubrey says that he intended to leave his library and papers to the Royal Society but as he had made no will at the time of his death the Society did not benefit by his intention.

Sir George Ent, M.D., was a distinguished classical scholar, who was educated at Sidney Sussex College, Cambridge, where he graduated in 1631; he was a close friend of William Harvey, whose discovery of the circulation of the blood he vindicated in a work published in 1641; he also wrote on respiration. He was President of the Royal College of Physicians in 1682 and 1684. He bequeathed his library to the Royal Society in 1689.

Dr F. Glisson, who graduated at Caius College, Cambridge, in 1624 and received his degree of M.D. at Cambridge in 1634, was a physician of distinction.

Samuel Foster, the mathematician, was educated at Emmanuel College, Cambridge, in 1623, and was Professor of Astronomy at Gresham College in 1636, and again from 1641 until his death in 1652. He was an active member of the group which became the Philosophical College, and their meetings often took place after his weekly lectures at Gresham College.

Theodore Haak, who is said to have been the first to suggest holding the meetings of the philosophers, was a German of Neuhausen who came to England at the age of twenty in 1625; he spent some time at each of the Universities and after a period of travel on the Continent returned to this country and became a commoner of Gloucester Hall, Oxford. He associated with Gabriel Plattes, the agricultural writer, John Pell the mathematician, Amos Comenius of Prague, Samuel Ward of Sidney Sussex College, Cambridge, and Samuel Hartlib, in their schemes for an international college.

Another man of influence was Dr Seth Ward, who was educated at Sidney Sussex College, Cambridge, and took his M.A. degree there in 1640; he was elected a Fellow in 1640 and to a lectureship in mathematics in 1643, being much esteemed for his learning in mathematics and philosophy. In 1644 he was summoned with others to appear before the Committee of Visitors, and was expelled from the post which he held and from the University. He then went to London where he studied mathematics with Rev. W. Oughtred, and made the acquaintance of those men of science who had begun to meet at Gresham College to discuss the new philosophy. In 1649 an invitation from Lord Wenman of Thame to be his chaplain took him to Oxford, where shortly afterwards the Visitation of that University took place, with the result that John Greaves, the Savilian Professor of Astronomy, was dismissed and, largely through the support and advocacy of Sir John Trevor, Dr Seth Ward was appointed in his place. He chose Wadham College for residence, being influenced in his choice by the reputation of Dr J. Wilkins, who was then the Warden.

Lawrence Rooke, a brilliant mathematician and astronomer, was educated at King's College, Cambridge, and was elected to a Fellowship in 1643. When Dr Seth Ward, whom he had known intimately at Cambridge, was ejected and went to Oxford in 1649, Rooke followed his example and became a fellow-commoner at Wadham College where he was an energetic member of the philosophical group which met in Dr Wilkins' rooms at that College. In 1652 he went to London, having been elected to the professorship of astronomy and later of geometry at Gresham College. He was modest, retiring and a man of few words except when with his intimate friends. Rooke was elected a Fellow of the Royal Society but died on 26 June 1662. Dr Seth Ward presented the Royal Society with a fine pendulum clock by Fromantel in memory of his friend; it is known to have been at Crane Court until 1755, but all trace of it has since been lost.

Sir Ralph Bathurst was a Fellow of Trinity College, Oxford, in 1640 and became an M.D. in 1654. Though a royalist, he was employed by Parliament as a physician, but gave up medicine at the Restoration and was appointed Chaplain to the King in 1663. He was a member of the philosophical group at Oxford and later became one of the original Fellows of the Royal Society.

Among those who also co-operated with the philosophers and attended their meetings in 1646 was Robert Boyle, who had been studying with his tutor, M. Marcombe, at Geneva and returned from the Continent in 1644. Samuel Hartlib was another who attended the meetings of the philosophers at Gresham College. He was of Polish origin and published works on agriculture for which he was granted a pension by Parliament. He endeavoured to organize a *bureau d'adresse* on the lines of that which Renaudot established in Paris, but without success. He was never a Fellow of the Royal Society.

William Petty, who joined the philosophers in 1646 at the same time as Boyle and Hartlib, was born in 1623, at Rumsey in Hampshire, a seat of the local woollen industry, where his father worked at his trade as a cloth weaver and dyed his own cloths. As a boy he was anxious to see the great world and at the age of fifteen he bound himself apprentice to the master of a vessel sailing for France. There he met with an accident and was landed near Caen with a broken leg. While he was in hospital he learned French, and on his recovery he entered a school at Caen to continue his education, but finding that better teaching was to be had at the Jesuit College there he accepted an offer from the Fathers to study with them. At twenty years of age, having now a good knowledge of Latin, Greek and French, as well as of astronomy, geometry, navigation, etc., he spent three years in France and the Netherlands at the schools of Utrecht, Leyden and the School of Anatomy in Paris. He also met in Paris Dr J. Pell, the English mathematician, and Thomas Hobbes, the philosopher, who recognized his ability and assisted him. Through Hobbes he became acquainted with the Marquis of Newcastle and Sir Charles Cavendish, royalist refugees, and with the Minorite Father, Marin Marsenne. On the death of his father he returned to England in 1646, and then joined the philosophers at their weekly meetings in London. Evelyn describes him as 'a public-spirited and ingenious person who had propagated many useful things and arts'; he also began to attend the philosophers' meetings at this time and later endeavoured to form a bureau on the French plan but was not successful. At the end of 1647 in a letter to Boyle he describes Petty as follows: '...one Petty of twenty-four years of age...a perfect Frenchman, and a good linguist in other vulgar languages besides Latin and Greek, a most rare and exact anatomist, and excelling in all mathematical and mechanical learning; of a sweet natural disposition and moral comportment'. Petty now decided to seek for a post in anatomy and with the support of his friends John Graunt the statistician, and Edmund Wylde, both adherents of the parliamentary party, as well as of Colonel Kelsey, who was in command of the garrison at Oxford, he left London to go there. He was successful in obtaining his doctor's degree in Physic in 1649; later, in 1650, he was elected a Fellow of Brasenose College and was appointed Deputy Professor of Anatomy, becoming Professor in 1651.

In 1651 the philosophers at Oxford and those who were in the habit of attending their meetings formed themselves into the Philosophical Society of Oxford, of which the minutes and a copy of its rules are preserved at the Ashmolean Museum. Its meetings took place at first weekly but later they occurred at irregular intervals until 1690 when they ceased altogether.

In 1656 Dr John Wilkins, the Warden of Wadham College and a prominent member of the group of philosophers, married Robina, the widow of Dr Peter Trench, a canon of Christ Church and the youngest sister of the Protector, a relationship through which he was able to exert some influence in favour of men of learning, the value of which was little understood or appreciated by many of the parliamentarians.

On 3 September 1651 the battle of Worcester was fought which resulted in the defeat of the royalist forces and gave undisputed control to the parliamentary party. Charles II became a fugitive and not until 16 October was he able to escape and land at Fécamp whence he went to Paris to undergo many hardships and privations for the next nine years.

Dr Petty, who had been appointed Assistant Professor in 1651 and later Professor of Anatomy at Oxford, was sent by the Government to Ireland as physician-general to the Army in Ireland and to the Commander-in-Chief; he arrived at Waterford in September 1652 to take up his duties. He set to work at once and reorganized the medical service so efficiently that he was recognized as being an administrator of exceptional ability; his services were therefore utilized by the Lord Deputy Fleetwood to carry out a survey of the lands of Ireland in order to provide a basis for the settlement on them of the officers and men of the army and others who had been given grants of the lands from which the Irish population had been driven out, and to settle these on the lands which had been reserved for them.

A scheme put forward by the Surveyor-General, Benjamin Worsley, was rejected as inadequate but an alternative one put forward by Dr Petty was accepted. He undertook to survey the country, and to map out the whole of the forfeited lands in thirteen months from an appointed day, 'if', as he said, 'the Lord give seasonable weather and due provision be made against Tories, and that my instruments be not found to stand still for want of bounders' who could point out the property boundaries. On 24 June 1657 'all the books with the respective mapps well draune and adorned, being fairly engrossed, bound up, indexed and distinguished were...delivered into the Exchequer'. Some additional surveys were carried out in 1658 and 1659.

Petty now returned to London where he rejoined the philosophers, who were now holding their meetings whenever practicable; but the Protector's death in 1658 threw everything into confusion.

In the autumn of 1653 Dr J. Wilkins wrote to Boyle to suggest that he should move to Oxford and take up his residence there; this he agreed to do and took rooms in High Street, where he lived until he went to reside with his sister, Lady Ranelagh, in London in 1668 as being more convenient.

In 1652 Joseph Glanvill entered Exeter College, Oxford, where he graduated in 1655. He was a man of very considerable ability and is remembered as a divine and as a philosopher. He was elected a Fellow of the Society in December 1664. In 1661 he attacked scholastic philosophy in his book *The Vanity of Dogmatizing*; afterwards he was a strong advocate of the new philosophy, though he defended the belief in witches and witchcraft in *Sadducismus Triumphatus* of 1666. Anthony à Wood tells us that Glanvill regretted in after years that he had not been at Cambridge where in 1652, when he went to Exeter College, the new philosophy was held in greater esteem than at Oxford.

About this time there was living partly in Scotland and partly in France an accomplished and versatile Scot, Sir Robert Moray, who was to render to the Royal Society a few years later most important services at and after the Restoration. He was born in 1608 and was educated in Scotland, though not at any of its Universities;[1] there and on various occasions

[1] A. Robertson, *The Life of Sir Robert Moray*, p. 2. London, 1922.

during his eventful life, he studied chemistry, mathematics and music; he was a friend of the learned, a recluse and yet a man of affairs and a most agreeable companion. In 1643 he was knighted by Charles I and served as a lieutenant-colonel in the Scottish Guards in France, but in November the French army was defeated and Moray was sent as a prisoner of war to Ingoldstadt in Bavaria, where he occupied his time in studying magnetism with Dr Kircher. In the following year he was ransomed and returned to England. In 1646 he arranged for the escape of Charles from Newcastle but at the last minute the king refused to attempt it, considering it to be too dangerous and impracticable. He then remained for some years in Scotland trying to raise levies for France but without much success, but waiting in readiness for any attempt that might be made to bring about the king's return. In 1654 he was denounced as being involved in a plot against Charles II, but the evidence adduced was proved to be false and he was released in the following year by the king's order. He then went to Paris but did not stay there long; by August 1656 he was in Bruges, and by June 1657 in Maastricht, where he remained until 1659 studying chemistry until the time for action arrived. Before long, more favourable reports arrived and in August he went to Paris hoping to obtain a grant of money from Richelieu to assist in the restoration of Charles, but in this he was unsuccessful. He stayed in Paris throughout the winter and with Lord Lauderdale urged the French Protestants to support Charles II.

In 1658 Charles II was living as an exile in Brussels when the news of the death of the Protector, Oliver Cromwell, reached him, and his agents at home were at once instructed to open negotiations with General Monk who was then in command of the army in Scotland. There was much unrest in the country and even in London where Matthew Wren, writing in October 1658, reported that on visiting Gresham College he found that the parliamentary forces had taken possession of the buildings and that the professors had been turned out. In 1659 Sir Richard Willis, Charles' principal agent, communicated the royalist plans to the parliamentary party, and the restoration of the monarchy seemed to be more remote than ever. Charles left Brussels on a visit to the Spanish Court in the hope of gaining some support there. There was a general feeling in the country that some change was imminent. Several of the philosophers who were at Oxford came to London and met those who had remained in the capital at Gresham College where they attended lectures by Christopher Wren on astronomy and by Lawrence Rooke on geometry, adjourning afterwards to dine at the Bull Head Tavern in Cheapside.

Oldenburg, who had been travelling on the continent with Richard Jones, afterwards Lord Ranelagh, was in Paris in 1659, and they were present at some of the meetings of the members of the Cabinet Montmor. Several years later these visits were quoted as evidence that Oldenburg was encouraged by what he saw and heard there to promote the establishment of the Royal Society. But at this time Oldenburg had no connection with the English philosophers and did not become one of the Society's Secretaries until 1662.

In May 1660 the Restoration took place, but Moray was too discreet to be one of those who hurried to London to secure a share in whatever gifts or appointments might be obtainable, and delayed his return until July when he renewed his acquaintance with some of the philosophers who warmly welcomed so influential a colleague. During the autumn he renewed his former friendship with many whom he had known at the Court and was received kindly by the king, so that by December, when the philosophers were forming their society he was able to explain to the king that their aims and the character of the institution which they were proposing to establish were in no way prejudicial to either Church or State.

This short sketch of the conditions under which a small group of enthusiastic men met together for about twenty years whenever practicable with the object of promoting the study of natural knowledge, and of accepting only such theories and explanations as were supported by verified observations or by carefully made experiments, led up to the incorporation of the Royal Society by Royal Charter in 1662. They were few in number, those who were most active hardly numbered twenty though the meetings were attended by their friends who were interested, and who sometimes did later enrol themselves in the Invisible College as Boyle called it. The only qualification demanded of those who joined this company was that they should be zealous supporters of the 'new philosophy'; political opinions carried no special weight—a few like Sir Robert Moray and Sir Ralph Bathurst were royalist in their sympathies, but most of the philosophers were supporters of the parliamentary party, and not a few came from dissenter families. This may have protected them from some degree of interference in the years between the royalist defeat in 1651 and the Restoration ten years later; but they were in these days a small body of men several of whom were well known for their learning and were extremely unlikely to be in any way a danger to the State. Socially they were drawn mainly from the professional classes and not infrequently from the skilled craftsmen of those days. They were one and all distinguished for zeal and energy in promoting natural knowledge such as were sadly lacking in many of their successors a century later. It was not until the Society had been formed and the expense of its administration was causing difficulties that members of the nobility and of the well-to-do classes were admitted in any number.

REFERENCES

Brown, Professor H. *Scientific Organizations in Seventeenth-century France.* Baltimore, 1934.

Clark, G. N. *History of the Seventeenth Century.* Oxford, 1929.

Clark, G. N. *Science and Social Welfare in the Age of Newton.* Oxford, 1937.

Dampier, Sir William. *A History of Science.* 3rd edn. Cambridge, 1942.

Fitzmaurice, Lord Edward. *Life of Sir William Petty.* London, 1895.

Robertson, A. J. *Life of Sir Robert Moray.* 1922.

Royal Society, Notes and Records of. 1939–41.

Royal Society, Record of. 4th ed. 1940.

Sprat, T. *History of the Royal Society.* London, 1667.

Wallis, W. *Defence of the Royal Society, an Answer to the Cavils of Dr William Holder.* 1678.

Weld, J. C. *History of the Royal Society.* Vol. 1. London, 1848.

CHAPTER II

THE FOUNDING OF THE SOCIETY: 1660–1670

ON 1 January 1660, Samuel Pepys began the Diary in which he was to record his acts and experiences during the next ten years. He describes the political situation as follows: 'The Rump was lately returned to sit again; Admiral Lawson lies still with the fleet in the river; Monk is with his army in Scotland. The new Common Council of the City do speak very high and had sent to Monk their Sword-bearer to acquaint him with their desires for a full and free parliament.'

On the same day General Monk commenced his march southwards from Coldstream on the Scottish border. He met with no resistance and all Yorkshire came out in arms to meet him. There was great work to be done, though still none dared speak of the Restoration as being at hand. In London the wildest rumours were current and everyone wondered what would happen next. On 3 February Monk and his troops entered London, and the next day the Rump offered him the Oath of Abjuration of the House of Stuart, which he refused to take. The Council of the City of London having resolved to pay no taxes until a Free Parliament had been called, the Rump on 8 February ordered Monk to march into the city and pull down the gates and portcullises; he marched into a city in which all were anxiously awaiting his next move. On the following night after a conference with his principal officers he wrote to the Rump demanding the issue of writs for filling up the House with the excluded members, and an early dissolution to make way for a free Parliament. Two days later he marched again into the city and told the Mayor and Aldermen what he had done, on which the wildest enthusiasm broke out on every side.[1] Pepys describes in his Diary on 11 February the spontaneous rejoicings of the inhabitants of the city:

In Cheapside there was a great many bonfires, and Bow Bells and all the bells in all the churches as we went home were a-ringing. Hence we went homewards, it being about ten at night. But the common joy that was everywhere to be seen! The number of bonfires, there being fourteen between St Dunstans and Temple Bar, and at Strand Bridge I could at one time tell thirty-one fires. In King Street seven or eight; and all along burning, and roasting and drinking for rumps. There being rumps tied on sticks and carried up and down. The butchers at the May Pole in the Strand rang a peal with their knives when they were going to sacrifice their rump. Indeed it was past imagination both the greatness and the suddenness of it.

[1] A. Bryant, *Samuel Pepys: The Man in the Making*, p. 81. Cambridge, 1933.

On 30 March Sir John Greville, bearing a letter from General Monk, reached King Charles at Brussels, and a month later Parliament received the king's reply. It was at once resolved that he should be invited to return and rule them.

The Restoration had become a fact without blood having been shed.

As the news of the king's Restoration spread throughout the length and breadth of the country, there arose everywhere a general feeling of relief at the ending of the restrictions and arbitrary acts to which all classes had been subjected for many years past. These had been modified to a certain extent after the civil war had come to an end, but now there was every prospect that greater freedom of action, thought and speech would be available to all. To none was this more welcome than to the philosophers, and the thoughts of many of them must have gone back to their discussions of some twenty years before when they had debated the possibility that one day a society might come into existence before which their views might be developed and made known to a far wider circle than their own. Now they were able once again to hold their meetings which had been discontinued during the occupation of Gresham College by the Commonwealth troops at the end of 1658. During 1659 several of their company from Oxford had come to London and rejoined those of their colleagues who had continued to hold their meetings at Gresham College so long as this was practicable; now that the Restoration had taken place and the repressions of recent years were at an end, these meetings could be revived; they were attended by a considerable number of those who were interested in the new philosophy and in the discussion of all that was related to it. No record exists of their meetings or of the consultations which took place between the end of May when the king arrived in London and the end of November when the meeting, at which the philosophers decided to take definite action, was held, but there must have been several. How many members would be likely to join such a society as they had in mind? What funds would be required and how were they to be obtained? Would the court or the State provide some such subvention as Cardinal Richelieu had obtained in 1635 for the French Academy? All these and many other matters had certainly been considered on numerous occasions during the twenty years which had passed since the small group of pioneer philosophers had first turned their thoughts to planning a society.

At this time a most valuable recruit joined them in the person of Sir Robert Moray, who had supported Charles I and had been in high favour with his son Charles II until 1654 when he was falsely accused of being involved in a plot against him; he was released in the following year, and now in August 1660 rejoined the philosophers to several of whom he was well known; soon afterwards he was restored to favour at court and in February 1660/1 he was made a Privy Councillor and later a Lord of the Exchequer. With his experience of home politics and his close connection

with the court his help and guidance were most valuable to the philosophers at this time when they were occupied with the formation of the
new society which they were establishing. It was on 28 November 1660,
after attending a lecture by Christopher Wren, the Professor of Astronomy
at Gresham College, that twelve of those who had been the most active in
promoting the new philosophy met in the rooms of Lawrence Rooke, the
Professor of Geometry, where the formation of a college for physico-
mathematical experimental studies was discussed. This is recorded in the
opening pages of the Society's Journal-book in the following words:

These persons following, according to the usual custom of most of them, mett
together at Gresham College to heare Mr Wren's lecture, viz. The Lord
Brouncker, Mr Boyle, Mr Bruce, Sir Robert Moray, Sir Paul Neile, Dr Wilkins,
Dr Goddard, Dr Petty, Mr Ball, Mr Rooke, Mr Wren, Mr Hill. And after the
lecture was ended, they did, according to the usual manner, withdrawe for
mutuall converse. Where amongst other matters that were discoursed of, something was offered about a designe of founding a Colledge for the promoting of
Physico-Mathematicall Experimentall Learning. And because they had these
frequent occasions of meeting with one another, it was proposed that some
course might be thought of, to improve this meeting to a more regular way of
debating things, and according to the manner in other countryes, where they
were voluntary associations of men in academies, for the advancement of various
parts of learning, so they might doe something answerable here for the promoting
of experimentall philosophy.
    In order to which, it was agreed that this Company would continue their
weekly meeting on Wednesday, at 3 of the clock in the tearme time, at Mr
Rooke's chamber at Gresham Colledge; in the vacation at Mr Ball's chamber in
the Temple. And towards the defraying of occasionall expenses, every one
should, at his first admission, pay downe ten shillings[1] and besides engage to pay
one shilling weekly, whether present or absent, whilest he shall please to keep
his relation to this Company. At this Meeting Dr Wilkins was appointed to the
chaire, Mr Ball to be Treasurer, and Mr Croone, though absent, was named for
Register.
    And to the end that they might the better be enabled to make a conjecture of
how many the elected number of this Society should consist, therefore it was
desired that a list might be taken of the names of such persons as were known to
those present, whom they judged willing and fit to joyne with them in their
designe, who, if they should desire it, might be admitted before any other.

Most of the philosophers who were present at this historic meeting had
been resident in London for some time; of the others Robert Boyle came
up to London from Oxford where he had been living since 1654, and
Dr Petty had recently returned from Ireland after completing the survey
and settlement of lands there; Sir Robert Moray had returned to London
from Paris in August; and Dr J. Wilkins had come to London after having
been deprived of his Mastership of Trinity College, Cambridge.

[1] This was increased to twenty shillings on 15 March 1661/2.

A list containing the names of the following forty persons was therefore prepared, and to each of them an invitation was sent urging them to become members of the new Society in addition to the twelve who had already resolved to hold regular weekly meetings. The response to this appeal was very satisfactory, for of those whose names appear on the list only five did not become Fellows of the Society. Of the remaining thirty-five candidates nineteen may be considered as men of science, while the other sixteen included statesmen, soldiers, antiquaries, administrators and one or two literary men. Medical men, of whom there were fourteen, formed the largest group. There was only one peer in the list, but on 12 December barons and all of higher rank were declared to be eligible for election as a special privilege on the same day that they were proposed.

LIST OF THOSE WHO WERE INVITED TO BECOME MEMBERS OF THE PHILOSOPHERS' SOCIETY

LORD HATTON, Statesman

HON. ROBERT BOYLE, Physicist and Chemist

MR RICHARD JONES, later EARL RANELAGH

*MR COVENTRY

MR WILLIAM BRERETON, later BARON BRERETON

SIR KENELM DIGBY, Courtier, Chemist

SIR ANTHONY MORGAN, Soldier

MR JOHN EVELYN, Diarist, Horticulturist

*MR RAWLINS

MR MATTHEW WREN, Secretary to Lord Clarendon

MR HENRY SLINGSBY, Warden of Mint

MR THOMAS HENSHAW, Historian

SIR JOHN DENHAM, Poet

MR THOMAS POVEY, Civil Servant

*MR WILDE

DR S. WARD, D.D., Bishop of Salisbury

DR JOHN WALLIS, D.D., Mathematician

DR FRANCIS GLISSON, M.D.

DR GEORGE BATE, M.D.

DR GEORGE ENT, M.D.

DR CHARLES SCARBURGH, M.D.

*DR PHRASIER

DR THOMAS COXE, M.D.

DR CHRISTOPHER MERRETT, M.D.

DR DANIEL WHISTLER, M.D.

DR TIMOTHY CLARKE, M.D.

DR RALPH BATHURST, M.D.

*DR A. COWLEY, Poet

DR THOMAS WILLIS, M.D.

DR NATHANIEL HENSHAW, M.D.

SIR JOHN FINCH, Physician

DR THOMAS BAYNE, M.D.

DR CHRISTOPHER WREN, Architect, Mathematician

DR GEORGE SMYTH, M.D.

MR ELIAS ASHMOLE, Antiquary

MR JOHN NEWBURGH

MR JOHN AUSTEN

MR HENRY OLDENBURG, Literary

SIR PETER PETT, Lawyer

DR WILLIAM CROONE, M.D.

* Did not become Fellows of the Society.

The way was now clear for those who had the matter in hand to press forward with their scheme for the formation of the Society which they had planned, and it was highly desirable that its aims and constitution should not arouse suspicions in the minds of the authorities. Here Sir Robert Moray was able to render invaluable service for he was not only well known to the king but was trusted by him; he was therefore a most suitable emissary to bring to the king's knowledge what the philosophers had done, and what they were proposing to do in organizing their Society; and this he did without delay. On 5 December, a week after the preparation of the preliminary list of additional members, another meeting was held, which is recorded as follows in the Journal-book:

Sir Robert Moray brought in word from the Court, that the King had been acquainted with the designe of the Meeting. And he did well approve of it, and would be ready to give encouragement to it.

The king's pleasure having been noted, the meeting turned to the business of the day which included the following:

It was ordered that Mr Wren be desired to prepare against the next meeting for the Pendulum Experiment.

That Mr Croone be desired to looke out for some discreet person skilled in short-hand writing, to be an amanuensis.

It was then agreed that the number be not increased, but by consent of the Society who have already subscribed their names: till such time as the orders for the constitution be settled.

That any three or more of this company (whose occasions will permit them) are desired to meete as a Committee, at 3 of the clock on Fryday, to consult about such orders in reference to the constitution, as they shall think fit to offer to the whole company, and so to adjourne *de die in diem*.

At the same meeting the following obligation was adopted:

Wee whose names are underwritten, doe consent and agree that wee will meet together weekeley (if not hindered by necessary occasions), to consult and debate concerning the promoting of experimentall learning. And that each of us will allowe one shilling weekely, towards the defraying of occasionall charges. Provided that if any one or more of us shall thinke fitt at any time to withdrawe, he or they shall, after notice thereof given to the Company at a meeting, be freed from this obligation for the future.

To this are attached the signatures of nearly all those persons comprised in the catalogue of names prepared at the meeting on 28 November, and also those of seventy-three others, who were subsequently elected into the Society, as recorded in the Journal-book.

On 12 December another meeting was held at which the following business was transacted and recorded in the Journal-book:

It was referred to my Lord Brouncker, Sir Robert Moray, Sir Paul Neile, Mr Matthew Wren, Dr Goddard, and Mr Christopher Wren, to consult about a convenient place for the weekly meeting of the Society.

It was then voted that no person shall be admitted into the Society without scrutiny, excepting only such as are of the degree of Barons or above.

Sir Kenelm Digby, Mr Austen and Dr Bate, were then by vote chosen into the Society.

That the stated number of this Society be five and fifty. That twenty-one of the stated number of this Society be the *quorum* for Elections.

That any person of the degree of Baron or above may be admitted as supernumerarys, if they shall desire it, and will conforme themselves to such orders as are or shall be established.

Whereas it was suggested at the Committee that the Colledge of Physitians would afford convenient accommodation for the meeting of this Society; uppon supposition that it be granted and accepted of, it was thought reasonable, that any of the Fellowes of the said Colledge, if they shall desire it, be likewise admitted as Supernumerarys, they submitting to the Lawes of the Society, both as to the pay at their admission, and the weekly allowance; as likewise the particular works or tasks that may be allotted to them.

That the Publick Professors of Mathematicks, Physick and Naturall Philosophy, of both Universitys, have the same priviledge with the Colledge of Physitians, they paying as others at their admission, and contributing their weekly allowance and assistance when their occasions do permitt them to be in London.

That the *quorum* of this Society be nine for all matters excepting the Businesse of Elections.

At the end of March 1661 it was agreed that the limit of fifty-five, which had been fixed for the membership, might be exceeded as the number of candidates for election had proved to be much in excess of what had been expected; in all seventy-three more were admitted. This augured well for the prospects of the young Philosophers' Society which had at last been formed and had received royal approval; more than fifteen years of strenuous effort on the part of the small body of men who had been the pioneers of the movement had now been crowned with success. The Royal Society, to call it by the name by which it was to be known a few months later, had now been founded, though the date of the sealing of the First Charter—15 July 1662—is more usually taken as being that of the Society's foundation.

The committee now proceeded to draft the regulations for the new Society, and though these only remained in force until the sealing of the First Charter of 15 July 1662, they are of interest as showing the procedure which it was proposed to adopt, and which formed the basis of the statutes of the Royal Society that were drafted in the autumn of 1662 and approved by the Council in January 1662/3. The need for such regulations and for their strict observance was fully realized by those who were familiar with the difficulties which the academies and bureaux of the brothers Dupuy, of Renaudot, of de Montmor and others in Paris had experienced. At their meetings it had been too often the practice of the members to make long speeches in support of theses or proposals though little evidence in support of them might have been brought forward; this led to vague discussions, altercations, and too often to accusations and personalities. The Philosophers' Society, and its successor the Royal Society, from the beginning set their faces resolutely against anything which would prejudice the orderly discussion of whatever was brought before them. This conduct of the Society's business at the Ordinary meetings greatly impressed foreign visitors and was referred to by Sorbière, the historiographer of Louis XIV, who attended several of the meetings in 1663. He wrote:[1]

Le President tient une petite masse de bois à la main, dont il frappe sur la table lors qu'il veut faire silence. On parle à luy découvert, jusques à ce qu'il fait signe de l'on se couvre; et l'on rapporte en peu de mots ce que l'on·trouve à propos de dire sur l'experience que le Secretaire a proposée. Personne ne se haste de parler, n'y ne se picque de parler longtemps, et de dire tout ce qu'il sçait. On n'interrompt jamais celuy qui parle, et les dissentimens ne se poussent pas bien avant, n'y d'un ton qui puisse des-obliger en aucune manière. Il ne se peut rien voir de plus civil, de plus respectueux, et de mieux conduit que cette Assemblée telle qu'elle me parut. S'il y a quelques entretiens particuliers qui se forment tandis que quelqu'on parle, ils se passent à l'oreille, et l'on s'arrete tout court au moindre signal que le President fait; de sorte que l'on n'achevé pas mesme de dire sa pensée.

Monconys, who attended the meeting which was held on 23 May in the same year, comments in similar terms on the procedure.

The regulations which were adopted on 12 December 1660 were as follows:

Concerning the Manner of Elections

That no man shall be elected the same day he is proposed.

That at the least twenty-one shall be present at each election.

That the Amanuensis doe provide severall little scrolles of paper of an equall length and breadth, in number double to the Society present. One halfe of them shall be marked with a crosse, and being roled up shall be lay'd in a heap on the table, the other halfe shall be marked with cyphers, and being roled up shall be lay'd in another heap. Every person coming in his order shall take from each heap a role, and throwe which he please privately into an urne, and the other into a boxe. Then the Director, and two others of the Society, openly numbering the crossed roles in the urne, shall accordingly pronounce the election.

That if two thirds of the present number do consent uppon any scrutiny, that election to be good, and not otherwise.

Concerning the Officers and Servants of the Society

The standing Officers of this Society to be three, that is to say, a President or Director, a Treasurer and a Register. The President to be chosen monthly.

The Treasurer to continue one yeare, as also the Register.

---

[1] *Relation d'un voyage en Angleterre*, par le Sieur Sorbière. Cologne, D.C. LXVII, pp. 64-5.

That there be likewise two servants belonging to this Society, an Amanuensis, and an Operator.

That the Treasurer doe every quarter give in an account of the Stock in his hand and all disbursements made, to the President or Director, and any three others to be appointed by the Society; who are to report it to the Society.

That any bill of charges brought in by the Amanuensis and Operator, and subscribed by the President and Register for any experiment made, and subscribed by the Curators of the experiment, or the major part of them, be a sufficient warrant to the Treasurer for the payment of that sum.

That the Register provide three bookes, one for the statutes and names of the Society, another for experiments and the result of debates: and a third for occasionall orders.

That the salary of the Amanuensis be £40 per annum, and his pay for particular business at the ordinary rate, either by the sheet or otherwise, as the President and Register can best agree with him.

That the salary of the Operator be foure pounds by the yeare, and for any other service, as the Curators who employ him shall judge reasonable.

That at every meeting, three or more of the Society be desired that they would please to be reporters for that meeting, to sitt at table with the Register and take notes of all that shall be materially offered to the Society and debated in it, who together may form a report against the next meeting to be filled by the Register.

When the admission money comes to £20, then to stop.

At a subsequent meeting, held 19 December 1660, it was 'ordered that the next meeting should be at Gresham Colledge, and so from weeke to weeke till further order'; the suggestion that the Society should meet at the Royal College of Physicians was not proceeded with. Wednesday was the day on which the meetings were held, and the weekly subscription was due on that day.

On 6 March 1661 Sir Robert Moray was chosen President of the new Society and on 10 April he was re-elected 'for another month'. The intention was that the President should be appointed monthly, but the names of those who occupied the chair have not been fully recorded. Moray served nine times between March 1661 and July 1662, Boyle and Lord Brouncker each served once, and Wilkins five times. Moray was one of the earliest and most active members of the Royal Society which benefited greatly by the free access to the king which was allowed to him. During its first decade he carried on a long and varied scientific correspondence in French with Huygens, who described him as being the 'Soul' of the Society, and with other scientific men in Paris. At his suggestion Charles II established a Chair of Mathematics at St Andrews University to which James Gregory, one of the most distinguished mathematicians of his day, was appointed in 1668.

Ever since he wrote in January and May 1660 to his friend Dr Alexander Bruce, afterwards Lord Kincardine, on his plans for the Society, Sir Robert Moray had been considering how the group of philosophers might best be converted into an organized institution with the necessary safeguards; he knew that bitter opposition was to be expected from both ecclesiastical and political bodies to any development of natural philosophy in this country. The only effective safeguard which it was within the philosophers' power to employ would be to obtain from the king their incorporation by means of a Royal Charter. Moray was extremely active throughout the year 1661 and this was recognized by his colleagues who elected him to the presidency of the new Society oftener than any other of its members. He undoubtedly had much to do with the preliminary drafting of the First Charter, for on 21 June 1661 he wrote hopefully to Huygens: 'Dans quelques jours nous espérons que notre Société sera établie de la bonne sorte.' Three weeks later Huygens wrote to Chapelain: 'Vous savez quel est le dessein de ces messieurs....Ils ont une personne entre autres qui travaille avec grand zèle à l'établissement de l'Académie, et qui en est comme l'âme: c'est le chevalier Moray.'

In 1661 Colonel Samuel Tuke, a recently elected member of the Society, was sent to Paris on a mission by the king to represent him at the funeral of Cardinal Mazarin who had died on 9 March. He was a cousin of John Evelyn, a poet and dramatist as well as being keenly interested in art, and was one of those who had been admitted to the Society when the membership was increased. The Society took advantage of his visit to Paris to learn as much of the activities of French men of science as he could gain knowledge of while he was there. An old acquaintance of his, M. de Roberval, a professor of mathematics, arranged that he should visit the home of M. de Montmor, where he was introduced by the Comte d'Albonne. After an experiment had been demonstrated and discussed, Tuke was invited to give an account of the Royal Society, which he did, describing the prosperity of England and praising King Charles as a patron of science and philosophy. At the end of the meeting it was agreed that the Academy and the Society should correspond regularly. A few months later Sorbière wrote to Tuke to explain that the correspondence which had been arranged had not begun because there had been illness in the Montmor family. He expressed the wish to visit England and two years later he did so, being present at one or more of the meetings of the Society. Tuke seems to have given the members of the Montmor Academy a somewhat exaggerated idea of the endowments which the Royal Society might expect from Charles II, and which led them to press actively for a royal establishment on similar lines in Paris.

By this time the Society was becoming well known in learned circles, and John Hoskins, who was elected President of the Society in 1683, wrote to his friend Aubrey, the antiquary, of Oxford, in July 1661: 'I wonder you tell mee nothing of the famous Academy of our philosophical scepticks that believe nothing not tryed.' They distrusted lengthy discussions and elaborate theories, preferring to concentrate first on the improvement of apparatus and instruments, being convinced that the close study of the realities of nature was the only sound way of advancing philosophy.

About November 1661 John Winthrop arrived in London from America; eighteen years had passed since his previous visit, but a number of those philosophers with whom he had then discussed the possibility of forming a scientific society were still living, and he had been in the habit of corresponding with them while he was in America. On this occasion he had come in a different capacity for he was now the Governor of the Connecticut colony and had been deputed by the colonists to present a loyal address on their behalf to the king; he was also to obtain if possible a Royal Charter for it. He had been provided with £500 with which to overcome any difficulties which might arise in the course of the negotiations. Sir Robert Moray and he had not met on his previous visit in 1642–4 since the former was then a prisoner of war in Bavaria, but now he was an influential member of the Royal Society and frequently presided at its meetings; he was also in high favour with the king and was therefore able to promote the objects of Winthrop's mission. On 18 December Winthrop was proposed for membership in the Society by Dr William Brereton, afterwards Lord Brereton, and was elected on 1 January 1661/2. In May 1663 he was elected an Original Fellow of the Society under the Second Charter.

From the time that he was elected a member of the Society Winthrop presented numerous communications at the meetings, most of which were concerned with the practical application of science and especially of chemistry. His friends among the philosophers had received him warmly, and certainly Sir Robert Moray was able to render him valuable assistance at court in the matter of the Charter for his colony at Connecticut. Winthrop was completely successful in this part of his mission and the terms of it were even more favourable than the colony had hoped to obtain. In August 1664 Oldenburg wrote to him by the order of the Council to 'inform John Winthrop that he was invited in a particular manner to take upon him the charge of being the Chief Correspondent of the Royal Society in the West, as Sir Philiberto Vernatti was in the East Indies'. Besides the experiments and the observations which he carried out he sent considerable collections of minerals and other objects of interest to the Society. He and Henry Oldenburg corresponded regularly, and from the latter he received the Society's publications, including the *History of the Royal Society* by Dr Thomas Sprat when it was published in 1667, as well as news of what was happening in scientific circles at home. Oldenburg also urged him to write an account of New England, but his many duties as Governor prevented him from doing this.

After his return to the colony he continued to correspond with many in England and in Europe; Dr Cromwell Mortimer, in a letter of dedication to his grandson of the same name (*Phil. Trans.* vol. 40), records the names of about eighty of these correspondents of whom thirty were or had been Fellows of the Royal Society. He continued to hold the post of Governor of the colony of Connecticut until his death, which took place in 1676.

During the years between 1645 and 1660, when the philosophers were holding their informal meetings in London and at Oxford, no record of the discussions which took place or of the various matters which were brought before them was kept; but now that they had constituted themselves a Society which had received the king's approval it became necessary that its business should be regularly recorded. The Journal-book which had been opened on 5 December 1660 has preserved the story of the launching of the Society on its career, and the transactions of its earlier years. Between the beginning of the year 1661 and the sealing of the First Charter in July 1662 the Society met on about fifty occasions at which many subjects were discussed and dealt with. A good many of these related to the administration of the Society, the admission of new members, the collection of fees and subscriptions which were due, and other business matters; a considerable number of curiosities, quaint happenings and so-called rarities were reported as had been done at the earlier informal meetings, and these were often referred to some member for his opinion or investigation; many local superstitions were also brought before them, and these were seriously considered and tested in order either to establish their validity or to refute them. One of the most valuable tasks which the Society set before itself from its earliest days was that of testing any legends or ill-supported statements, which were reported to it, by a critical verification of the facts. This procedure did much to discredit the belief in witchcraft, demoniac possession and supernatural intervention in human affairs which was even then beginning to lose its influence over the minds of many in the educated classes; but there were still some even in the ranks of the philosophers who clung to their belief in magical and occult powers. Sir Kenelm Digby, despite his interest in chemistry, or rather alchemy, firmly believed in cures by means of sympathetic powder and such remedies. Evelyn describes him as an 'errant mountebank'. The king referred several matters to the Society for consideration, sending at one time two lodestones for experiment, and at another glass balls for examination; he also desired the Society's opinion on sensitive plants. When he asked that the Society should undertake the construction of a model of the moon the task was allotted to Christopher Wren, who duly carried it out. The model was accepted by the king, who placed it in his collection of rarities.

On the evening of 3 May 1661, King Charles was shown Saturn's rings and Jupiter and his satellites 'through His Majesty's great telescope, drawing thirty-five feet; on which were divers discourses'. Eleven days later Evelyn records in his Diary: 'May 14. His Majesty was pleased to discourse with me concerning several particulars relating to our Society, and the planet Saturn, &c., as he sate at Supper in the withdrawing-room to his bedroom.'

As soon as the Society had been formed the meetings and discussions were resumed on much the same lines as during the preceding years.

Dr William Petty, Dr Goddard, Mr Christopher Wren and Sir Kenelm Digby were appointed as a committee on ship construction, a matter in which the king was especially interested; Lord Brouncker was desired to continue his experiments on the recoil of guns, and Robert Boyle was invited to show the air-pump which he had designed, as well as his experiments on the air; Dr Goddard was asked to show his experiments on colours, and Dr Merrett to bring his history of refining. The first volume of the Register-book opened with a list of twenty-two questions which Lord Brouncker and Robert Boyle had drafted in order that they might be sent to Tenerife for study by one of their acquaintances there and for replies to be furnished in due course. Evelyn was asked to bring what he had written on engraving and etching; his book on this subject was published later by the Society. He was also invited to communicate his observations on the anatomy of trees. Sir Kenelm Digby, who gave a discourse on the vegetation of plants, was invited to put the same in writing. Dr Petty, who had worked with his father as a clothier and dyer in his youth, was asked to report on this trade, and Henry Slingsby, an official of the Mint, to discourse on metallurgy and the work of the Mint. Later, Lord Brouncker gave an account of experiments made at the Tower of London on the increase of weight of bodies when strongly heated. A diving-bell which was under construction was to be tried shortly at Deptford. Such matters would fall naturally within the scope of the Society's aims, but there were also many rumours, reports and strange occurrences which came before them for explanation or enquiry; among these were the supposed virtues of May-dew collected before sunrise; Dr Dee's skill in weather forecasting reported by Sir K. Digby; the petrification of wood in Ireland; the production of young vipers from the powdered livers and lungs of vipers; a poison that turned a man's blood to jelly; the composition of a 'sympathetic' powder which would cure a wounded man if the weapon or his blood were treated with it; divining by means of hazel rods; whether a spider set within a circle of powdered unicorn's horn could cross it: that it could not do so was refuted by an experiment; the tale of a city in northern Africa where all the inhabitants had been petrified was vouched for by a friend of Sir K. Digby.

At the meeting of the philosophers on 5 February 1661, Dr D. Whistler, a physician, of Merton College, Oxford, and Professor of Geometry at Gresham College in 1648, brought in a book, *Natural and Political Observations on the Bills of Mortality* from Mr John Graunt, and read a letter of dedication to Sir Robert Moray who was then President. Fifty copies of the book were presented by the author for distribution among the Fellows of the Society. The thanks of the Society were returned to him, and he was proposed as a candidate. Later, on 20 May 1663, he was included in the list of the Original Fellows of the Royal Society. He is best known for this statistical work though he also carried on successfully a commercial business in the city of London. Dr T. Sprat, in his *History*, refers to Graunt's admission and adds that the king highly commended the action of the Society: 'It was so farr from being a prejudice that he was a shopkeeper of London that his Majesty gave this particular charge to his Society that if they found any more such tradesmen, they should be sure to admit them all without any more ado.' But at no time was there any attempt on the part of the Society, either then or later, to limit admission to the Fellowship to any social class, or to hinder the entry of skilled craftsmen or any candidate of middle-class parentage so long as he possessed adequate knowledge and the wish to promote the aims of the Society. George Graham, the instrument maker and watch- and clockmaker, was elected a Fellow in 1720; John Harrison, a carpenter's son who became a celebrated horologist, received the Copley Medal in 1749 for the improvements which he introduced in the design of chronometers; he ultimately won the prize offered by the Board of Longitude for one which could be used in the determination of longitude at sea. Sir William Petty, whose father was a working clothier and dyer, educated himself and joined the philosophers in 1646; he was knighted in 1662 and elected an Original Fellow in 1663. Sir Isaac Newton was the son of a yeoman farmer, and George Gregory, an eminent mathematician of the end of the seventeenth century, was a son of the manse from near Aberdeen. For many years the Society did admit peers and statesmen as a privileged class in the hope that they might advance the aims of the Society; but this was rarely the case, and the prestige of the Society has always been built up and maintained by men who owed nothing to their social position or connections but were prepared to devote their genius and zeal to improving natural knowledge. Sir Robert Moray, who accepted Graunt's gift and dedication, was re-elected to preside in the following month and on several subsequent occasions. In a letter addressed to M. de Montmor on 22 July 1661, he subscribes himself as 'Societatis ad Tempus Praeses'.

At this time the activity of the de Montmor Academy was diminishing, and Sorbière with others who were interested in maintaining meetings and discussions on natural philosophy were endeavouring to induce the minister Colbert to support the institution of an Academy of Science.

The name by which the Society was to be known was doubtless the subject of much discussion at this time. That which was ultimately chosen was first publicly applied to it by Evelyn in the dedication to the Earl of Clarendon of his translation of Gabriel Naudé's *Avis pour dresser une Bibliothèque*, where he praised that nobleman for his services 'in the promoting and encouraging of the Royal Society'. The book appeared about the middle of November and by that time Evelyn had accustomed himself to think, and even in his Diary to write, of the company of philosophers as the 'Royal Society'. His colleagues lost no time in expressing their gratification to him for his public recognition of them by this title, and in his Diary on 3 December he writes: 'By universal suffrage of our philosophic assembly, an order was made and registered, that I should receive their public thanks for the honourable mention I made of them by the name of the Royal Society, in my Epistle Dedicatory to the Lord Chancellor before my traduction of Naudaeus. Too great an honour for a trifle.' We may believe that during the frequent conversations which Evelyn had with the king at this time, when many subjects w re discussed, the important matter of the Society's name was considered. Charles was now genuinely interested in the work of the philosophers, and if he did not propose the title himself he no doubt readily approved of it when it was suggested by his esteemed courtier.

The selection of a suitable name had become urgent, for the grant of a Royal Charter of Incorporation for the Society was discussed by the members during the summer of 1661, with the result that on 18 September of that year a petition to the king for a royal grant of incorporation was read at a meeting of the Society. At the request of the Society, Christopher Wren prepared a preamble to the draft Charter which reads as follows:

Whereas amongst our Royal hereditary titles, to which, by Divine Providence and the loyalty of our good subjects, we are now happily restored, nothing appears to us more august or more suitable to our pious disposition, than that of father of our country, a name of indulgence as well as of dominion, wherein we would imitate the benignity of heaven, which in the same shower yields thunder and violets, and no sooner shakes the cedars, but dissolving the clouds drops fatness: We, therefore, out of paternal care for our people, resolve, together with those laws which tend to the well administration of government and the people's allegiance to us, inseparably to join the supreme law of Salus Populi, that obedience may be manifest by not only the public but the private felicity of every subject and the great concern of his satisfactions and enjoyments in this life. The way to so happy a government we are sensible is in no manner more facilitated than by promoting of useful arts and sciences, which upon mature inspection are found to be the basis of civil communities and free governments, and which gather multitudes by an Orphean charm into cities, and connect them in companies; that so by laying in a stock as it were of several arts and methods of industry, the whole body may be supplied by a mutual commerce of each other's peculiar faculties, and consequently, that the various miseries and toils of this frail life may be, by as many various expedients ready at hand, remedied or alleviated, and wealth and plenty diffused in just proportion to every one's industry, that is to every one's deserts.

And whereas we are well informed that a competent number of persons, of eminent learning, ingenuity, and honour, concording in their inclinations and studies towards this employment, have for some time accustomed themselves to meet weekly, and orderly, to confer about the hidden causes of things, with a design to establish certain and correct uncertain theories in philosophy, and by their labours in the disquisition of nature, to prove themselves real benefactors to mankind; and that they have already made a considerable progress by divers useful and remarkable discoveries, inventions and experiments in the improvement of mathematics, mechanics, astronomy, navigation, physic, and chemistry, we have determined to grant our Royal favour, patronage, and all due encouragement to this illustrious assembly, and so beneficial and laudable an enterprize.

In this preamble the promotion of useful arts and sciences as being the basis of free governments, and the collection of information relating to the several arts and methods of industry to improve the lot of all, are given prominence as being of special importance and according with the king's known preferences. Physical and mathematical study and research, as well as chemistry, are mentioned later, for they could not be expected to make so wide and popular an appeal.

The style of this preamble follows that of the writers of the early and middle part of the seventeenth century rather than the severer and plainer form adopted by Wallis, Wilkins, Sprat and other pioneers of the Royal Society; but Wren's preamble, though it was not incorporated in the Charters, is valuable as a description of the Society's aims as understood and described by one of the most able of the small group of men who founded it at Gresham College at the end of 1660.

Probably in large measure owing to Sir Robert Moray's influence a favourable response to the request for a Charter of Incorporation was not long delayed, for at a meeting held on 16 October, Sir Robert Moray acquainted the Society that 'hee and Sr. Paul Neile kiss'd the King's hands in the Company's Name, and is intreated by them to return most humble thancks to His Majesty for the Reference he was pleased to graunt of their Petition: and to this favour and honour hee was pleased to offer of him selfe to bee enter'd one of the Society'.

This Charter of Incorporation passed the Great Seal on 15 July 1662, which has therefore been taken as being the date of the foundation of the Royal Society. The Charter was read before the Society on 13 August of the same year, and a fortnight later the President, Council, and Fellows went to Whitehall and returned their thanks to His Majesty.

Those members of the Society who were appointed to be members of Council under the First Charter were

VISCOUNT BROUNCKER, President

SIR ROBERT MORAY	WILLIAM AERSKINE
ROBERT BOYLE	JONATHAN GODDARD, M.D. Prof. of
WILLIAM BRERETON	Medicine, Gresham College
SIR KENELM DIGBY	CHRISTOPHER WREN, M.D., Prof. of
SIR PAUL NEILE	Astronomy, Oxford
HENRY SLINGSBY	WILLIAM BALL, Treasurer
SIR WILLIAM PETTY	MATTHEW WREN
JOHN WALLIS, D.D.	JOHN EVELYN
TIMOTHY CLARKE, M.D.	THOMAS HENSHAW
JOHN WILKINS, D.D., Secretary	DUDLEY PALMER
GEORGE ENT, M.D.	HENRY OLDENBURG, Secretary

All these, except Aerskine, Wren, Henshaw, Palmer and Oldenburg, had a knowledge of some branch of science or of technical industry.

Viscount Brouncker, who was appointed President of the Society in the First Charter, was then forty-two years of age. He was the son of Sir William Brouncker, who was gentleman of the Privy Chamber to Charles I, and Vice-Chancellor to Charles II when he was Prince of Wales: he was created Baron Brouncker of Newcastle and Viscount Brouncker of Castle Lyons, Co. Cork, in 1645. The future President was educated at Oxford where he studied mathematics, natural philosophy and medicine. He was well known both at home and abroad for his mathematical knowledge, to which he devoted himself and in which he attained high distinction. After the Restoration he was appointed in 1662 to be Chancellor to the Queen Consort and Keeper of her Great Seal; he was also one of the Commissioners for executing the office of the Lord High Admiral.

Robert Boyle was at this time living at Oxford, where since 1654 he had been devoting his time to research in chemistry; he served on the Council in 1663-4, 1665-6 and in 1667-8, and three times while he was resident in London after 1669.

Sir William Brereton was educated at Breda and studied mathematics under John Pell, as did Sir W. Petty, becoming an accomplished algebraist. He was a close friend of John Winthrop of New England.

John Evelyn had become a member of the Society in 1661 and later was one of the Original Fellows of 1663; he was twice invited to be President but declined. He was a member of both the first and second Councils and served on fifteen others between 1668 and 1700.

Sir Paul Neile was also a member of the first two Councils, and served as a councillor in each of the ten following years; he was keenly interested in astronomy.

Timothy Clarke, who was educated at Balliol College, Oxford, took his degree in 1652. He was a member of the first two Councils and of four others; he died in 1672. He was physician to Charles II and was a friend of Samuel Pepys. It will be seen that the Society chose its early councillors with care and they served it well at a time when such support was all important to it.

Now that the Society had received its Charter of Incorporation it became necessary to prepare for the coming session and a most important item was the appointment of a competent curator whose duty would be 'to furnish the Society every day they mett, with three or four considerable experiments, expecting no recompense till the Society gett a stock enabling them to give it'. It was essential that a suitable man should be chosen and the Society was very fortunate in securing the services of Robert Hooke, whose appointment was proposed by Sir Robert Moray in 1662. Hooke was twenty-seven years of age at the time, and had for several years assisted Robert Boyle in his experiments at Oxford and elsewhere, as well as in the construction of his pneumatic engine. Boyle rendered a great service to the Society by releasing Hooke from his engagement with him, for he thereby enabled him to carry out for forty years valuable scientific work at first in the employment of the Society and afterwards as one of its Fellows.

Robert Hooke, the son of a country parson, was born in 1635, and was educated at Westminster School under Dr Busby. In 1653 he went to Oxford as a servitor or chorister at Christ Church. A small deformed man, greatly handicapped by ill-health, he grew up melancholy in temperament, jealous of other workers and disposed to claim as his own their improvements and advances whenever they published them. His active mind and his very original outlook soon attracted the notice of Dr J. Wilkins, through whom he came to know many of the philosophers. With the mathematics and astronomy which he studied under Dr Seth Ward, and the chemistry which he learned in the laboratory of Dr Thomas Willis who recommended him to Boyle as his personal assistant, together with eight years' personal contact with Robert Boyle in his laboratory, Hooke, at the time when he was offered the post of Curator of the Royal Society, had acquired an exceptional knowledge of the science of the day, and his own genius suggested to him many ways in which it could be utilized. He demonstrated discoveries and inventions of his own before the Society, as well as illustrating the communications which had been received from the Fellows; and he was ready to devise and explain any apparatus which was wanted to clarify any point under discussion. It was an inestimable boon for the Society to have so brilliant and learned a member on its staff who was at the same time a skilful experimenter in physics and chemistry.

That no salary could be paid to him at that time by the Society may not have been so much to his disadvantage as it would appear. His Diary shows that he maintained a close contact with all the most competent and skilful artisans in London at that time, such as instrument makers, clock- and watchmakers, metal and wood workers and many others from whom he learnt much from their practical competency, and to whom he could give valuable information and advice in matters in which they were engaged. In this way Hooke no doubt received more than the Society would have offered him. In his later years his technical and scientific knowledge enabled him to set aside a very considerable sum as the proceeds of a busy and lucrative advisory practice, as a skilled mechanic, surveyor and also as an architect.

It was not until January 1673 that the Council provided him with any assistance; they then appointed Henry Hunt as a boy-assistant to him. For more than forty years Hunt continued to be a member of the Society's staff and filled several posts; in November 1676 he replaced Richard Short-grave as 'Operator' at £20 per annum; in 1696 he was promoted to be Keeper of the Library and the Repository at a salary of £40 per annum. He also engraved plates for the *Philosophical Transactions* and other works.

Though Hooke was known as being very difficult to work with, he always treated Hunt with great kindness and consideration, more as though he were his son than a subordinate. Hunt died in 1713, ten years after Hooke, after having been a generous benefactor to the Society on more than one occasion (p. 143).

Although the king was unable to provide the Society with any endowment in its early years, a record appears in the Register-book on 17 September 1662 of his having written to the Duke of Ormonde, the Lord Lieutenant of Ireland, recommending the Society 'for a liberall contribution from the adventurers and officers of Ireland for the better encouragement of them in their designes'. On 3 January 1662/3 the President, Lord Brouncker, wrote to the Duke of Ormonde referring to the king's letter and begging that a suitable grant of lands might be made for the benefit of the Society. No action of any kind was taken by the duke, and in the following year Sir William Petty explained to his friend Sir Robert Southwell that there was no prospect whatever of any grant being made.

By the time that the First Charter had passed the Great Seal and the Council had given it their careful consideration, several of the Fellows had come to the conclusion that there were some additional privileges which they would wish to see included in it while retaining all the clauses of incorporation: they therefore very wisely decided to press for an amending Charter while the royal interest in the Society and its work was still active.

At the end of 1662 and in the early part of 1663 the Council devoted several meetings to drafting such alterations as were desired. Sir Robert Moray, who had been thanked by the Council in August 1662 for the important part that he had played in obtaining the grant of the First Charter, was naturally chosen to take up negotiations for the Second, and on 25 March 1663 he was commissioned to hand the revised draft to Sir Henry Bennet, the Secretary of State, who had been made Keeper of the Privy Purse to the King at the Restoration, and in 1662 was made Secretary of State, being raised to the peerage as Earl of Arlington in the following year.[1] The draft of a Second Charter was accompanied by a petition praying for a new Charter to which the king consented and this, the Second Charter, passed the Great Seal on 22 April 1663, and was read before the Society on 13 May following. During the ten months which had elapsed between the sealing of the First and Second Charters no Council meetings were held, elections as well as any other business being transacted at the Ordinary meetings of the Society. The Anniversary meeting of 1662, which should have been held on 30 November, did not take place, and no new members of Council were then elected, but those appointed by the First Charter remained in office until the Second Charter came into operation.

[1] *Royal Society Journal*, Book 1.

On 1 June 1663, soon after the Second Charter had passed the Great Seal, Huygens wrote to congratulate Sir Robert Moray on the Incorporation: 'Je sais que ce vous sera une grande satisfaction de voir bien réussir ce à quoi vous avez travaillé avec tant de zèle et de constance.'[1] On 13 August 1662 the Society had resolved that 'Sir Robert Moray should be thanked for his concern and care in promoting the constitution of the Society into a Corporation'.[2] It may seem strange that he was not chosen as the Society's first President, but it was not his habit to seek his own advancement, and he was probably well content to see Lord Brouncker appointed to this post by the Charter, Moray being one of the members of the Council in both the First and Second Charters. At this time there was no one more competent than he to take a leading part in guiding the young Society's affairs, and he was re-elected to the Council in every year except 1667 until his death in 1673. The Society was already an object of suspicion to many, and it was only the enthusiasm and determination of its more energetic members that kept it alive; some years later when these flagged its vitality seriously diminished.

The Council and officers nominated in the First Charter were re-nominated in the Second, except that the names of Sir George Talbot and Abraham Hill were substituted for those of Dr Wallis and Dr Christopher Wren; William Ball was again appointed Treasurer and Dr John Wilkins and Henry Oldenburg were retained as Secretaries. The Second Charter contained a provision that all persons whom the President and Council should receive into the Society within two months from the date of the Charter should be named Fellows of the Royal Society. In accordance with this provision, at a meeting held on 20 May 1663, one hundred and fifteen persons were declared to be members, and they, together with four members who were elected subsequently at a meeting held on 22 June, constituted the Original Fellows of the Society; eighteen ordinary Fellows were also elected before the Anniversary meeting on 30 November. A special interest attaches to this list, inasmuch as it gives an indication of the lines on which the early founders of the Royal Society chose the men with whom they wished to be associated in one common fellowship for the furtherance of Natural Knowledge.

On 13 May 1663 the Council of the Royal Society met for the first time when the Second Charter was read, and the new members of the Council were sworn in before the President. On 13 May one Michael Wicks was appointed Clerk of the Society; he had previously been employed by Dr Jonathan Goddard at Gresham College, and according to Aubrey worked as an assistant in the laboratory there to look after the stills used by Goddard. He it was no doubt who recommended the Society to appoint Wicks to the post of Clerk. His work included making fair copies of the minutes of the Council meetings and the Ordinary meetings; the

[1] Huygens, Corr. III, Nos. 869, 873.   [2] *Royal Society Journal*, Book I.

communications received by the Society had also to be transcribed into the Register. He was in this way the assistant of Oldenburg, who complained of receiving too little pay, which meant probably that he really needed someone having higher qualifications than a copyist. In the same year Richard Shortgrave was appointed Operator to the Society and was paid £20 per annum for preparing and demonstrating experiments.

At the same meeting of the Council the following resolutions were adopted:

Ordered that all persons that have been elected or admitted into the Royal Society doe pay their whole arrears unto this day according to their subscription; and that the Treasurer, or collector by him appointed, do repair to every such person and demand the said arrears, showing unto him this order, together with the form of the subscription hereunto annexed.

Form of the Subscription

Wee whose names are underwritten do consent and agree that we will meet together weekly (if not hindered by necessary occasions), to consult and debate concerning the promoting of Experimental Learning, and that each of us will allow one shilling weekly towards the defraying of occasional charges: Provided, that if any one, or more of us, shall think fit at any time to withdraw, he or they shall after notice given thereof to the Company at a meeting, be freed from this obligation for the future.

On 27 May 1663 the Council resolved, and on 17 June confirmed, their decision, that

the following obligation should be signed by every Fellow of the Society, and that anyone refusing to do so shall be ejected from the Society; and if any person elected shall refuse to subscribe to it his election shall be void:

'We who have hereunto subscribed, do hereby promise each for himself, that we will endeavour to promote the good of the Royal Society of London for improving Natural Knowledge, and to pursue the ends for which the same was founded: that we will be present at the meetings of the Society, as often as conveniently we can, especially at the anniversary elections, and upon extraordinary occasions; and that we will observe the Statutes and Orders of the said Society: Provided that whensoever any of us shall signify to the President under his hand, that he desireth to withdraw from the Society, he shall be free from this Obligation for the future.'

This obligation is still in force and is inscribed at the head of each page of the Charter-book in which every elected Fellow is required to sign his name before being admitted.

The Council also resolved that

the Ordinary Meetings of the Society should be held every Wednesday, at 3 o'clock p.m., and continue until 6, unless the major part of the Fellows present shall resolve to rise sooner or sit longer, and no Fellow shall depart without giving notice to the President:

and that the President, when in the Chair, is to be covered, notwith-
standing the Fellows of the Society be uncovered. The custom that the
President should be covered when presiding was observed for many years;
Dr A. B. Granville, when describing the procedure of the Society early in
the nineteenth century, says that in 1830 it was still in force, and the edition
of the statutes which was printed in 1840 includes a rule to this effect; but
Weld, writing eight years later, says that it was then no longer the practice.
Thus it had been discontinued at some date between 1840 and 1848 during
the presidency of Lord Northampton, though the regulation reappears
in the statutes which were printed in 1847 after those governing the
election of candidates had been revised.

In addition to some other minor alterations the principal changes made
by the Second Charter were that the name of the Society was changed
from 'The Royal Society' to 'The Royal Society of London for improving
Natural Knowledge'; that the king declared himself to be the Patron of
the Society; that the Society was granted the following blazons of honour:
in the dexter corner of a silver shield our three Lions of England, and
for Crest a helm adorned with a crown studded with florets, surmounted
by an eagle of proper colour holding in one foot a shield charged with
our lions: Supporters, two white hounds gorged with crowns.[1] The Latin
texts, together with English translations of these two Charters, as well as
of the Third Charter which was granted on 8 April 1669, are printed in
the *Record of the Royal Society of London* (4th edition, 1940, pp. 215–84).

When the Society was preparing its petition for the grant of a Charter,
the design for the arms of the new institution came up for consideration,
as well as the wording of a suitable motto. The grant to the Society of the
right to use as its arms a shield bearing the three Lions of England with
Supporters and a crest—a helm surmounted by an eagle—which was made
by the Second Charter settled the first point, and a series of designs for
these which had been proposed by Evelyn were no longer required. The
Society did however accept the motto which he had suggested—*Nullius
in Verba*—and this was adopted as indicating the Society's determination
to withstand the domination of authority and to test all statements by
appeal to facts which had been carefully verified by it. The words are taken
from a passage of Horace in which the poet compares himself to a gladiator
who, having earned peace and retirement by a long and successful career,
is now free from the control of those who had hitherto directed his actions.[2]

---

[1] Appendix I.

[2]   'Ac ne forte roges, quo me duce, quo lare tuter,
          Nullius addictus iurare in verba magistri.'

                                            Horace, *Ep.* I, i, ll. 13, 14.

Sir Cecil Carr has contributed the following translation:

              'You shall not ask for whom I fight
                  Nor in what school my peace I find;
              I say no master has the right
                  To swear me to obedience blind.'

In a similar manner the Fellows of the Society claim to have freed themselves from the restrictions laid down by authorities in the past, and are willing in future to accept only what they have satisfied themselves by examination and experiment to be reliable and true.

The First Charter, which is engrossed on four skins of vellum, was drawn by Sir Robert Sawyer, who was chamber-fellow with Samuel Pepys at Magdalene College, Cambridge; he was first Craven Scholar in 1649 and became the Attorney-General in 1681. It is remarkable for its clearness and legal terseness. It is not known by whom the English translations of the Charters were made, but it is thought that they were prepared about 1848, perhaps at the Record Office.

In it and the other Charters the members of the Society are described as 'Fellows', a term which was used by Bacon in his *New Atlantis* whence it is said to have been adopted by the Society. For some years the terms 'Member' and 'Fellow' were used almost indiscriminately in the minutes and the Journal-book, but later the term 'Fellow' became general, 'Member' being applied to the foreign members only. It is of special interest that the word 'sodalis' should have been used as the Latin equivalent of 'Fellow' thereby providing a permanent record of the informal dinners at which the early philosophers carried on their discussions, and which have been held almost uninterruptedly at some inn, tavern or restaurant, until the present day. Cicero's description of such meetings of friends in his day so closely reproduces those of the Fellows of the Society that it is worth quoting. In *De Senectute*, XIII, he writes: 'I have always had my club companions, and so I used to attend the dinners with my fellow-members (*cum sodalibus*); it was always done in decent moderation; all the same there was a certain enthusiasm appropriate to one's age. It was not only the physical sensation of pleasure which one counted up at these gatherings; there was just as much the enjoyment of meeting friends and of good talk.' This applies equally well to the early meetings of the philosophers, and later to those of the Fellows of the Society at which they discussed scientific questions of common interest after they had attended the lectures of Foster, Rooke or Wren at Gresham College, or after the Ordinary meetings of the Society from the time when it was formally established in 1662 up to the present day.

The Charters were kept at first in the Treasurer's chest which Ball, the first Treasurer, presented to the Society. Later, when the Society had a house of its own, they were kept in as safe a place as could be found at the Society's rooms. In 1935 the Council decided to have the three Charters examined and where necessary repaired; this was done by the permission of the trustees in the laboratory of the British Museum. The sheets of vellum were found to be in good condition and the only repairs needed were to some of the seals which were cracked. The Charters have now been mounted flat in glazed frames, instead of being rolled up as they had been for many years, and the frames are kept in a specially made box in the strong room of the Society.

The object for which the Society was founded is defined in the Charter as being for 'the improving Natural Knowledge by experiment'. The word 'natural' is here used as excluding all that is 'supernatural'. Sprat speaks of 'Experiments of natural things as not darkening our eyes, nor deceiving our minds, nor depraving our hearts'; and elsewhere he describes the Society as 'following the great precept of the apostle, of trying all things, in order to separate superstition from truth'. Belief in witchcraft and divination as well as superstitions of all kinds were rife in the seventeenth century and instances, for which irrefutable evidence was claimed, were constantly being brought before the early meetings of the Society to be examined in order to see whether any reasonable explanation for them could be found. Sir Walter Scott, in his *Demonology*, records his conviction that the belief in witchcraft decreased materially after the Royal Society began to investigate critically this and other reputed supernatural manifestations.

The task which the Society set before itself in order to carry out its objects is not described in detail in any of its journals or publications, but a very comprehensive statement by Robert Hooke occurs in a document of which the original is included in a volume of papers written by him and is now in the British Museum. It bears the date 1663 and was probably drawn up very soon after the Second Charter was granted:

The business and design of the Royal Society is:

To improve the knowledge of naturall things, and all useful Arts, Manufactures, Mechanick practises, Engines and Inventions by Experiments—(not meddling with Divinity, Metaphysics, Moralls, Politicks, Grammar, Rhetoric or Logick).

To attempt the recovery of such allowable arts and inventions as are lost.

To examine all systems, theories, principles, hypotheses, elements, histories, and experiments of things naturall, mathematicall and mechanicall, invented, recorded, or practised by any considerable authors ancient or modern. In order to the compiling of a complete system of solid philosophy for explicating all phenomena produced by nature or art, and recording a rationall account of the causes of things.

In the meantime this Society will not own any hypothesis, system or doctrine of the principles of natural philosophy, proposed or mentioned by any philosopher ancient or modern, nor the explication of any phenomena whose recourse must be had to originall causes (as not being explicable by heat, cold, weight, figure and the like as effects produced thereby); nor dogmatically define nor fix axioms of scientificall things, but will question and canvass all opinions, adopting nor adhering to none, till by mature debate and clear arguments, chiefly such as are deduced from legitimate experiments, the truth of such experiments be demonstrated invincibly.

And till there be a sufficient collection made of experiments, histories, and observations there are no debates to be held at the weekly meetings of the Society concerning any hypothesis or principal of philosophy, nor any discourse made for explicating any phenomena, except by the special appointment of the Society, or allowance of the President. But the time of the assembly is to be employed in proposing and making experiments, discoursing of the truth, manner, grounds, and use thereof, reading and discoursing upon letters, reports and other papers concerning philosophicall and mechanical matters, viewing and discoursing of curiosities of nature and art, and doing such other things as the Council or President shall appoint.

The Fellows of the Society had now been granted a Charter which gave them all that they required; they were authorized to make all such laws, statutes, acts and ordinances as might seem to them desirable or necessary for the better government of the Society so long as these 'be reasonable and not repugnant or contrary to the laws, customs or statutes of this our Realm of England'; the king had declared himself their patron, and they were assured of his favour towards them. The Fellows themselves could decide of how many the Fellowship should consist and could lay down what qualifications a candidate should possess to justify them in admitting him to their Society. The President and Council were given 'full authority power and faculty from time to time to draw up, constitute, ordain, make and establish such laws, statutes, acts, ordinances and constitutions as may seem to them, or to the major part of them, to be good, wholesome, useful, honourable and necessary, according to their sound discretions, for the better government, regulation and direction of the Royal Society'. This account of the Society's activities is an account of how the Fellows and the councillors have made use of the powers conferred upon them, how they have dealt with such difficulties as have arisen and how they have modified their by-laws and statutes from time to time as new conditions have arisen in the course of the past two hundred and eighty years.

From time to time it has been proposed that the Charter should be amended, or replaced by another differently worded, but careful consideration has always shown that the objects desired could be as effectively and more readily obtained by a revision of the statutes then existing, or by the enactment of new ones; such alterations it has always been within the powers of the Council to carry out.

It will be useful therefore to describe what the position of the Society was at the end of 1663 when it was about to make use of the very wide powers which had been conferred upon it. In August 1663 King Charles presented the Society with the Mace which is always placed before the President at each meeting of the Council or of the Society. The Council's Minute-book records that on 3 August 1663 the President (Lord Brouncker) informed the Society that Sir Gilbert Talbot, Master of the Jewell House, had sent to him, without taking any fees, the Mace bestowed by His Majesty on the Society, and that he, the President, had, in the book of

His Majesty's Jewell House, acknowledged the receipt thereof for the Society. The Mace is made of silver richly gilt, and weighs 190 oz. avoirdupois (150 oz. troy weight). It consists of a stem, handsomely chased, with a running pattern of the thistle terminated at the upper end by an urn-shaped head surmounted by a crown, ball and cross (*Record of the Royal Society*, 1940, Plate III). On the head are embossed figures of a rose, harp, thistle and fleur-de-lys, on each side of which are the letters C.R. Under the crown, and at the top of the head are the royal arms richly chased; and at the other end are two shields, one bearing the arms of the Society and the other the following inscription:

EX MUNIFICENTIA
AUGUSTISSIMI MONARCHAE
CAROLI II
DEI GRA. MAG. BRIT. FRANC. ET HIB.
REGIS ETC.
SOCIETATIS REGALIS AD SCIENTIAM
NATURALEM PROMOUENDA INSTITUTAE
FUNDATORIS ET PATRONI
AN. DNI. 1663.

For many years it was believed that this Mace was the one which was turned out of the House of Commons by Oliver Cromwell when he dissolved the Long Parliament. As this seemed to be highly improbable, Mr C. R. Weld, the Assistant Secretary, investigated the history of the Society's Mace in 1846, and was so fortunate as to find in the archives of the Lord Chamberlain's office the original warrant, ordering a Mace to be made for the Royal Society in the following terms:

A Warrant to prepare and deliver to the Rt. Hon. William Lord Viscount Brouncker, President of the Royall Society of London, for the improving of Natural Knowledge by experiment; one guilt Mace, of one hundred and fifty oz., being a gift from His Majesty to the said Society.

The warrant is among those issued in 1663; previous ones bear dates from January to April, and others entered subsequently are dated from May to July, so that the warrant which refers to the Society's Mace was issued in April or May which agrees with the Society having received the Mace from the Master of the Jewell House in August of the same year. The arms of the Society and the inscription were engraved on the Mace by the direction of the Society in 1663. In 1756 the President, Lord Macclesfield, being of the opinion that the Mace needed regilding, gave orders that this should be done at his expense; the work was carried out by Wyckes and Netherton, silversmiths of Panton Street. In 1828 it was again regilt and also in 1912.

On 30 November 1663 the Society held the first Anniversary meeting to elect the officers and members of the Council for the coming twelvemonth as required by the Charter; ten members of the old Council had to retire and were replaced; as the Treasurer appointed in the Charter, William Ball, did not seek re-election, his place was taken by Abraham Hill who held it for the next two years and again from 1679 to 1700. Lord Brouncker was re-elected President and Dr J. Wilkins and H. Oldenburg as Secretaries. The retiring Treasurer had to report that £158 was due to the Society from Fellows who had not paid their subscriptions; two years later arrears had risen to four times this amount, and it was many years before negligence on the part of its Fellows ceased to handicap the activity of the Society.

The definite instruction in the Charters, that the annual election of the Council and officers should take place on 30 November, St Andrew's Day, in each year, has given rise to speculation why that particular day should have been chosen; by some it has been taken for granted that the date was selected with some reference to Scotland and her patron saint. But there was perhaps no such intention; when the philosophers resumed their meetings after the restoration of the monarchy, the one at which the first steps were taken towards the formal organization of what became a year and a half later the Royal Society was held on 28 November 1660. That date might therefore be appropriately held to mark the anniversary of the Society, but as it was only two days from St Andrew's Day, which is an important date in the ecclesiastical year, we may suppose that 30 November was chosen as being more familiar and notable; once chosen, the belief that St Andrew was to be the patron saint of the Society would readily follow. The influence of that eminent Scotsman, Sir Robert Moray, who had been so active in procuring the grant of the Charters, may well have been contributory to the choice. Whether this suggested explanation is well founded or not, it was generally accepted at the time that the Royal Society had some special relation to the patron saint of Scotland, for on the first Anniversary after the granting of the Second Charter Evelyn made the following entry in his Diary: 'The first anniversary of our Society for the choice of new officers, according to the tenour of our patent and institution. It being St Andrew's Day, who was our patron, each Fellow wore a St Andrew's Cross of ribbon on the crown of his hat.' Samuel Pepys, in his Diary, refers to the practice of wearing a St Andrew's Cross in their hats at the Anniversary meeting in his day, and notes that he had to pay two shillings for his in 1668. This custom does not seem to have lasted for more than twenty years, and it is said by Aubrey to have fallen into disuse about 1683, when Robert Plot was one of the Secretaries. The antiquary, John Aubrey, who was an Original Fellow of the Society, has preserved the following note of a conversation with Sir William Petty, another Fellow:

I remember one St Andrew's Day—which is the day of the generall meeting of the Royall Society for annual elections—I says 'Methought 'twas not so well that we should pitch upon the Patron Saint of Scotland's Day; we should rather have taken St George or St Isidore (a philosopher who was canonized)'. 'No', says Sir William, 'I had rather have had it been on St Thomas's Day, for he would not believe till he had seen and putt his fingers into the holes; according to the motto *Nullius in Verba*.'

At this first Anniversary meeting fifty-seven or fifty-eight Fellows are said to have been present, almost half the membership of the Society at that time. Evelyn further notes that on this St Andrew's Day the Fellows dined together after the election at the Bull Head Inn, Cheapside, His Majesty having sent them a gift of venison. A. Hill, the new Treasurer, in a letter written about this time says that only nineteen were present on this day, but he may have been referring to the attendance at the dinner, though this number would have been unusually small for such an occasion.

Robert Hooke, in his Diary, says that the Anniversary dinners of the seven years 1673 to 1679 were held at Thomas Blagrove's Crown Tavern, Threadneedle Street, but at two of them he was not present. He adds the interesting information that the charge for each dinner in 1676 was 7s. 6d. and 5s. 6d., though he does not explain why two prices are quoted; in 1677 and 1679 the charge was 5s. per head, and the total reckoning in the former year amounted to £11. 15s., which would correspond to forty-seven diners, a likely number. Hooke was elected Secretary in November 1677, so his information is probably reliable. By dining together in the afternoon after the Anniversary meeting had been held and the officers and councillors for the coming twelvemonth had been elected, the Society observed on this day the old custom by which after the ordinary weekly meetings the Fellows and their friends adjourned to a tavern near by to dine together and to continue their discussions. These informal dinners, which date from about 1640–5, were attended more regularly when the Society had been established, and the one which was held on the day of the Anniversary meeting became at once the natural conclusion to that day's celebrations. Evelyn also refers to the Anniversary dinners which were held in 1680, 1683, 1688, 1693, and in 1696, but does not suggest that dinners were not held on the other anniversaries. In his account of the dinner of 1696 he says: 'after the Anniversary meeting and its elections we dined at Pontac's as usual'. This tavern was in Abchurch Lane in the city, and so was conveniently near Gresham College where the Society was still holding its meetings. The election of the officers and Council, the address delivered by the President and the Fellows' dinner afterwards thus became the normal items in the procedure at this annual festival.

A well-planned administration is the first need of any new institution if it is to deal efficiently with its finances and membership, to record the communications which it receives, and the correspondence which it maintains with other bodies of like interests. With aims so wide in scope as the 'Improvement of Natural Knowledge' the amount of routine work was bound to be large even in the Society's early days and to increase rapidly as its membership grew, but the provision of an adequate administrative staff to deal with it was for a good many years more than the Royal Society's meagre resources could afford. Its officers, as well as the salaried staff, were overworked, and it was not until the early part of the eighteenth century that a satisfactory system had been gradually built up.

Oldenburg had only been Secretary for two years when he protested against the amount of work which was required from him, and for which he should have received an honorarium of £40, but the amount was often less. He wrote:

The business of the Sec. of the R.S. He attends constantly the Meetings both of the Society and Council, noteth the observables said & done there, digesteth them in private, takes care to have them entered in the Journal and Register-books, reads over and corrects all entrys, sollicites the performances of taskes, recommended and undertaken, writes all letters abroad, and answers the returns made to them, entertaining a correspondence with at least fifty persons, employs a great deal of time, and takes much pains in satisfying forreign demands about philosophical matters, disperseth far and neare, store of directions and enquiries for the Society's purpose, and sees them recommended.

Query: Whether such a person ought to be left unassisted?

The Council granted Oldenburg the sum of £40 in December 1666, which shows that they recognized his protest to be reasonable; but it was not until 1669 that they approved an increase of £40 to his salary on the proposal of Robert Boyle. The Society's resources were evidently slender at this time and expenditure had to be kept within very narrow limits; the Council issued a letter to the public in 1667 asking for financial assistance, but nothing came of it. For two centuries two-thirds of the Fellows were men who had no scientific knowledge or any real interest in the advancement of science, so that for many years, in fact until after 1830, more than half the members of the Councils belonged to this category. It was left for the most part to the officers, if they were scientific men, to see that the claims of science were not overlooked.

THE OFFICERS

In the Charter it is laid down that the President, the Treasurer and the two Secretaries are the officers of the Society, and that they are to be elected by the Fellows at each Anniversary meeting when the Council for the coming twelvemonth is chosen. To them are entrusted the execution of the Society's policy and such action as may be decided upon by the Council from time to time, or by the Fellows at their meetings. They had therefore to keep in close touch with the current business of the Society, to report upon it to the Council and to assist that body in arriving at their decisions. The Council might delegate to them power to deal with various matters, and occasions arose from time to time when they had to act to the best of their own judgment, reporting to the Council at their next meeting how such situations had been dealt with. Seeing that until early in the nineteenth century the non-scientific Fellows constituted about two-thirds of the membership, it is not surprising that the officers were not selected solely from those who had scientific training and inclinations, but in part from those whose interests were mainly literary. The extent to which an officer could influence the Council's policy would depend on his personal zeal and energy, on his knowledge of the scientific aims of the Society and his interest in them; also on the energy with which he carried on the work and administration of the Society. Such matters are rarely touched upon in their biographies, and the Minute-books seldom provide any information about the men, their personal qualifications or their characters, but the interest taken by a President or by the Secretaries in the Society's work and advancement may be estimated from the number of Council meetings which were held and by the regularity with which the officers and councillors attended them. Whenever matters of exceptional importance had to be dealt with, such as the financial crises which occurred from time to time, the revision of statutes, or the change of the Society's headquarters from one house to another, the number of Council meetings increased notably and the attendance at them improved for the time being. Such information as we possess about the officers of the Royal Society during the first two centuries of its existence is not as full as could be wished and a good deal more may exist in letters, documents and diaries lying unstudied in libraries, institutions and not improbably in private possession also; if this could be made available, a much fuller and clearer picture of the Society's administration in the past would be at our disposal.

Biographies of many of the Presidents are available, but these usually deal more fully with their lives and the appointments which they held than with the influence which they exercised on the Society and its policy. Our information relating to the other officers is even more meagre, and this is much to be regretted since the Secretaries were more closely associated with the administration of the Society than either the President or the Treasurer, for they dealt with the correspondence which was received, and drafted or approved the replies which were sent; for this reason the knowledge which they possessed of the matters which came up for discussion by the Council was more complete and more informative than that which was in the possession of the other members, and their influence on its decisions would consequently be greater.

Whether the President was a scientific man or not he could from his position exercise very considerable influence on the decisions of the Council when he was present, and so we find that presidents who attended the meetings fairly regularly, and were capable administrators, were able to do much to influence the work of the Society in the way of gaining support for their projects or of improving the organization. The Society was exceptionally fortunate in having for its first President Lord Brouncker, who occupied that position from 1662 to 1677; he was not only recognized both in this country and abroad as an able mathematician, but was appointed Chancellor to Queen Catherine in 1662, and a Commissioner of the Admiralty in 1664; he was therefore in a position to advocate the claims of the Society at court and in official circles. He may not have had so much influence with the king as Sir Robert Moray or John Evelyn, but he devoted his energies to promoting the welfare of the young Society. From the first there was much to be done; statutes had to be drafted and adopted, new members enrolled and the business at the weekly meetings dealt with. In the twelve months between 1 December 1663 and 30 November 1664 the Council met thirty-six times and at all but one of these meetings the President presided, as he did also at most of the Ordinary meetings. During the fourteen years of his presidency he presided at 148 out of the 171 Council meetings which were held, a regularity of attendance which was only surpassed by Newton and Banks. Evelyn says in his Diary: 'Lord Brouncker, Boyle and Sir Robert Moray were above all others the persons to whom the world stands obliged for the promoting of that generous and real knowledge which gave the ferment which has ever since obtained, and surmounted all those many discouragements which it at first encountered.'

The Treasurer ranks next to the President in the Society and has since 1699 almost always been nominated by him as one of the Vice-Presidents. His duties are mainly concerned with the control and management of the Society's finances, but as he should be present at all the Council meetings he can, if he is energetic, render helpful service to the other officers and to the Council. He was charged with 'paying and disbursing all moneys, and with keeping particular Accompts of all such receipts and payments'. Sums under five pounds were to be paid by order of the President, and those above that amount by order of the Council. His duties at first cannot have been heavy, but the neglect of many Fellows to pay the fees and subscriptions due from them to the Society soon caused him much anxiety, and often made it very difficult for him to meet the Society's obligations, or even to pay the salaries of its staff. The Council lost no time in organizing the control of the Society's funds, and the first auditing committee was nominated on 11 November 1663 so that their report might be ready before the Anniversary meeting at the end of the month. The statutes required that 'the Treasurer's accounts should be audited and examined quarterly by a committee of five members of the Council and annually by five Fellows of the Society not members of the Council nominated by the President in accordance with the consent of the Fellows as given by ballot at an Ordinary meeting'. This statute remained in force until 1891 when the number of members of the auditing committee was reduced to three nominated by the Council and three others not of the Council who were nominated by an Ordinary meeting of the Fellows.

William Ball, the first Treasurer, was an astronomer and had been one of the philosophers who were at Oxford in 1658; on his retirement in favour of Abraham Hill in 1663 he presented the Society with an iron-bound wooden treasure-chest with three locks in which cash and documents of value could be kept. This is still at the rooms of the Society (see *Record of the Royal Society*, Plate XII). He also gave the sum of £100 towards the expenses of the Society. Hill, who succeeded him, was a Commissioner of trade in 1689; he was elected Treasurer in 1664 and 1665, and again from 1679 to 1700 when the Society's financial position had become very precarious. Colwall, who was Treasurer from 1665 to 1679, was a citizen of London and a benefactor of Christ's Hospital. He inaugurated the Society's Museum in 1668, having in 1666 contributed £100 towards it which the Council expended in the purchase of the collection of rarities which had been the property of Mr Hubbard. Some of the Fellows had presented the Society with a few objects of interest already, but as soon as its Museum was definitely established the number which were received increased very rapidly. The habit of collecting what were called 'rarities' was very fashionable at this time and various private collections were to be found, in different parts of the country, wherein, besides valuable and interesting specimens in natural history and antiquities, all sorts of curiosities, natural and artificial, were gathered together. There was at that time no public institution to which zoological, botanical, geological, or mineralogical specimens could be sent for examination or preservation. The Society, therefore, very properly undertook the task of collecting, arranging, and cataloguing specimens in all departments of natural science, doubtless in those days including much that might be curious, but had no real scientific value. Some care, however, was exercised to prevent the inclusion of useless or undesirable objects. In a letter from Oldenburg to Boyle of 18 January 1667/8, it is mentioned that persons, not Fellows, who desired to present specimens to the Society were obliged to show them first to the President 'for fear of lodging unknownly ballads and buffooneries in these scoffing times'.

In his history of the early years of the Royal Society, Sprat remarks: 'All places and corners are now busy and warm about this work; and we find many noble rarities to be every day given in, not only by the hands of learned and professed philosophers; but from the shops of mechanics, from the voyages of merchants, from the ploughs of husbandmen, from the sports, the fish-ponds, the parks, the gardens of gentlemen.'

But besides receiving the various objects that were presented to it, the Society at one time employed an emissary to travel over the country for the purpose of collecting other specimens, more particularly in natural history. Thus, in the Journal-book under date 21 October 1669 it is recorded that 'Thomas Willisell the botanick Traveller, entertained by the Society, brought in his collection of plants gathered in several parts of England and Scotland, together with some rare Scottish fowl and fish'. It is added that Dr Merret 'digested these plants', and Mr Hooke was instructed to 'take the whole collection into his custody, for the Repository, making first an Inventory of them and producing that before the Society'.

In less than twenty years the Repository had increased so much that a folio volume of nearly 500 pages was published as a catalogue and description of its contents by Nehemiah Grew, M.D., Fellow of the Royal Society and of the 'Colledge of Physitians: London, 1681'. Oldenburg was most active in explaining to his numerous correspondents that the Society would be grateful for specimens illustrating natural knowledge. Fellows of the Society residing abroad contributed many examples, and some, like the Winthrops of New England, sent considerable collections of botanical, mineralogical and zoological specimens.

In the statutes which were approved by the king and by the Council in January 1663 the Secretaries of the Society are charged with the care of its books, documents and correspondence. They are to attend all meetings of the Society and Council, and take notes of the orders and material passages of the meeting; they shall also draw up all letters written in the name of the Society. Their duties as laid down in these statutes remain in force to-day with but little change. It is worth noting that although the Philosophers' Society of 1660 had but one secretary, both the First and Second Charters give the Royal Society two, and those who were to serve until the next Anniversary meeting are mentioned in them by name; they were: Dr John Wilkins, D.D., and Mr Henry Oldenburg. The reason for the change is nowhere mentioned, but it seems possible that the draft of the Charter of Incorporation of the Society, with which the king had so closely identified himself by becoming its patron, intentionally followed to some extent the constitution of the Privy Council. Here, according to Professor G. N. Clark, 'the king was the guiding and driving power. A number of committees, some permanent, others temporary, were needed to deal with special matters and certain Ministers had acquired a special importance as links between the Privy Council and the various executive officers or bodies with whom it corresponded.... The chief co-ordinating Ministers were the two secretaries of state.... They had, besides their own executive work, extensive functions in setting the other departments of state in motion.' The Society's organization was much simpler, but as the years went by and it became better adapted to the work required of it, the provision of two Secretaries has proved to be most useful in dealing with the wide range of subjects to be considered. At first the division seems to have been a matter of private arrangement, but by the nineteenth century it had become customary for one of them to take the mathematical and physical business and the other to take the biological as is the practice to-day.

Whether Presidents were active in carrying out their duties or did no more than was absolutely necessary, the administrative work of the Secretaries was always heavy. Notes had to be taken of all that passed at the Council meetings and at the Ordinary meetings of the Society, and these had to be written out in the Council minutes and Journal-books by the Secretary or the Clerk; Council minutes were not as a rule lengthy, but the record of announcements made, of communications received and of the discussions which took place at the Ordinary meetings, often ran into many pages and must have taken a considerable time to copy. As the Society had no house of its own yet, the Council minutes and Journal-books, etc. were kept by the First Secretary at his own house so that they could be written up by him in his spare time; the Society's Clerk took them over from the Secretary before each meeting in case they should be required for reference and returned them to him afterwards. When a Secretary retired from office it was his duty to hand them over to his successor. Besides all this, the exchange of correspondence which was maintained between the Society and foreign men of science was considerable, and was of much value in maintaining contact with workers in other countries. The editing and publication of the *Philosophical Transactions*, when they were instituted in March 1664/5, also fell to the lot of one of the Secretaries. Apparently it was not easy at this time to find suitable men who were willing to undertake the onerous work attached to a secretaryship, and to hold the position for several years. Between 1663 and the end of the century fourteen Fellows were appointed Secretary; of these Oldenburg served for fourteen years, Gale for eight, Wilkins for six and Henshaw for four years; nine others served for one or two years only, and not until 1687, when R. Waller was appointed, and 1694, when Hans Sloane joined him, did longer terms of office become usual. The constant change of Secretaries after only two or three years' service, and before they could possibly have made themselves familiar with the needs and procedure of the Society, must have seriously hampered good administration, even though they had the assistance of a clerk, M. Wicks, who had been appointed in December 1663.

COUNCILS

Soon after the Anniversary meeting of 1663 the Council began their meetings. In accordance with the Charter ten of the councillors who had been appointed in the Charter were replaced by other Fellows, and of the total twenty-one nine were in this year men of science and twelve were not. There was much to be done and thirty-six meetings were held, at all of which except one Lord Brouncker, the President, was present. The average number of councillors who attended the meetings was 12·2. The number of Council meetings held in each year would depend on the amount of business there was to transact, and in the seven years 1664 to 1670 it varied from thirty-six to six, the average number being seventeen; for the same period the average number of scientific members was 9·2 out of twenty-one; and the average attendance of members at a meeting was 9·2. When the difficulties of communication and the fact that only a minority took any active interest in the aims of the Society are taken into account, the attendance was perhaps as good as could be expected, but the fact remains that the Society's progress was left to the efforts of the officers and a few councillors (see Appendix II C).

THE FELLOWSHIP

For many years the membership of the Royal Society was not limited to men of science; in the seventeenth century such men were comparatively few in number and many of these did not live in or near London. It was also clear that the Society in its early days was very unlikely to receive any financial assistance from the Crown or the State, so it was essential to secure the support of some of those holding high official positions if the young Society was to survive the difficulties which lay before it. The policy of recruiting a proportion of the Fellows from statesmen and from those of the educated and well-to-do classes of society who might be interested in, or at least well disposed towards the promotion of, the new learning was therefore adopted. Bishop Sprat states this very definitely when he says:

As for what belongs to the Members themselves they are to constitute the Society: it is to be noted that they have freely admitted Men of different Religions, Countries, and Professions of Life....But though the Society entertains very many men of particular Professions; yet the farr greater Number are Gentlemen, free and unconfin'd. By the help of this there was hopefull Provision made against two corruptions of Learning, which have been long complained of, but never removed: The one, that Knowledge still degenerates to consult present profit too soon; the other that Philosophers have bin always Masters and Scholars, some imposing, and all the others submitting; and not as equal observers without dependence.

Thus it came about that in the early days of the Society about one-third only of the Fellows consisted of scientific men of eminence and merit, the remainder being made up of those who might be interested in the new philosophy and its aims, but who did not devote themselves seriously to the advancement of natural knowledge. The Society therefore started with a membership which was composed in part of scientific men, and in part of those who may be taken as representing the ordinary intellectual and cultivated society of the day, but who for the most part were but little interested in scientific investigation. If we are to estimate correctly the influence which the Society and the social or industrial factors of the time exerted on one another, we should know the respective proportions of the scientific men and the non-scientific men who had been admitted to the Fellowship.

The number of Fellows who had been admitted up to 30 November 1663, according to the records of the Society, was 137, which includes the Original Fellows together with a few who had been elected in the autumn of that year. A manuscript in the British Museum gives the number of Fellows in the register of the Society at the Anniversary meeting on 30 November 1663 as 131, which included 18 noblemen, 22 baronets and knights, 47 esquires, 32 doctors, 2 bachelors of divinity, 2 masters of arts and 8 foreigners. For the first ten or fifteen years the number rose till it reached about 215, but by then the novelty of the movement was beginning to wear off, and political conditions were more disturbed, both of which factors affected the membership. At this time candidates could be elected at any meeting of the Society; on the average nine Fellows per annum were being added, but this was not enough to ensure a steady growth of the Society as resignations became more numerous. These were due in part to the discussions becoming more technical, but mostly to the disinclination or the inability of many to pay their annual subscriptions, and ten years later the total sum owed by Fellows to the Society had reached twelve times what it had been in 1663. Many years were to pass before the Treasurer was able to count with any certainty on receiving the greater part of what was due to the Society from the Fellows without unreasonable delay.

Lists of the Fellows of the Society were prepared for the Anniversary meetings, and copies of all these, with very few exceptions, are filed in the archives of the Society, but the profession or occupation of each Fellow is not given. It is therefore necessary to collect this information from other sources, and much of it is to be found in the *Dictionary of National Biography*. This has been extracted for the period 1663 to 1860 and has been supplemented by any additional details that could be gleaned from other sources. There were also a number for whom details are not available or are inadequate. The information about those who have been classed as scientific is more satisfactory, for in each case the profession or occupation of the Fellow is definitely stated (see Appendix II, A and B).

In 1664 a special register, the Charter-book, was provided and brought into use. It is a handsome volume bound in crimson velvet with gold clasps and corners; on one side there is a gold plate on which is engraved the shield of the Society, and on the other a similar plate showing the Society's crest—an eagle holding a shield bearing the three lions of England. It includes a number of sheets of fine vellum, on the first of which are emblazoned the arms of England, and on the next those of the Society; a copy of the Second Charter follows occupying seventeen pages, and one of the Third Charter which takes up sixteen pages. Statutes of various dates take up seventy-six pages and eleven more are blank. Then follow the autograph signatures of the Patrons and Fellows of the Society. Charles II signed his name on the first sheet on 9 January 1664/5, and James

Duke of York and Prince Rupert added theirs. Samuel Pepys writes in his Diary on this date: 'I saw the Royal Society bring their new book, wherein is nobly writ their Charter and Laws; and comes to be signed by the Duke as a Fellow; and all the Fellows' hands are to be entered there, and lie as a monument; and the King hath put his with the word Founder.' At the top of each page is inscribed the obligation which was adopted by the Council on 27 May 1663, and the signatures of the Fellows are written below it. This volume is still in use, and in 1912, on the occasion of the two hundred and fiftieth anniversary of the grant to the Society of its First Charter, a facsimile reproduction of the Charter-book was published. This was brought up to date in 1935 by the reproduction of fourteen additional pages bearing the signatures of those admitted between 1912 and 1935.

Before long several of our leading philosophers had felt the need of a clear and more precise style of expression than was usual at the time. However dignified the prose of Milton, Sir Thomas Browne and others might be there was a general desire among those who founded the Royal Society that simplicity, correctness, lucidity and precision in speech and writing should be cultivated. As early as 1646 Dr J. Wilkins had recommended in his book *Ecclesiastes, or the Gift of Preaching* that the style of preaching should be plain and without rhetorical flourishes, and he continually urged the philosophers to profit by his precepts. Sprat, whose prose style was highly commended by his contemporaries, insists in his *History* on the importance of this essential reform:

There is one thing more about which the Society has been most solicitous; and that is the manner of their Discourses: which unless they have been only watchful to keep in due temper, the whole spirit and vigour of their design had soon been eaten out by the luxury and redundance of speech....And in few words I dare say that, of all the studies of men, nothing may be sooner obtain'd than this vicious abundance of Phrase, this trick of Metaphor, this volubility of Tongue, which makes so great a noise in the world....It will suffice my present purpose to point out what has been done by the Royal Society towards the correcting of excesses in Natural Philosophy to which it is of all others a most profest enemy. They have therefore been most vigorous in putting into execution the only Remedy that can be found for this extravagance, and that has been a constant Resolution to reject all amplification, digressions, and swellings of style; to return to the primitive purity and shortness, when men deliver'd so many things almost in an equal number of words. They have exacted from all their members a close, naked, natural way of speaking, positive expressions, clear senses, a native easiness, bringing all things as near the Mathematical plainness as they can, and preferring the language of Artizans, Countrymen, and Merchants before that of Wits and Scholars.

Sir William Petty criticized a speaker at a meeting of the Society in 1683 for using the expression 'considerably bigger' as being too indefinite. We may see the change in progress if we compare Joseph Glanvill's earlier publication *The Vanity of Dogmatizing* with his later one of *Sadducismus Triumphatus* where his style conforms more closely to the ideal described by Sprat.

When so much stress was being laid on the need for a terse and precise prose style, it is not surprising that a special committee should have been appointed to study the matter. On 2 December 1664 'it being suggested that there were several persons of the society, whose genius was very proper and inclined to improve the English tongue, and particularly for philosophical purposes; it was voted that there be a committee for improving the English language; and that they meet at Sir Peter Wyche's lodgings in Grays Inn once or twice a month, and give an account of their proceedings to the society when called upon'. Twenty-one members besides Sir Peter Wyche, the chairman, were appointed; they included: Sir Robert Atkyns, a legal writer; John Dryden, the poet and playwright; John Evelyn; Bishop Sprat, the historian of the Society; Sir Robert Southwell, and Sir Joseph Williamson, Charles II's librarian who in 1665 founded the *Oxford Gazette* which in 1666 became the *London Gazette*. Nothing definite seems to have resulted from the consultations of this committee. The movement arose from the desire of the early Fellows to encourage clear and precise prose rather than from any wish to emulate the Académie française which had been founded in 1635 by Richelieu for the special purpose of safeguarding the purity of the French language. Seeing that for two centuries two-thirds of the Fellowship of the Royal Society were not scientifically minded, and that a considerable proportion of these were interested in literature, history, etc., it is strange that the attempt to develop the Society's activities in this direction was not repeated whether by forming a special branch of it for this purpose or in any other way.

Until 1664 the Society had taken no steps to make the work that it was doing more widely known. Sir Robert Moray, Robert Boyle, Oldenburg and some others of its Fellows corresponded with learned circles in France and elsewhere, so that news of what was being done in London was gradually spreading on the Continent. As early as September 1661 Sir Robert Moray, when writing to Huygens, describes his correspondence as making great demands on his time and energy; he anticipates that before long 'we shall print what passes among ourselves, at least everything that may be published. Then you shall have copies among the first, and if there is something withheld from publication, it will be much easier for me to communicate it to you than to have to send you word of everything by letter.' At that time the active members of the Society were too fully occupied in drafting the Charters and preparing the statutes for the Society's guidance as soon as it should be incorporated for them to take up the question of publishing the communications and reports which were being received.

In August 1664 M. de Sallo, a lawyer living in Paris, received authorization to publish a *Journal des Sçavans*. This was registered on 30 December 1664 and the first number was published in Paris on 5 January 1664/5. It was proposed to review in it books of importance which were printed in Europe; to print obituary notices of distinguished authors; and to communicate accounts of physical experiments which had been performed, newly invented machines, astronomical observations and medical matters. In February Sir Robert Moray wrote to Huygens to tell him of Oldenburg's plan for a scientific journal which would avoid the contentious matters of the kind that in March 1665 led to the suppression for the time being of the *Journal des Sçavans*:

As for the Gazette des Sçavans we have seen a sample of it, but already we have found things to criticize in it. You say very well that the thing can be useful provided it be not spoilt. Mr Oldenburg has shown us a sample of similar plan, much more philosophical, and we hope to get him to begin it, if it can be done. He will not interfere with legal or theological matters, but in addition to philosophical matters which come from abroad, he will publish the experiments, at least the most important, performed here.

The matter was discussed by the officers of the Society at the end of February and on 1 March 1664/5 the Council resolved

That the Philosophical Transactions, to be composed by Mr Oldenburg, be printed on the first Monday of every month, if he have sufficient matter for it, and that the tract be licensed under the Charter by the Council of the Society, being first reviewed by some members of the same; and that the President be now desired to license the first papers thereof, being written in four sheets in folio, to be printed by John Martyn and James Allestree,

the printers to the Society. The first number of the *Transactions* therefore appeared on Monday, 6 March 1664/5, and consisted of sixteen quarto pages, at the end of which are the words, 'Printed with license'. It opened with an introduction written by Oldenburg, which was followed by a list of the articles in it, namely:

An accompt of the Improvement of Optick Glasses at Rome: Of the Observation made in England of a Spot in one of the Belts of the Planet Jupiter: Of the Motion of the late Comet predicted: The heads of many new Observations and Experiments in order to an Experimental History of Cold, together with some thermometrical discourses and experiments: A relation of a very odd monstrous Calf: Of a peculiar Lead Ore in Germany very useful for essays: Of an Hungarian Bolus of the same effect with the Bolus Armenus: Of the new American Whale-fishing about the Bermudas: A Narrative concerning the success of the Pendulum-Watches at sea for the Longitude: and the grant of a Patent thereupon: A catalogue of the Philosophical Books published by Monsieur le Fermet, Connsellour at Toulouse, lately dead.

This table of contents shows the class of contribution which Oldenburg was able to obtain for his publication; two of them treated of physics, two of astronomy, two deal with zoology, one with metallurgy and one with navigation and the measurement of time. Oldenburg continued to edit and publish the *Philosophical Transactions* until his death in 1677; after that date they were published by Hooke, who replaced Oldenburg as Secretary, for the next few years under the title of 'Philosophical Collections'. In January 1683 the publication of the *Philosophical Transactions* was resumed, the 143rd number being published under the editorship of Dr Robert Plot, who replaced Hooke as Secretary; Plot was succeeded by Halley in 1686 on his appointment as Clerk.

In the same year that the *Philosophical Transactions* appeared the Plague broke out in London. Many of the Fellows left for Oxford and other places, and the weekly meetings had to be discontinued. This interfered greatly with the sale of the new periodical and Oldenburg wrote to Boyle in December saying that only three hundred copies had been sold which would scarcely pay for the paper; but the situation soon improved and the *Transactions* found a ready sale to men of science abroad. It was not long before some of them sent accounts of their own work and discoveries with a request that the Society would publish them. In 1669 M. Malpighi, Professor of Medicine at Messina, sent his 'Dissertatio Epistolica de Bombyce', and the Society accepted it for publication.

Those who were devoting their energies to organizing the new Society realized the importance of publishing as soon as possible an authoritative description of the Society and its aims in order to counteract the inaccurate and exaggerated reports which were being circulated. In 1664, therefore, the Council invited Dr Thomas Sprat (1635–1713) to undertake this task, and appointed a committee to select papers which would be suitable for inclusion as an appendix to the *History*. The choice was a good one, for Sprat was a well-educated man having been admitted a Scholar of Wadham College in 1651 and elected to a Fellowship in 1657. He was made Dean of Westminster in 1683 and Bishop of Rochester in 1684. He was distinguished in literary circles for his excellent prose style.

In his *History* he attributes the origin of the Royal Society to the meetings of the philosophers and others which were held at Wadham College under Dr J. Wilkins, the Warden, from 1649 onwards, being then unaware of the informal meetings which the philosophers had been holding at Gresham College from 1645 at least, as Dr J. Wallis who took part in them has recorded. Sprat was only sixteen years of age when he first went to Wadham College, and Dr J. Wallis was then Professor of Geometry at Oxford; but not until twelve years later was he invited to compile an account of the Society. His ignorance of the informal meetings at Gresham College which had been held twenty years before was quite reasonable under the circumstances. The *History* was published in 1667 and a second edition was printed in 1683, and others subsequently; it was widely read and circulated on the Continent both in the English editions and in translations. If it did not wholly check ill-natured and irresponsible criticisms and calumnies, it rendered the Society valuable and notable service by describing its constitution and the objects which its members sought to attain.

The foundation of the Royal Society was hailed as a notable event by some of the most eminent literary men of those days. Dryden extolled the achievements of Bacon and the work of his successors in the paths of experimental philosophy, and Cowley composed a laudatory ode on the Royal Society which appeared in 1667 as a prefix to Sprat's *History* of the Society. But these early appreciations were soon succeeded by criticisms of a wholly different tone. The general community was not yet prepared to welcome so novel an experiment as the association of a company of learned men not for the purpose of political intrigue or of literary or antiquarian co-operation, but of devoting themselves to the earnest investigation of nature. It is not surprising that the Society, which came so quickly into public notice, was supported by the king, and declined to be bound by the decree or injunction of any authority unable to produce convincing proof of the validity of its assertions, should be attacked by many who for one reason or another disliked its procedure. The Society had hardly taken definite shape before it was abused and disparaged; for about a century and a half it continued to be a mark for the shafts of ridicule launched by some of the foremost men of letters in each successive generation. Sprat's *History* which contained Cowley's complimentary poem also included evidence that already, within less than four years from its foundation, the institution of the Royal Society and the doings of its members had roused the antagonism of two classes of opponents. We learn that, on the one hand, 'some over-zealous Divines do reprobate Natural Philosophy as a carnal knowledge, and a too much minding worldly things,' while on the other side 'the men of the world and business esteem it meerly as an idle matter of fancy and as that which disables us from taking right measures in humane affairs'. 'The greatest part of men, if they can bring inquirers into Experimental Philosophy under the scornful titles of Philosophers, or Schollars or Virtuosi, it is enough: they presently conclude them to be men of another world, only fit companions for the shadow of their own melancholy whimsies.'

That the aims and pursuits of the Society should have been looked upon as tending to the subversion of religion seems strange when it is remembered that one of the most notable among the early Fellows was Robert Boyle, who, besides being the most illustrious physicist of his day, was a generous supporter of religious activity, distinguished for his piety and benevolence, for his active efforts to circulate translations of the Bible in the East, as well as in the colonies in New England, and for his institution of the Boyle lectures in defence of Christianity which are still delivered annually. He wrote a remarkable treatise which, under the title of 'The Christian Virtuoso', reveals the dignity and the magnanimity of his character; in it he remarks that some had thought it 'very strange that I, whom they are pleased to look upon as a diligent cultivator of experimental philosophy, should be a concerned embracer of the Christian religion'. He adduces many proofs of how much more the virtuosi see than others can 'of the diverse excellencies displayed in the fabric and conduct of the universe, and of the creatures it consists of'. He will not admit that his intercourse with men of science in any degree disposes him to atheism, and he thinks that there are not so many speculative atheists as men are wont to imagine. Having had a tolerably wide familiarity with naturalists, not only of this but of foreign countries, he declares that he has met with 'so few true atheists that I am very apt to think that men's want of due information or their uncharitable zeal has made them mistake or misrepresent many for deniers of God, that are thought such, chiefly because they take uncommon methods in studying His works, and have other sentiments of them than those of vulgar philosophers'. Notwithstanding Boyle's earnest and eloquent vindication, the charges against science and scientific men which he so well contested continued to be brought against them for many years. Depreciatory comments on the aims and objects of the Royal Society were heard even within the walls of the Universities. On 9 July 1669, at the Oxford Encaenia, as Evelyn records, 'Dr South, the University Orator, made an eloquent speech, which was very long, and not without some malicious and indecent reflections on the Royal Society, as underminers of the University, which was very foolish and untrue, as well as unseasonable'. There must have been many who listened with surprise to this attack, when they remembered the notable share that the Warden and some of the Fellows of Wadham College had taken only twenty years before in fostering science and promoting the study of the new philosophy.

It is not difficult to understand how easily this misconception arose and why it lasted so long. That a company of intelligent men should think it worth their while to devote themselves to enquiries into the most ordinary phenomena; that they should meet together to encourage each other in such a pursuit, and with infinite labour and at no small cost should organize experiments to prove what nobody cared about or thought of disputing; that they should give up valuable time to the study of such unattractive things as 'beasts, fishes, birds, snails, caterpillars, flies'; that they should collect and arrange all manner of 'curiosities' which were not worth house-room and appeared to have no practical use or sensible interest for anybody—all this seemed to be eccentric behaviour on which it was widely felt that no men with serious duties in life ought to waste their time. Moreover, the general name of 'virtuosi', which was then in common use, included not only true men of science sincerely anxious for the discovery of truth in every department of nature, but also men of culture and lovers of all manner of 'articles of vertu', as well as mere collectors who had a passion for gathering together whatever was ancient, uncommon or odd. The term was at first employed in a complimentary sense, but before long, on account of the vagaries of these indiscriminate collectors of 'rarities', it acquired a more or less contemptuous meaning. As it was not possible for the ignorant public to discriminate between the true seeker after science and the mere curio-hunter, the literary critic took advantage of his opportunity and classed the whole confraternity together, putting them all into one common pillory as objects of his sarcasm and ridicule.

Among the experiments conducted at the early meetings of the Royal Society were those shown by Hooke with the lately perfected air-pump, or what was called 'Mr Boyle's engine', whereby some fundamental laws in the physics of the atmosphere were demonstrated. To the ordinary mind, however, the occupation of weighing the air seemed incredibly futile, and appeared so even to Charles II, who, as Pepys records, 'mightily laughed at Gresham College for spending time only in weighing of ayre and doing nothing else since they sat'. When His Majesty, who took so friendly an interest in the Society's success, could not resist making fun of what had been only one of the numerous subjects that had engaged its attention, there were sure to be many others ready to ridicule the philosophers.

Robert Crosse, the Vicar of Chew Magna in Somerset, had maintained in a controversy with Joseph Glanvill in 1655 that the philosophers had done nothing to advance science, and that their store of knowledge was far less than that which Aristotle possessed. Glanvill answered him in his book *Plus Ultra*, declaring that 'the impertinent taunts of those who accused the Society of doing nothing to advance knowledge were no more to be regarded than the little chat of idiots and children'. Evelyn, in thanking Glanvill for a copy of his book, says: 'I do not conceive why the Royal Society should any more concern themselves for malicious and empty cavells of these delators after what you have said...the Society every day emerges, and her good genius will raise up one or other to judge and defend her.'

A certain Dr Henry Stubbs (1632–76), who is described by Anthony à Wood as being 'the most noted Latinist and Grecian of this age...and a singular mathematician', also attacked the Society vehemently, accusing its members of atheism, disingenuity, of attempting to displace the Universities as teaching bodies and of promoting revolutionary ideas—in short, of stirring up distrust in professional and official circles generally. His attacks were not due to any philosophical or religious differences of opinion, but rather to personal spite or self-interest. A manuscript written by the nephew of Dr Baldwin Hamey (1660–76), who was a talented physician and a benefactor of the college, has been preserved in the library of the Royal College of Physicians;[1] it records that Dr Hamey was very jealous of the Royal Society, which, he feared, would before long extend its activities to medicine, anatomy and surgery; these, he considered, should be more properly left to the College. The writer continues:

> Dr Hamey therefore found out a person of his own profession but a Country Practiser, one Dr Henry Stubbs, a man of as much Acrimony as Wit, with as knowing a head, as he had an able hand, and that wanted no ill nature to compleat ye Satirist in him, and this man he generously retayn'd for his Champion against the Royal Society; Stubbs then drew his Pen with great virulence, and lay'd about him most furiously indeed, and was well gratified by Dr Hamey for it, who meant onely to keep this Leviathan in its proper element, if he could have done it, tho ye attempt was vain and to no purpose.

Stubbs was educated at Westminster School where he attracted the notice of Sir Henry Vane who became his patron; in 1656 he went to Christ Church, Oxford, where his learning was highly esteemed, but he was expelled from there in 1659 for writing against the clergy and the Universities.

On the stage the Society and its Fellows were held up to ridicule by Shadwell in his comedy of *The Virtuoso* (1676). This dramatist, who, according to Dryden, 'never deviated into sense' must have perused with some diligence the early numbers of the *Transactions* in order to gather material for his farcical travesty. Samuel Butler indulged his caustic humour on the same subject, satirizing the Society in his *Elephant in the Moon*, and enumerating

> Their learned speculations
> And all their constant occupations,
> To measure wind, and weigh the air,
> And turn a circle to a square.

Among the wits of Queen Anne's reign it continued to be the practice to disparage the virtuosi in general and the Fellows of the Royal Society in particular. Addison, for instance, in the *Spectator* for 31 December 1711, wrote:

> Among those advantages which the public may reap from this paper, it is not the least that it draws men's minds off from the bitterness of party, and furnishes them with subjects of discourse that may be treated without warmth or passion. This is said to have been the first design of those gentlemen who set on foot the Royal Society; and had then a very good effect, as it turned many of the greatest geniuses of that age to the dispositions of natural knowledge, who, if they had engaged in politics with the same parts and application might have set their country in a flame. The air-pump, the barometer, the quadrant, and the like inventions were thrown out to those busy spirits as tubs and barrels are to a whale, that he may let the ship sail on without disturbance while he diverts himself with those innocent amusements.

[1] Quoted by permission of the Council of the College.

As we have seen, 'those busy spirits' retired of their own accord from the political troubles of the time to enjoy a freer and calmer air in the study of nature. In another paper, the class of men who had nothing to do is said to include 'all contemplative tradesmen, titular physicians, Fellows of the Royal Society, Templars that are not given to be contentious, and statesmen that are out of business' (*Spectator*, 12 March 1710/11).

There was sometimes a tone of singular bitterness in the invective, as in a paper in the *Tatler* (No. 236, 12 October 1710), which has been attributed to Steele:

There is no study more becoming a rational creature than that of Natural Philosophy; but, as several of our modern virtuosi manage it, their speculations do not so much tend to open and enlarge the mind as to contract and fix it upon trifles. This in England is in a great measure owing to the worthy elections that are so frequently made in our Royal Society. They seem to be in a confederacy against men of polite genius, noble thought and diffusive learning; and choose into their assemblies such as have no pretence to wisdom, but want of wit, or to natural knowledge, but ignorance of everything else. I have made some observations in this matter so long, that when I meet with a young fellow that is an humble admirer of these sciences, but more dull than the rest of the company, I conclude him to be a Fellow of the Royal Society.

Swift, at greater length and with more laboured sarcasm, caricatured the philosophers in his *Voyage to Laputa*. Pope, too, assailed them, but with a lighter touch. Assembling them to receive 'titles and degrees' from the Queen of Dulness, he placed them in her presence among the

> More distinguished sort
> Who study Shakespeare at the Inns of Court,
> Impale a Glow-worm, or Vertu profess,
> Shine in the dignity of F.R.S.   (*Dunciad*, Bk IV, 567.)

Early in 1664 the Council turned their attention to their scientific responsibilities; administrative matters still took up much of their time and the small proportion of scientific men among their members could not provide the expert knowledge needed to deal with the wide range of problems which in science and technology came up for consideration; they therefore adopted the very natural solution of appointing committees each of which would deal with a special field of enquiry, and this was done in March when eight of them were formed, namely:

i. Mechanical (69 members)

ii. Astronomical and Optical (15 members)

iii. Anatomical (with Boyle, Hooke, Wilkins and all the physicians of the Society)

iv. Chemical (with all the physicians of the Society and seven other Fellows as members)

v. Georgical (Husbandry) (32 members)

vi. Histories of Trades (35 members)

vii. Natural Phenomena (21 members)

viii. Correspondence (20 members)

One result of this activity was that the Fellows were urged to plant potatoes as being a valuable food crop, and to induce their friends to do the same.

These committees made their own arrangements to carry on their work and from time to time sent their reports to the Society; a number of these are preserved in the archives. The Husbandry (Georgical) Committee was one of the most active and R. V. Lennard has recently discussed its operations.[1] They had correspondents in many parts of the country from whom they collected information about local practices of cropping, working the land, grazing and much else. What remains of the other reports has not yet been as critically examined. To provide the number of members allotted to these committees was a task of great difficulty when scientific Fellows were few in number, and frequent meetings of the committees must have been impracticable; they are not often referred to and in 1680 only three of them, the anatomical, cosmographical, and georgical, were re-appointed.

In 1665 and the following year the life of the population of London was seriously disorganized by two great calamities; the first was the Plague which became serious early in the summer of 1665 with the result that many of the Fellows retired into the country, being 'exhorted by the President to bear in mind the several tasks laid upon them, that they might give a good account of them on their return'. Oldenburg, however, the Secretary, remained at his house in Pall Mall during the whole period, and after cataloguing and packing up the Society's papers and books so that they could be removed if necessary, carried on a correspondence with Boyle and others on scientific matters. Several of the Fellows found themselves at Oxford where Boyle was still living, and they at once made it their practice to meet at his lodgings where discussions were held and experiments were carried out: he mentions Sir R. Moray, Sir P. Neile, Sir W. Petty, Dr Wallis, Dr Coke, Captain Graunt and Mr Williamson as being of their number, so that the Society's continuity was well maintained. The seventh and eighth numbers of the *Philosophical Transactions* had to be printed at Oxford in consequence of the lack of printers remaining in London. Wilkins, Petty and Hooke stayed with John Evelyn at Durdans and there carried out experiments which would otherwise have been made in London. By the following February a sufficient number of Fellows had returned to London for the meetings of Council to be resumed, and Hooke in a letter to Boyle, speaking hopefully of prosecuting experiments and observations rigorously and perhaps of setting up an observatory equipped with instruments, looks forward to an early resumption of the Ordinary meetings; in fact the first one was held on 14 March.

At a meeting held three weeks earlier the Council had decided to use Mr Colwall's gift of £100 in purchasing Mr Hubbard's collection of

[1] *Economic History Review*, vol. IV (1931).

rarities which was to form the basis of the Society's own Museum. This was for many years the only important institution of the kind in London, where it attracted much attention. The number of exhibits increased rapidly by gifts, including some important collections sent by members of the Winthrop family from the New England colonies, by objects obtained from abroad, and by what the Society's own collectors were able to acquire. Its arrangement in a hall at Gresham College and the elimination of faked or worthless objects occupied much of the Society's time and attention.

The Anniversary meeting which should have been held on 30 November 1665 was postponed on account of the Plague until the following April, when the Treasurer had to report that arrears of subscriptions had increased to £678.

Four months later the Great Fire of London, which destroyed a large part of the city, again interrupted the Society's work, and though it did not extend so far as Gresham College or beyond the end of Bishopsgate Street, the College buildings were required for the use of the Lord Mayor and of the merchants. For a while therefore the Society held its meetings at Dr W. Pope's lodgings in the College, but very soon Mr H. Howard, who later became the sixth Duke of Norfolk, came to the Society's rescue and offered to provide the necessary accommodation in Arundel House. This was gratefully accepted, and on 19 September the Council recorded their thanks to Mr Howard for 'his great respect and civility to the Society'. At the meeting of the Society held on the same day Hooke showed a model of his design for rebuilding the burnt area of the city which it was decided to submit to His Majesty.

James Gregory, a brilliant mathematician who became Professor of Mathematics at St Andrews in 1668 and later at Edinburgh, published his book *Optica Promota* in 1663, in which he described his well-known reflecting telescope. He never constructed it, but in the following year visited London to discuss the matter with Rieve, a well-known instrument maker. For technical reasons the requisite parabolic object-mirror could not be satisfactorily made so Gregory gave it up and went abroad to Italy. He was elected a Fellow of the Society in June 1668, and died at Edinburgh in 1675.

On 9 January 1667 the Society's meetings were first held at Arundel House and the President then availed himself of the opportunity to thank their benefactor for his hospitality as well as for the gift of the greater portion of the library which his grandfather and earlier ancestors had collected. This he had done at the instigation of John Evelyn who, seeing that many persons were availing themselves of the opportunity to take whatever they fancied, urged on the owner the claims of the Society. The collection which was thus acquired was a most valuable one and included 3287 printed books and 544 volumes of manuscripts; a committee was appointed at once to prepare a complete list of all that the gift included, but it was not until 1678 that the books, etc. could be transferred to the rooms which the Society was occupying in Gresham College.

It was at this time that the Council began to discuss the provision of a suitable place for the Society's meetings in the future and a committee was appointed to consider what might be done. The Council had already realized that Gresham College would not be available for them indefinitely, nor were its buildings wholly suitable; some of the Fellows were already complaining that Gresham College was too far from their places of residence. Arundel House solved the question for the time being, and there was now a good prospect of Chelsea Hospital and the lands adjoining being transferred to the Society. Nothing could be done at the time but the question often came before the Council in subsequent years and, as will be seen, before long the trustees of Gresham College themselves began to impress on the Society the desirability of finding other quarters. A suitable house was not acquired until 1710.

The arrest and committal of Oldenburg to the Tower in the course of the summer of 1667 gave the Council an unpleasant shock, since they were authorized by the Charter to correspond with foreigners on philosophical matters. No details are given, but it is possible that the letters complained of had not been licensed by the President, or that the subjects treated of in the letters, went, in the opinion of the authorities, beyond what was permissible under the Charter. He was liberated after two and a half months' detention in the Tower.

In the autumn of this year Evelyn received orders to hand over Chelsea Hospital to the Society, but though the buildings were in a very dilapidated condition no repairs could be carried out until the Third Charter making the grant of them to the Society had passed the Great Seal; this was not until April 1669. The Council however resolved that subscriptions should be invited to a fund for building a college 'as the most probable way of the Society's establishment'. A committee which was appointed to solicit contributions to the building fund made every effort to obtain donations from members of both houses of Parliament and wealthy landowners who were Fellows of the Society, but with very little success. Oldenburg wrote to Boyle and many others, and the President used his influence with high officials who might be in a position to assist the Society in one way or another. Mr Henry Howard gave the Society 4000 square feet of ground in Arundel Gardens for a site, and he as well as Christopher Wren and Robert Hooke prepared designs for the building. There is in the Society's archives a register entitled 'Contributions towards Building the College' in which the names of Fellows who contributed are entered as well as the amount given; the total sum collected amounted to £1075 contributed by about twenty-five Fellows out of the total Fellowship of about two hundred, which was but a meagre response to the Council's appeal. Pepys writes in his Diary on 8 April 1668: 'With Lord Brouncker to the Royall Society when they had just done: but I was forced to sub-scribe to the building of a college and did give £40; several others did subscribe, some greater and some less sums; but several I saw hang off; and I doubt it will spoil the Society, for it breeds faction and ill-will, and becomes burdensome to some that cannot or would not do it.' It had been decided that when £1000 had been collected the building should be put in hand, but obstacles arose regarding the conveyance of the land, and on 10 August 1668 the operations were postponed until the following spring.

The Society lost no time in taking up the question of Chelsea Hospital and the lands belonging to it, which the king had granted to the Society, but many difficulties were met with. Not only were the buildings in a bad state of repair, but various parties claimed rights over certain parts of the estate, and lengthy legal disputes seemed to be inevitable. The committee charged with investigating these matters met frequently and at last agreed to a proposal by John Evelyn that it should be let for £100 per annum as a prison-house during the war; there seemed also to be some prospect of a Fellow of the Society being willing to undertake the management of the house and to plant and cultivate the lands belonging to it. But neither of these schemes came to anything, nor were the many attempts which were made to find a tenant any more successful. It therefore remained in the hands of the Society as an unprofitable burden until, as explained later, the property was surrendered to the king at his desire in 1682 on the pay-ment of £1300 to the Society as compensation.

Those who were occupied in advancing knowledge by research in such subjects as made a special appeal to them, were much disturbed in these times by the possibility of their results being appropriated by other workers, or by accusations which might be made against them of appro-priating without acknowledgment the work of others. To guard against such charges some recorded their proofs and their results in cipher, others made use of anagrams, or deposited copies of their work in places where they could not be consulted without their knowledge and permission. The matter was brought before Council on 16 November 1667 when

Mention being made, that a security might be provided for such inventions or notions, as ingenious persons might have, and desired to secure them from usurpation, or from being excluded as having a share in them, if they should be lighted upon by others; it was thought good, if anything of that nature should be brought in, and desired to be lodged with the Society, that, if the authors were not of their body, they should be obliged to shew it first to the President, and then it should be sealed up both by the small seal of the Society, and by the seal of the proposer; but if they were of the Society, then they should not be obliged to shew it first to the President, but only to declare to him the general heads of the matter, and it then should be sealed up as mentioned before.

At a meeting of the Society on 6 February 1678 Mr Oldenburg proposed a paper from Mr R. Boyle, sealed up, which had been sent to him to be deposited with the Society, containing some notion or invention of Mr Boyle not yet perfected. Mr Oldenburg was desired to deliver it to the President, that he might lay it up according to an order made by Council on 16 November 1667 concerning the depositing of such papers with the Society. On 18 February Oldenburg, in a letter to Boyle, mentions that the President not being at the meeting of the Society on the 13th Mr Boyle's paper was ordered to be delivered to him, and that the President on receiving it, two to three days after, put it in Mr Oldenburg's presence in a box by itself, after he had written on it the day and year of his receiving it, as Mr Oldenburg the same minute it came to his hands. In the same way Boyle deposited sealed packets with the Royal Society in 1668, 1680, 1683 and 1684; these were opened at his death in 1692 and handed to his legal representatives. The continuance of this practice in the eighteenth century is uncertain, for no references to it occur, but it was employed in the early part of the nineteenth when the acceptance of such manuscripts in 1834 is recorded; for some years after this the number of documents which had been committed to the Society's custody was reported by the Treasurer to the Council from time to time.

In 1668 two important changes took place among the leading men of the Society. Boyle left Oxford and took up his residence in London with his sister, Lady Ranelagh, at her house in Pall Mall, where he remained for the rest of his life. His health had never been good and his chemical studies were often as much as he could cope with; his enthusiasm for the advancement of religion and the distribution in many countries of translations of the Bible took up much of his time and energy. It was not to be expected, therefore, that he could do much towards the administration of the Society. He did however serve on the Council in 1662, 1663, 1665–6 and 1667–8 while he was living at Oxford; and after he came to London he was elected a councillor in 1671–2, 1673–4 and 1680–1. After this he spent more of his time than before at home among his books and on his personal interests. His advice was often sought by his colleagues and was highly esteemed, so much so that Evelyn placed him among the three Fellows of the Society who had, in his opinion, done most for it. This change of residence from Oxford to London brought him into closer contact with the Society, which was able to consult him and thereby to benefit by his advice and assistance in the years to come when it was to be sorely in need of them.

The other was the appointment of Dr J. Wilkins, one of the Secretaries of the Society, to the bishopric of Chester. From the time he joined the group of philosophers at Gresham College, about 1645, and later as Warden of Wadham College he had done all in his power to advance their aims, and this they had recognized by appointing him their Chairman when the Philosophers' Society was formed on 28 November 1660. A man of great administrative ability, moderation and geniality he rendered them valuable service, and was appointed a Secretary of the Royal Society in the Charters of 1662 and 1663. Oldenburg, the other Secretary, a good linguist and a man of method, took charge of the correspondence, the recording of the business transacted at the various meetings and of the publication of the *Philosophical Transactions* of the Society, but Wilkins provided much of the energy and driving power which was so urgently needed by an institution which had to make its way in the face of many difficulties and much opposition, as well as to establish a reputation for the wise accomplishment of its aims. Mr T. Henshaw, a barrister and historian, was elected Secretary in Dr Wilkins' place. He had been a member of the Philosophers' Society from February 1661 and served on the Society's Council in 1662 and 1663. He was sent to Denmark as Envoy Extraordinary 1672-5 but was again on the Council in 1675, being a Secretary in the next two years, 1676 and 1677.

It will now be of interest to turn to France and see what progress was being made there with the promotion of experimental research during these ten years. At the end of 1659 Huygens visited Paris and attended meetings of the Montmor Academy which was then the principal meeting place of the learned men in the capital. He notes that there were few good telescopes there and makes the comment that exact observation was not popular. He left for London in the following spring where he attended assemblies of the philosophers for experiments at Gresham College and saw experiments illustrating papers by Goddard, Lord Brouncker and others. Huygens' accounts of his visit to London and his conversations there with those whom he had met, as coming from an impartial observer, had much effect in Paris; and so did the impressions which he had gained of the support which the Royal Society was receiving from the nobility and wealthy classes, and also from ministers of State, though he greatly overstated this.

Thévenot and Petit were among the most active in working for the improvement of the conditions of science in Paris, and early in 1663 steps were being taken to reorganize the Montmor Academy. Sorbière delivered an address on the Academy and its aims, as well as the improvements which might be introduced. A copy of this, with a letter, was sent to Colbert, Louis XIV's minister, asking for his protection and encouragement for the scheme. Towards the end of 1665 Sir Robert Moray wrote to Oldenburg that 'Colbert intends to sett up a Society lyke ours and make Huygens Director of the design'; and about the middle of the following year Colbert's plans began to take more definite shape. Justel, on 26 May, wrote to Oldenburg saying that steps were being taken to establish an academy composed of men selected from all sorts of professions; Huygens and Auzout were mentioned as being probable members. By the end of 1666 the Académie des Sciences had been formed and began its regular meetings in December.

We have already traced the various attempts made by men of learning in Paris to organize the study of natural philosophy on an effective and permanent basis, but they had not achieved any permanent success although much useful work had been done by the academies and bureaux of Dupuy, Renaudot, Mersenne and Montmor. Though the Montmor Academy included among its members men of brilliant attainments, its meetings were marred at times by quarrels and disagreements which were not limited to the subjects under discussion but included personal insults and discourteous behaviour. Steps had to be taken to put an end to this bickering if the Academy's work was to continue, and in February 1658 Sorbière, the secretary of the Academy, sent to Hobbes a draft of some new regulations which had been drawn up by him and du Prat at the request of Montmor. They were far more formal than the unwritten code with which the Club of the Philosophers had successfully carried on their meetings for some twenty years without any difficulties being experienced, and out of them the statutes of the Royal Society had developed in due course. In France, however, there was no one who exercised sufficient authority to maintain the necessary discipline, and before long the discussions relapsed into their former ineffectiveness; both Sorbière and Thévenot did what they could to control them, but they failed.

In June 1664 Huygens, in a letter to Sir Robert Moray, says that the Montmor Academy had come to an end, but that some of his friends in Paris cherished hopes that its place might be taken by some other institution. Appeals to the minister Colbert for support and guidance were unavailing, and nothing was done until the Abbé d'Aubignac petitioned the king to countenance an organization of scientific men which would assuredly, if properly administered, be of the greatest benefit to the State. Auzout, a mathematician, in a letter to the king which was printed in his *Ephémérides du Comète de* 1664, followed a similar line in praying for an exercise of the royal patronage to this end, and emphasizing the benefits which would certainly result from it. Auzout had prepared a draft scheme for his project from which he prudently excluded all discussion of religion and State affairs just as the pioneers of the Royal Society had done twenty years earlier. The political situation had improved, and in the following year Colbert consulted an advisory committee on the subject; soon after he definitely committed himself to some such scheme.

Huygens went to Paris in 1666, but it was not until June that Colbert founded the Académie des Sciences, four years after the Royal Society had received its First Charter from Charles II. By the end of the year the Académie and the Royal Society were exchanging correspondence, and Auzout and Oldenburg were writing freely to one another.

An observatory was built as the headquarters of the members of the Académie, salaries were provided for them, and funds were forthcoming to meet the expenses which were necessarily incurred in carrying on their researches and in making their experiments. In this respect they enjoyed advantages which the Fellows of the Royal Society lacked but were unable to obtain. The latter however possessed the priceless boon of complete independence; they were responsible only to their colleagues, no minister could direct their work or control their discussions. The price which they paid for their freedom may, in the years of their financial difficulties, have seemed to them to be high, but the worth of it has been fully proved. Even in comparatively recent times the Royal Society has been able to take the lead in promoting innovations in international scientific relationships and has even been pressed by corresponding institutions in other countries to do so, because it was free from all State control.

Experiments on the transfusion of blood which had been made by Dr Lower of Christ Church, Oxford, in 1667 were resumed in 1668 with some success but, when death had resulted in several cases from experiments of a similar kind made in some other countries, public opinion turned against the practice and the Society did not pursue it further.

In October 1669 Dr J. Wilkins, now Bishop of Chester, told the Society that the king had expressed the wish that the measurement of a degree of latitude should be carried out in this country, he being desirous of emulating the work of this character which had recently been carried out in France under Cassini, the astronomer, with the support of Louis XIV, and in this he desired the assistance of the Society. He did not however offer to provide the necessary funds, and as the Treasurer's balance this year was but £70 and the arrears of unpaid subscriptions were steadily mounting up nothing could be done.

During the ten years which had passed since the philosophers first decided to form a Society for the improving of Natural Knowledge much had been accomplished: the king's favour had been gained; a Charter had been conferred on them which granted to them almost complete freedom in the management of their own affairs; the membership had already increased to two hundred Fellows; arrangements had been made for the regular publication of their proceedings which were already highly esteemed by scientific men in all parts of the world.

There were however two serious obstacles to the Society's advancement which were to hinder it for many years. The first of these was a lack of funds; no subvention was to be obtained from the State, and the contributions which Bishop Sprat hoped that the Society would receive from its Fellows when they saw how successful it was in achieving its aims amounted to very little. This absence of gifts and bequests was aggravated by the neglect of many of the Fellows to pay their subscriptions to the Society regularly. Even before the granting of the First Charter these arrears were mounting up and efforts were made in March 1662 by the Treasurer to collect them. At the first Anniversary meeting on 30 November 1663 the Treasurer had to report that the sum of £158 was owing to the Society from the Fellows and this amount continued to increase steadily until by the end of 1670 it had risen to £1475. From time to time persistent defaulters were removed from the list of Fellows but the practice continued, and it was not until about a century later that the yearly total of unpaid subscriptions ceased to cause anxiety to the Treasurer of the day. The other obstruction which proved to be far more difficult to overcome was the existence of a majority of the Fellowship made up of men who had no knowledge of scientific matters nor any interest in them, such as Bishop Sprat described in his classification of the members as 'gentlemen free and unconfined'. This large group, which included about two-thirds of the Fellows and was represented by more than half the members on the Council, could not but hinder the Society's activity in the promotion of science. Two centuries were to pass before this difficulty was successfully overcome.

REFERENCES

BIRCH, T.   *History of the Royal Society*. London, 1660–1687.

BROWN, Professor H.   *Scientific Organizations in Seventeenth-century France*. Baltimore, 1934.

BRYANT, A.   *Samuel Pepys: The Years of Peril*. 1935.

FITZMAURICE, Lord EDWARD.   *Life of Sir William Petty*. London, 1895.

GEORGE, A. J.   'The Genesis of the Academy of Sciences, Paris.' *Annals of Science*. Vol. III. 1938.

HOOKE, R.   *Diary*. Edited by W. H. ROBINSON and W. ADAMS.

MASSON, F.   *Robert Boyle*. London, 1914.

PEPYS, S.   *Diary*. 1660–1670.

ROBERTSON, A. J.   *Life of Sir Robert Moray*. 1922.

Royal Society, Notes and Records of. 1940.

Royal Society, Record of. 4th ed. 1940.

SCOTT, J. J.   *The Mathematical Work of John Wallis*. 1938.

SPRAT, T.   *History of the Royal Society*. London, 1667.

TURNBULL, Professor H. W.   *Tercentenary Memorial Volume on James Gregory*. Edinburgh.

WELD, J. C.   *History of the Royal Society*. Vol. I. London, 1848.

CHAPTER III

DIFFICULT YEARS: 1671–1700

URING the first ten years of the Society's existence much had been accomplished. An effective administration had been evolved, the membership had risen to about 200, and the more important of the communications which were received were being published regularly in the *Philosophical Transactions*, which was widely recognized as being the most important scientific periodical of those days. Most of those pioneers to whose zeal and untiring energy the formation of the Society had been due were still in its ranks, and their influence did much to maintain its reputation. The next thirty years were to be the most critical in its history. The king's interest in it was beginning to wane, the nobility and officers of State did not understand the Society's aims and methods, wealthy men who had hoped for its assistance in developing their properties and under-takings were disappointed that results of value were not produced as rapidly as they had expected, those to whom Bishop Sprat had looked for generous support had done nothing. After 1675 the number of Fellows began to decrease until by 1693 there were only 113 Fellows left; 194 candidates were elected in the last thirty years of the century, besides a few foreign members, but in spite of this the number in 1698 was more than sixty less than it had been in 1671; the meetings were often poorly at-tended, and in June 1680 Evelyn was writing to Samuel Pepys begging him for 'one half-hour of your presence and assistance toward the most material concern of a Society which ought not to be dissolved for want of an redress....I do assure you we shall want one of your courage and address to encourage and carry on this affair. You know we do not usually fall on business till pretty late in expectation of a fuller company, and therefore if you decently could fall in amongst us by 6 or 7 it would, I am sure, infinitely oblige...the whole Society'.[1] The arrears of subscriptions due to the Society from its Fellows exceeded £1600 in 1671 and there seemed to be little likelihood of their ever being paid. Robert Hooke in his Diary notes that at the Anniversary meeting of 1676 no audit of the accounts or any report on the finances was presented, and none is recorded in the minutes of Council after this year until 1716; by 1685 the names of more than sixty Fellows had been removed from the Society's list for having been in default for several years in the payment of their subscriptions; the position could hardly have been worse. The Society's funds did not even suffice for purchasing the instruments and apparatus which were needed

[1] A. Bryant, *Samuel Pepys: The Years of Peril*, p. 337. Cambridge, 1935.

for experiments, and only Robert Boyle's generosity in lending those which he possessed enabled this branch of the Society's work to continue.

Men of resource and ability were not wanting, and hitherto the Society had been very successful in finding among its Fellows men of exceptional administrative skill, but within the next seven years Lord Brouncker retired from the office of President and Dr J. Wilkins, Sir Robert Moray and Henry Oldenburg died; such men could ill be spared. The effect of the loss of these men, and of others whom the Society was to lose before long, inevitably affected the efficiency of the young Society and increased the difficulties caused by the financial stringency. The activity of the Council, which was so prominent a feature for the first few years, was now becoming less marked, and for this the frequent changes of the officers occupying the various administrative posts must have been largely responsible. The uncertainty of the political situation, and the increasing difficulty of the Society's financial position had a considerable effect, but the absence of experienced officials to direct the administration for periods of several years without being replaced or superseded greatly hindered the development of administrative efficiency.

Various factors may influence the development of institutions and among those which have had most effect on the Royal Society during its long history have been: firstly, the zeal and energy shown by the officers and the Councils; secondly, the number and the composition of the Fellowship; and thirdly, its financial resources. These have always been operative though the effects of each have varied from time to time; their influence can therefore be best estimated by examining them over a period of years, and that between 1660 and 1700 is specially suitable for such an analysis since in the course of it the Society at first increased considerably in numbers and influence, but later a decline both in membership and resources followed which lasted until the closing years of the century.

THE OFFICERS

The duties which fell to the officers of the Society to carry out were arduous, and occupied much of their spare time, for they included not only the supervision of the administrative staff, the execution of resolutions adopted by the Council and decisions taken by the Society at its meetings, but also keeping a careful record of all this in the journals and registers of the Society. The Society's correspondence both at home and abroad was increasing and the regular publication of the *Philosophical Transactions* had added very considerably to the Secretaries' work. There were also the weekly meetings of the Council and the Society to be attended and the minutes of their business to be recorded. It is not surprising that Oldenburg should have protested in 1664 at receiving no assistance in his task after two years' experience of what was expected from him. For all this work to be correctly and efficiently carried out it was essential that officers should hold their posts sufficiently long to gain a thorough knowledge of their duties, but during the twenty-six years which elapsed between the retirement of Lord Brouncker in 1677 and the election of Sir Isaac Newton in 1703 the office of President was held by ten different men and only one of these, Sir Christopher Wren, was a scientific man by training and profession; the others included Sir Joseph Williamson, Sir John Hoskins, Sir Cyril Wyche, Samuel Pepys, John, Earl of Carbery, Thomas, Earl of Pembroke, Sir Robert Southwell, Charles Montagu, afterwards Earl of Halifax, and John, Lord Somers, all of whom were eminent statesmen or administrators but knew little or nothing of any branch of Natural Knowledge, and consequently have left little mark on the history of the Society.

At this time the Secretaries received no fixed allowance to recompense them for devoting a large portion of their spare time to the Society's work; but they were supposed to receive an honorarium which was voted annually by the Council, and which might be £60 or some lesser sum. It was not until 1720 that the Council, on a motion by the President, Sir Isaac Newton, fixed the annual remuneration of each of the two Secretaries at £50. Under such conditions it is not surprising that the Council should have found difficulty in securing suitable men to undertake so arduous and thankless a task; in 1663 there were about forty-four scientific Fellows in the Society, in 1671 there were about forty-seven and in 1698 not more than thirty-six, so it was unlikely that a suitable officer from these groups, and living in or near London, would always be forthcoming. In the forty years which elapsed between the beginning of Lord Brouncker's presidency and that of Newton, the post of Secretary was filled by sixteen Fellows, of whom Dr J. Wilkins, Nathaniel Grew (botanist), Robert Hooke (physicist), Dr Hans Sloane (physician) and Richard Waller (zoologist) alone could be classed as scientific men. Most of the others who were appointed held the post for two or three years only; of those who served for longer periods Dr Wilkins retired after five years on being appointed Bishop of Chester; Oldenburg served for fifteen years until his death in 1677; Robert Hooke was Secretary for five years; Thomas Gale, who was also charged with the foreign correspondence, acted for eleven years but he was an historian and classical scholar and did not take an active interest in the Society's scientific work. The situation was greatly improved when Richard Waller, a zoologist, was elected in 1687 and held the post for twenty-seven years, until his death in January 1715; for two-thirds of this period, from the end of 1694 until 1713, with the exception of one year, he had as his colleague Dr Hans Sloane. Too frequent a change of Secretaries was not in the best interests of the Society, and indeed was most prejudicial to them, so that Waller and Sloane, who took up the post of Secretary in 1687 and 1694 respectively, are deserving of the Society's

grateful recognition for having undertaken the laborious task of reorganization which resulted in a noticeable improvement of the administration within the next twenty years.

The Secretary, who kept the Minute- and Journal-books, attended
Council meetings and took note of all that passed, is sometimes referred
to as the First Secretary, but his duties as distinct from those of his colleague
are nowhere defined. Francis Aston became a First Secretary on taking
over the Registers from Robert Hooke in 1682, and so did Waller on
taking them over from Sir John Hoskins at the end of 1687; his colleague,
Dr T. Gale, only attended three Council meetings in eleven years. Dr Hans
Sloane replaced Gale at the end of 1694 and at once began to attend all
Council meetings while Waller rarely did so from 1695 to 1711; Sloane
had apparently become the First Secretary either by arrangement or on
account of his energetic character. For example, in 1698 he arranged that
fifteen meetings of Council should be held to discuss the financial position
of the Society, which was then very serious, and of these he attended
eleven; the President, Lord Somers, attended none of them, nor did
Waller the other Secretary; Sir John Hoskins, a Vice-President, presided
at most of them.

COUNCILS

Since under the Charter the administration of the Society and the making
of its statutes and laws were entrusted to the President and Council, any
exceptional activity should be reflected in the records of its meetings and
in the attendance of its members. At first, when there was much to be
done, meetings were frequent, 117 having been held during the years 1665
to 1670, but not regularly; in one year thirty-six meetings took place and
in another only six, the number being determined by the amount of
business to be transacted. By 1671 they were being held more regularly,
but the number of them decreased towards the end of the century (cf.
Appendix II C).

The scientific members of Council seldom numbered more than a third
of the twenty-one councillors at this time. There were several periods of
the eighteenth century when the attendance at Councils was not much
better, and until the scientific Fellows were in a majority on the Council no
marked improvement could be expected except when an energetic
President like Lord Macclesfield or Lord Morton brought about a temporary improvement. But even then there was but little change in the
number of councillors who attended the meetings, and it was apparently
regarded as sufficient if the officers and five or six councillors were present
to deal with the Society's business, the others being content to leave the
management of the Society to them. Communications were slow and
difficult so that regular attendance at Council meetings could only be
expected from those who were living in or near London. In 1670 the

carriage of a letter from London to St Andrews took nearly three weeks, and in March 1671, when William Collins sent a second copy of Part 79 of the *Philosophical Transactions* to James Gregory at St Andrews instead of Part 80 which contained Newton's discovery of the refractive nature of coloured light, Gregory, who had at once recognized the importance of Newton's discovery, had to wait many months before he saw an account of it in print.[1]

THE FELLOWSHIP

By the end of 1663 the Fellows on the register of the Society numbered between 130 and 140, which included a few foreign members, but these were not shown separately on the annual lists until 1681; this number increased steadily until about 1670 by which time it reached 225; it then fell off until 1693 when it was only 113. The number of admissions, including both British and foreign candidates, to the Fellowship between January 1664 and 1700 amounted to 384, or between ten and eleven in each year, but deaths, resignations and ejections for non-payment of fees and subscriptions outnumbered the accessions, a state of things which did not augur well for the future of the Society.

Although the aim of the Society was to improve Natural Knowledge it had been decided from the first that a number of educated persons who were neither learned in science nor particularly interested in its advancement should be admitted to the Fellowship; scientific men were as yet comparatively few in number, and if the Society was to receive adequate financial support from its members their numbers had to be sufficient to provide this. In order to form an estimate of the number of scientific and non-scientific Fellows belonging to the Society at various periods of its history the membership lists for the years 1663, 1671, 1698, in the seventeenth century when conditions were changing rapidly, and also for the years 1740, 1770, 1800, 1830 and 1860, when its development had become more uniform, have been analysed so far as is possible. The proportion of scientific to the non-scientific Fellows in the Society shown in Appendix II A has been taken from the lists which were prepared for these Anniversary meetings. The information about the professions and the occupations of the various Fellows has been obtained mainly from the *Dictionary of National Biography*, and this has been supplemented from the annual lists of the Fellows and from such other sources as seemed to be sufficiently definite and reliable; but there were always a number of Fellows whose occupations or interests have been described in the records of the Society in such indefinite terms as 'interested in natural philosophy' or 'in mathematics', an estimate which is probably of little or no value; these have therefore been included in the non-scientific group.

At first hopes had been entertained that the close connection of the

[1] H. W. Turnbull, *Royal Society Notes and Records*, 1940.

Society with the court and with the higher classes of statesmen and land-owners would lead to their being well represented in its membership, but this was not the case. The Fellows who belonged to the peerage numbered sixteen in 1663, eleven in 1671 and ten in 1698, or stated in percentages of the whole Fellowship 11·0 per cent in 1663 and 1671 and 8·4 per cent in 1698; but the nobility and the wealthy classes of the day for the most part took little interest in the Society and gave it no active support.

The scientific Fellowship in these years was distributed over few professions as the analysis in Appendix II B shows. At this time most of the scientific Fellows were either mathematicians or physicians, the biological sciences and technology claimed but few. The greatest number belonged to the medical profession for which the subjects that concerned its members were studied more thoroughly and by larger numbers than in other branches of science; and this continued to be the case for many years.

FINANCE

Most probably when the king declared his willingness to be a Fellow and the Patron of the Society it was generally hoped and believed that a subvention of some kind would be provided. It may not have been realized by the Fellows how difficult his position was at that time, though it must have been well known to such men as Lord Brouncker, Sir Robert Moray, John Evelyn and some others that the king would not be able to make any grant to relieve the Society from its financial difficulties. Documents of this period which have been studied in recent years show that at the Restoration the country's accumulated debt was over two million pounds and the annual deficit was about one million. The cash at the Exchequer at the time of Charles' return was but £11. 2s. 10d. Though an annual sum of £1,200,000 was guaranteed to the king by Parliament in no year did he receive so much, and in the first year only £70,000 was provided for ordinary expenditure.[1] It was therefore recognized very soon that the Society would have to depend on such resources as it could command, and the Council discussed at length various means whereby they might be increased. Bishop Sprat wrote in his *History*:

> But besides this, there is one thing more, that persuades me that the Royal Society will be *Immortal*, and that is that if their stock should still continue narrow, yet even upon that, they will be able to free themselves from all difficulties and to make a constant increase of it, by their managing. There is scarce anything has more hindered the True Philosophy; than a vain opinion, that men have taken up, that nothing could be done in it, to any purpose, but upon a vast charge, and by a mighty Revenue.

His sturdy optimism was shared by the Council, who realized that it rested with them to strain every nerve in order to put the Society's

[1] A. Bryant, *King Charles II.*

finances on a more satisfactory basis. In 1665, 500 forms were printed demanding from the Fellows the prompt payment of all arrears; in 1667 a circular letter was sent out asking for assistance from any of the public who might be interested; a list of Fellows who were in default with their subscriptions was compiled in 1672; a form of undertaking to pay off arrears in not less than six months was prepared in 1673, and at the end of December, after the Anniversary meeting of that year, it was resolved that anyone who did not pay any arrears which might be due from him at Michaelmas 1673, or give his bond for doing so within six months, should be ejected from the Society and sued for what he owed. In 1682 the Council ruled that anyone who had not paid his subscription up to date should be ineligible to serve on it as a member. In July 1685 a list of forty-seven Fellows who were considerably in arrear with their payments was brought before the Council and it was resolved that their names should be omitted from the list of the Society which would be printed for the Anniversary meeting of that year unless they had paid them in full before then. It cannot be said therefore that the Council were not doing their best to deal with the situation. They had also looked about for other means of increasing their funds. In 1664 it was proposed to solicit a grant from the king of such lands as were left by the sea; or alternatively that the king might be petitioned to confer such offices in the Courts of Justice, or the Custom House, as were in His Majesty's grant upon some members of the Society for the use of the whole. Finally a petition had been addressed to the king praying His Majesty to grant Chelsea College, and the lands belonging to it, to the Royal Society.

As early as September 1662 the king had addressed a letter, with his own hand, to the Duke of Ormonde, the Lord Lieutenant of Ireland, recommending the Royal Society 'for a liberal contribution from the adventurers and officers of Ireland for the better encouragement of them in their designs'. The President, Lord Brouncker, also wrote in January 1663 to the duke appealing for a grant of land to the Society, but without any effect. Sir William Petty, a Fellow of the Society, who was then in Ireland, after having surveyed the lands to be appropriated, was asked to send an estimate of the value of the lands granted by the king to the Society. He made a rough calculation but did not transmit it to the Society seeing that the lands referred to had already passed into other hands. Later, however, at the request of several of his personal friends among the Fellows, he communicated it to Sir Robert Southwell whose letter to Oldenburg dated from Dublin on 15 May 1663 is in the archives of the Society. No attempt at an exact estimate of the lands was made, for, as Sir William Petty says,

if the odd money and odd measure be understood in an unlimited way, then, I say it will amount to a great matter but I know not what....Those who told you that thirteen millions of acres were yet to be disposed of did not calculate well for I cannot imagine that there is one; and the better to confirm you, I am assured that all the profitable land in the whole kingdom of Ireland exceeds not nine millions; all the lands let out to adventurers and soldiers not much above two millions. Nor does all the forfeited lands, intended to be disposed of by Act of Settlement extend to three millions, and much of what was intended will fall short and return to the Irish.

It was quite clear that the Society would have to depend on its own exertions and resources. Many of the Fellows were no doubt disappointed that the Crown should not provide pensions and subventions for the Society's work on a scale such as that with which Louis XIV supported the Académie des Sciences in Paris when it was founded in 1666, but this was out of the question; the academy's members were few in number while no limit had been laid down in the Charters for the Fellowship of the Royal Society. When the Society was first formed in November 1660 the subscription of one shilling a week may have seemed very reasonable to those who fixed it, but to pay £2. 12s. 0d. per annum was perhaps considered by many to be too much for what they received in return. Early in 1661 members of the Philosophers' Society had been told that their subscriptions and arrears must be paid at the Wednesday meetings, and in January 1663 an order was made for payment of all arrears forthwith, in order that the financial position should show no outstanding liabilities when the Second Charter was sealed, but these warnings had little effect. At the end of 1663 the total amount due to the Society from its Fellows was £158, and by 1676 it had reached £2000. By 1700 it was probably considerably more, but no record of the amount then owed exists.

Information on the finances for many years is scanty for no detailed yearly statement of income and expenditure was rendered by the Treasurer to the Fellows until 1830, and the most that was done until then was to report to the Society at the Anniversary meeting the total sum that had been received during the preceding twelvemonth, and the total amount that had been expended; the difference remained in the hands of the Treasurer as a balance with which to begin the following year; no other details were given. Fuller details may have been laid before the Councils, for they certainly existed in the Society's ledger, but they are not recorded in the minutes or elsewhere. This very meagre information was given for each year from 1663 to 1676 after which no information of any kind appears in the Council minutes or the Journal-book of the Society until 1716; in the four years 1710 to 1713 the accounts are said to be 'True and correct' or 'Very exact and just', but no figures are given. The books of account which were in use in the early years no longer exist so that financial details of those days cannot now be recorded. The arrears were made up of subscriptions which were due to the Society from Fellows who either by negligence or from intention had omitted to pay them; there were not a few who considered that on account of their social or official standing they should not be called on to contribute; also sums were owing from the estates of deceased Fellows, and these the executors often did not regard as debts, and did not consider themselves justified in dealing with them as such. So the tale of arrears went on mounting up, and only a small proportion of them was ever recovered. Under the statutes then in force Fellows who had defaulted in the payment of their subscriptions could be removed from the list of Fellows, but Treasurers were inclined to remind defaulters of their obligations in the hope that some payments would be made sooner or later, rather than to ask the Council that such Fellows should be ejected. The report of a committee stated in 1673 that of 149 Fellows 56 paid well, 79 did not pay and 14 were absent; the total number of Fellows at that time is recorded as being 215 so there were 66 other subscribers who had not been accounted for. The subscriptions which should have been paid by 215 Fellows would have amounted to £559 besides the admission fees of £2 each from twelve candidates who were elected; but the sum actually received by the Treasurer was only £152. His expenditure had amounted to £146, thus leaving him but £6 with which to meet any salaries or other charges which might be out-standing. It is not surprising therefore that when the king expressed the wish that a degree of latitude should be measured in England, or that any other investigations involving expenditure should be undertaken, the Society was wholly unable to carry out the work from its own resources. In 1674 and 1675 the Council held numerous meetings to discuss what steps should be taken to enforce the payment of Fellows' dues, and it was 'ordered that there should be prepared a forme of a legal subscription for paying fifty-two shillings a year. That as many of the Fellows as are willing to further the business of the Society shall be desired to advance a year's weekly contribution for carrying on the work thereof with more vigour than hitherto.' The Attorney-General therefore, at the request of the Council, drew up the following Obligation, which was approved and ordered to be signed by every newly elected Fellow:

I.........................do grant and agree to and with the President, Council and Fellows of the Royal Society of London for improving Natural Knowledge, That, so long as I shall continue a Fellow of the said Society, I will pay to the Treasurer of the said Society for the time being, or to his Deputy, the sum of Fifty-two shillings per annum, by four equal Quarterly payments, at the four usuall days of payment: that is to say, the Feast of the Nativity of Our Lord: the Annunciation of the Blessed Virgin Mary: the Feast of St John the Baptist: and the Feast of St Michael the Archangel; the first payment to be made upon the ....next ensuing the Date of these Presents; and I will pay in proportion, viz. One shilling per week for any lesser time after any the said days of payment, that I shall continue Fellow of the said Society. For the true payment whereof I bind my Self and my Heirs in the penal sum of Twenty Pounds. In witness whereof I have hereunto set my Hand and Seal, this......... day of........................

(Witness' signature.)   (Signature of Fellow.)

The same Obligation was also signed by a good many of the Fellows who had been elected before it was adopted, for instance by John Evelyn whose bond is reproduced in the *Record* of the Society (Plate IX). A good many years passed before defaulters were sued for not fulfilling their liabilities under this Obligation, but in 1728 when Sir Hans Sloane was President the Society's rights were successfully enforced by legal process in a number of cases.

In June 1684 the Council decided that in order that E. Halley should undertake the measurement of a degree of latitude he should receive the sum of £50 for the expenses of the work, or fifty copies of Willoughby's *De Historia Piscium*; the work was not carried out, for the sum allotted for it was quite inadequate. In July 1687 the Council ordered that Halley should have fifty copies of the same book, instead of the sum of £50 which was due to him for his salary, and twenty other copies in settlement of arrears of salary due to him for the previous year, which offer Halley accepted. A similar arrangement was proposed to Hooke, but he asked for six months' time to consider the offer which he was not inclined to accept; later he was paid in cash.

Although the financial position was so difficult, some Fellows were from time to time excused payment of the whole or a part of their annual subscription; the Council evidently had to balance their policy between bringing pressure to bear on those who might reasonably be expected to pay and were merely negligent, and showing leniency to others who were in positions which justified it lest their services should be lost to the Society. When the yearly honorarium of a secretary could not be paid, this expedient was sometimes adopted as in the case of F. Aston in 1683 and 1687. Some of the arrears due to the Society must have been paid from time to time, for in 1676 the Treasurer held £200 of Africa Company Stock and £300 of East India Company Stock; but the general situation did not improve, and in 1687 the Society was on the point of having to sell its holding of East India Company Stock; this was however avoided by help which was given by a Fellow of the Society who became surety for the amount.

In 1694 Dr Hans Sloane, having replaced T. Gale as Secretary, at once gave his serious attention to the financial question, and achieved some success, for by 1697 the Society was able to increase its investments by £250 of East India Company Stock and £800 of Africa Company Stock. In the following year he arranged for fifteen meetings of the Council to be held at which the principal matters under discussion were the financial position of the Society and plans for its improvement. It would be interesting to know more of the expedients which the Council adopted under Sir William Hoskins' and Dr Hans Sloane's guidance to extricate the Society from its difficulties. Well-organized accounting was already in use by many financial houses and a treatise on the subject by Richard

Dafforne was published in 1635 and was reprinted three times before the end of the century. Account books the Society certainly had at this time; a cash book was a necessity, as well as a register in which ledger accounts were kept of the sums due from and paid by Fellows for their admission fees and their annual subscriptions which at this time were supposed to be paid quarterly. They provided the only means of ascertaining the arrears due to the Society, and this information was frequently required by the Council. What was eventually done about the accumulation of unpaid arrears is not known, but probably the greater part had to be written off as irrecoverable. In addition to these various factors which influenced the development of the Society in a greater or less degree as the years rolled by there were always a number of occurrences of minor importance which often throw an interesting light on its history and on the part which some of its Fellows played in it. These have come down to us from many sources and, though they may differ in their accuracy, they form a valuable and interesting addition to the Society's own records. It may well be that considerable additions may be made to them from such collections of letters, diaries, etc. as still exist in private possession.

Notwithstanding the difficulties which the young Society had to face during the latter part of the seventeenth century, its annals record a very considerable amount of activity in many fields. Most of this is recorded in the *Philosophical Transactions*, but there is much interesting information which was not important enough to be included in these though it has a considerable historical value, and was included in the minutes of Council.

About 1668 Newton had planned his reflecting telescope without knowing of James Gregory's earlier work in the same field, and constructed one with his own hands which he presented to the Society early in 1671; another of a similar type was made at Cambridge, and a description of it entitled 'An Accompt of a New Catadioptrical Telescope, invented by Mr Newton' was published in the 81st number of the *Philosophical Transactions*. In March 1671 Newton wrote to Oldenburg:

With the telescope which I made, I have sometimes seen remote objects, and particularly the moon, very distinct, in those parts of it which were neare the sides of the visible angle. And at other times, when it hath been otherwise put together, it hath exhibited things not without some confusion; which difference I attributed chiefly to some imperfection that might possibly be either in the figures of the metalls or eye-glasse; and once I found it caused by a little tarnishing of the metal in four or five days of moist weather.

On 8 February 1671/2 an entry in the Journal-book records the receipt of a communication from Newton

concerning his discovery about the nature of light, refractions and colours, importing that light was not a similar, but a heterogeneous thing, consisting of difform rays, which had essentially different refractions, abstracted from bodies they pass through, and that colours are produced from such and such rays, whereof some in their own nature are disposed to produce red, others green, others blue, others purple, etc., and that whiteness is nothing but a mixture of all sorts of colours, or that 'tis produced by all sorts of colours blended together.

It was

Ordered that the Author be solemnly thanked, in the name of the Society, for this very ingenious discourse, and be made acquainted that the Society think very fit, if he consent, to have it forthwith published, as well for the greater conveniency of having it well considered by philosophers, as for securing the considerable notions thereof to the Author, against the arrogations of others.

This paper was published in No. 80 of the *Philosophical Transactions* in 1672, the experiments which are described having been made in 1666. By providing the means of publishing his work promptly in the *Philosophical Transactions*, which were already widely known and studied, the Society rendered to Newton an important service, for at this time the London booksellers were 'much averse to publishing mathematical books'. John Collins, a Fellow of the Society, who corresponded frequently with Newton, James Gregory and other leading Fellows of the Society on their work and that of others, in a letter to Newton says, 'our Latin booksellers have no vent for mathematical works; and so when such a copy is offered instead of rewarding the author, they rather expect a dowry with the treatise'. The Society gave £5 with the copy of Horrox's *Opera Posthuma* to encourage a bookseller to print it. In 1671 John Collins, when writing to James Vernon, who was then at Oxford and afterwards became a Secretary of State, says: 'Here Printing and Bookselling being a trade (as they are not in Holland) books are much dearer than in either France or Holland. Dr Barrow's books are not to be had for money, lying pawned and the bookseller unable to redeem them.'

On 21 December 1671 it is recorded in the Journal-book that 'the Lord Bishop of Sarum [Dr Seth Ward] proposed for candidate Mr Isaac Newton, Professor of Mathematicks at Cambridge'. Two years earlier Dr Barrow, a Fellow of the Society, who was then Lucasian Professor of Mathematics at Cambridge, had resigned the chair in favour of Newton who was at this time achieving some of the most brilliant successes of his long and distinguished career. On 11 January 1671/2 Newton was elected a Fellow, and for nearly half a century his work notably advanced the cause of science; his genius conferred great distinction on the Society, in whose publications his most important work appeared. Most of this was done before the end of the seventeenth century when he was appointed an official of the Royal Mint.

In November 1672, by the death of Dr John Wilkins, Bishop of Chester, the Society lost one of its earliest and most active supporters. He had joined the group of philosophers in London about thirty years before and had devoted his energies to the advancement of the new philosophy and later to the formation of the Society. He was its first Secretary, only resigning the post in 1668 when he became Bishop of Chester. In his will he bequeathed to the Society the sum of £400, the first bequest that it had received, and this the Council decided to invest. On 17 December 1674 therefore a committee was appointed consisting of Sir John Lowther, Sir John Bankes, Mr Samuel Pepys and Dr J. Goddard, to consider 'whether the four hundred pounds legacy might not be best laid out upon fee-farm rents; and they were desired to ripen this business for the 17th of January, and make their report to the council'. Sir John Bankes, Bt., was a wealthy East India merchant; he had 'a noble mansion at Lincoln's Inn Fields, and an estate, Oakleigh Towers, in Kent'. He was elected a Fellow of the Society in 1668, having been proposed by Dr J. Goddard, and was elected a member of Council on 30 November 1674. He was a close friend of Samuel Pepys and gave evidence in his favour in the House of Commons when Pepys was accused of 'being a Papist'.[1] On 14 January 1675 the committee reported that they had found upon the books three fee-farm rents from Lewes in Sussex (whereof one was from the estate of the Earl Marshal) amounting in all to twenty-four pounds per annum, and on 21 January, after receiving a full report from Sir John Bankes concerning these three fee-farm rents, the Council very prudently accepted the proposal of their Committee. On 28 January Sir John Bankes produced the conveyance of the twenty-four pounds fee-farm rents yearly payable from Lewes in Sussex; and on 25 February he was requested to extract out of his conveyance the particular parcels of the lands out of which these fee-farm rents were payable. The lands had formerly been the property of the Abbot and Priory of Lewes. Before long, difficulties arose about the payment of the rent, and by 1703 the sum of £450 was owing to the Society. This was due mainly to an uncertainty about the proportion of the rent which was said to be due from two or three owners of property in that part of Sussex.

At the end of 1672, Thomas Henshaw, who had succeeded Bishop Wilkins as Secretary four years earlier, did not stand for re-election since he had been appointed Envoy Extraordinary to Denmark. He was succeeded by his intimate friend, John Evelyn, who only held the post for one year, being then replaced by Abraham Hill, who occupied it until Henshaw's return from abroad in 1675. Henshaw then resumed his work as Secretary with Oldenburg as his colleague until the end of 1677, when they were replaced by Dr N. Grew and Robert Hooke. On Oldenburg's death in 1677 Dr Grew continued the publication of the *Philosophical Transactions* until the end of 1678, but so keen was the interest which was now taken in them and the desire of philosophers generally that they should be revived that Hooke was desired by the Council in February

[1] A. Bryant, *Samuel Pepys: The Years of Peril*, p. 114. Cambridge, 1935.

1679 to 'publish a sheet or two every fortnight of such philosophical matters as he shall meet with from his correspondents; not making use of anything contained in the Register-books without the leave of the Council and author'. It was later agreed that, in consideration of propositions made by Hooke to improve these *Philosophical Collections*, his salary should be increased by forty pounds per annum. These *Philosophical Collections* consisted of seven numbers which included papers by various contributors and gave 'accounts of Physical, Anatomical, Chymical, Mechanical, Astronomical, Optical, and other Mathematical and Philosophical Experiments and Observations'. The second and third numbers appeared in 1681 and the next four in the early part of 1682. In January 1683 the publication of the *Philosophical Transactions* was resumed and the 143rd number was issued under the editorship of Dr R. Plot, who had replaced Hooke as Secretary at the Anniversary meeting of 1682. The preface to the first number contained the following statement: 'Although the writing of these Transactions is not to be looked upon as the business of the Royal Society, yet in regard they are a specimen of many things which lie before them, contain a great variety of useful matter, are a convenient Register for the bringing in and preserving many experiments, which, not enough for a book, would else be lost, and have proved a very good ferment for the setting of uncommon thoughts in all parts a-work.' In spite of this and other disclaimers, the scientific world continued to regard the *Philosophical Transactions* as the Council's publication; it was not until some years later, in 1753, that the Council definitely undertook their regular publication, and accepted the responsibility for them (see p. 179).

In March 1672/3 Newton wrote to Oldenburg as follows: 'Sir, I desire that you will procure that I may be put out from being any longer a Fellow of the Society: for though I honour that body, yet since I see I shall neither profit them, nor (by reason of the distance) can partake of the advantage of their assemblies, I desire to withdraw. If you please to do me this favour you will oblige.'[1] Oldenburg was much disturbed by this letter, and in his reply expressed surprise 'at his resigning for no other cause than his distance, which he knew as well at the time of his election'. 'Offering withal my endeavour to take from him the trouble of sending hither his quarterly payments without any reflection.' Newton also confides to Collins his troubles with the Royal Society in a manner which in the opinion of L. T. More makes it certain that the payment of dues was but an excuse for the true reason. He wrote: 'Concerning the expenses of being a member of the R.S. I suppose there hath been done me no unkindness, for I met with nothing of that kind besides my expectations. But I would wish I had met with no rudeness in some other things. And therefore I hope you will think it strange, if, to prevent accidents of that nature for the future, I decline that conversation which hath occasioned

[1] L. T. More, *Isaac Newton, a Biography*, p. 154.

what is past. I hope this, whatever it may make me appear to others, will not diminish your friendship to me.'[1] Oldenburg's action did not always help matters, for when objections were received from professors at Liège on the controversy on light he failed to realize that they were not critics of the same standing as Hooke and Huygens but continually pressed Newton to reply to their criticisms. It was Hooke's mistrustful and jealous disposition, and his suspicions that others were appropriating the results of his work, that made intercourse with him so intolerable to Newton, who desired to avoid service on the Council so long as Hooke was alive. Weld quotes the Council minutes of 28 January 1673/4 to the effect that: 'It was mentioned by the Secretary that Mr Newton had intimated his being now in such circumstances that he desired to be excused from the weekly payments; it was agreed by the Council that he should be dispensed with as others were.' He had mentioned the subject earlier to Oldenburg in a letter dated Cambridge, 23 June 1673, which was apparently in reply to Oldenburg's letter already quoted: 'For your proffer about my quarterly payments I thank you; but I would not have you trouble yourself to get them excused if you have not done so already.' The Society has always been ready to exempt those of its Fellows to whom the payment of the subscription was a hardship so long as they were among those who were active in promoting the scientific aims of the Society.

In July 1673 Sir Robert Moray died suddenly at the age of sixty-five; he was still working actively for the Society where his tact in composing such differences as arose from time to time between the Fellows or between them and their foreign correspondents was always welcome; he had been a member of its Council since 1662, except during 1667. Not only was his influence with Charles II of much value to the Society but his frequent correspondence with men of science in Paris and elsewhere abroad, especially with Huygens, usefully supplemented Oldenburg's letters to his regular correspondents. His loss to the Society so early in its career was irreparable. From the time when he returned to London in August 1660 he had devoted himself wholeheartedly to making the Society the important institution that in his opinion it should be. During 1661 and the first half of 1662 he was virtually President of the Society though the title was not formally used by anyone until the Charter had been sealed, and a Council elected. He succeeded in obtaining its incorporation by Royal Charter and thereby assured it of a position which its enemies failed to ruin. Unfortunately the Society does not possess a portrait of him.

For six years the Society had been meeting at Arundel House by the permission of the Duke of Norfolk, since Gresham House was occupied by the Lord Mayor and the Merchants of London after the Great Fire. But in April 1673 representatives of the Gresham Committee, the City of London, and the Mercers' Company waited on the President to invite the

Society to resume their meetings at the College as formerly, and on this being reported to the Council 'they thought good to have their hearty thanks returned to the Committee of Gresham College for their kindness and respect; yet without saying anything to them of acceptance or not acceptance; only in case they should give occasion of saying more, that then it might be intimated that this business was under consideration'. The Council had already in mind the desirability of either building or acquiring a house of their own as soon as their resources permitted them to do so. At their next meeting the Council informed the Duke of Norfolk of the offer, and begged that he would continue his hospitality to the Society at Arundel House at all events for the time being; this he agreed to do. However, as the south and west galleries of the College were not yet in a state to accommodate the library and the repository, it was not until November of the following year that meetings at Gresham College could be resumed.

In 1674, besides requiring every Fellow to give a bond that he would pay punctually his subscription when it fell due, the Council ordered that 'such of the Fellows as regard the welfare of the Society should be desired to oblige themselves to entertain the Society, either *per se* or *per alios*, once a year at least, with a philosophical discourse grounded upon experiments made or to be made; and in case of failure to forfeit £5'. A circular letter asking for discourses from selected Fellows was sent to those who seemed to be most likely to co-operate, and Oldenburg was active in pressing many to help the Society at this time when the number of suitable communications was regrettably few.

In 1675 the Royal Observatory at Greenwich was built by the order of Charles II, and Flamsteed, who was elected a Fellow in February 1677 (N.S.), was appointed Astronomical Observator with a salary of £100 per annum. He gives the following account of its foundation. In 1675 a Frenchman, Le Sieur de St Pierre, who had some knowledge of astronomy, proposed to the king that facilities should be provided for him to study the better determination of longitude. His proposal was referred to Lord Brouncker, Dr S. Ward, Sir Christopher Wren, Sir Charles Scarborough, Sir Jonas Moore, Colonel Titus, Dr J. Pell, Sir Robert Moray, and Mr Hooke, all of whom were Fellows of the Royal Society, as well as to some others. Sir Jonas Moore, who was then the Surveyor-General of Ordnance, took Flamsteed to one of their meetings at which the proposals were read. Flamsteed criticized the scheme and pointed out that the places of the fixed stars were not truly given. On learning this the king ordered the building of an observatory to amend and improve this field of astronomical knowledge and appointed Flamsteed to take charge of it. The Royal Observatory has never been under the direction or control of the Royal Society, but an intimate co-operation has always existed between them. Flamsteed had made his first contribution to the Society in November 1669 and soon afterwards came to London where he made the acquaintance of Sir Jonas

Moore, F.R.S., at the weekly meetings. He was interested in the young astronomer and used his influence to forward the scheme for an observatory at Greenwich, a site recommended by Sir Christopher Wren. The king gave £500 and a quantity of building materials towards the execution of the project, Sir Christopher Wren being desired to prepare the plan. No provision was made for instruments which the observer was expected to provide for himself; but Sir Jonas Moore gave a sextant, two clocks, one of which was by T. Tompion of London, the most celebrated clockmaker of his day, a telescope and some books. In January 1677 the Society lent such instruments as it possessed to Flamsteed until they should be required by it again, but they were recalled two years later when the Society wished to equip its own observatory at Gresham College. This greatly annoyed Flamsteed who, perhaps unreasonably, attributed the action to ill-will on the part of Robert Hooke.

Early in 1674 Oldenburg brought before the Society a letter from Huygens describing a new pocket watch of a type which Hooke claimed to have invented several years earlier. Hooke then appealed to the Society to have the minutes of the earlier meeting examined to substantiate his statement; the Minute-book was produced but little bearing on the matter was to be found; Hooke then accused Oldenburg of not entering the minutes correctly; he also called him a 'trafficker in intelligence', and was ordered by the Council to offer an apology which was printed in the *Philosophical Transactions*. There the matter has remained until recently, when an article by R. D. Waller on Lorenzo Magalotti's visit to England between 1660 and 1669, has thrown a new light on the incident. Magalotti and Falconieri of the Florentine Academy visited England in 1668, and were present at a meeting of the Royal Society on 20 February at the invitation of Oldenburg. In his account of Hooke's demonstrations at the meeting Magalotti says: 'We also saw a pocket watch with a new pendulum invention. You might call it with a bridle, the time being regulated by a little spring of tempered wire which at one end is attached to the balance-wheel, and at the other to the body of the watch. This works in such a way that if the movements of the balance-wheel are unequal, and if some irregularity of the toothed movement tends to increase the inequality, the wire keeps it in check, obliging it always to make the same journey'.[1] Nothing of all this appears in the minutes; it would seem therefore that Hooke was correct in his statement and that his demonstrations had not been correctly recorded in the minutes. Professor E. Andrade,[2] in referring to this dispute between Huygens and Hooke, says: 'When it is remembered that Oldenburg had a financial interest in Huygens' watch, since Huygens had assigned to him the patent rights, and that his petition for a patent is on record, it is perhaps possible to believe that he was not completely unprejudiced.'

[1] *Royal Society Notes and Records*, 1937, pp. 92–4.   [2] *Nature*, 1935, p. 359.

On 30 November 1677 Lord Brouncker resigned the presidency at the Anniversary meeting after having held it for fifteen and a half years. He had steered the young institution through its early difficulties and had contributed largely to the esteem in which it was held both at home and abroad; he had also presided with exceptional regularity at the meetings of the Council and at the Ordinary meetings of the Society. The Council were now faced with the difficult task of finding a suitable successor to one who not only was an able mathematician but also held the position of the Queen's Controller, and was one of the Commissioners of the Navy; he could therefore exercise considerable influence among the statesmen of the day as well as being in touch with the king and court circles. He was not perhaps able to do so much for the Society at the beginning of its career as did Sir Robert Moray or John Evelyn, but this may have been due to his personality being a less attractive one than theirs. Lord Brouncker as President with D. Colwall as Treasurer and J. Wilkins and H. Oldenburg as Secretaries had carried the Society through its first ten years with conspicuous success; but there was a great deal to be done by the officers, and the work which fell to the share of the Secretaries was increasing. It was not easy to find men of energy and methodical habit, possessing sufficient scientific knowledge to guide the Society through the coming years which were certain to be difficult and in which even its continued existence was at times doubtful. Evelyn attributes Lord Brouncker's decision not to be nominated again to a very praiseworthy desire that a lengthy tenure of the presidency should not be considered as prejudicing the selection of a successor if that seemed to be desirable; nothing had been decided as yet about the period during which officers should hold their appointments, but the Fellows were left to decide each case at the Anniversary meetings.

As his successor the Council recommended Sir Joseph Williamson, and the Fellows elected him to be President in November 1677. He had been educated at Westminster and at Queen's College, Oxford, where he showed great ability. After the Restoration he was appointed a secretary to the Secretary of State and Keeper of the State Papers Office. He established the *Oxford Gazette* at Oxford in 1665 [1] where the king and court had gone on account of the Plague. He was knighted in 1667 and became a Secretary of State in 1674. He fell under suspicion at the time of the Popish Plot in 1678 and was committed to the Tower by the House of Commons but was released on the king's order. Evelyn had not a high opinion of him, describing him as 'Lord Arlington's creature and ungrateful enough'. He was a man of very considerable ability but had no understanding of the scientific aims of the Society. In the first two years of his presidency he attended none of the eighteen Council meetings which were held, but was present at most of those in the third year when he was no longer holding a State appointment. He did not stand for re-election at the end of 1680.

[1] *London Gazette*, 1, in 1666.

He presented a number of objects to the Society's Museum but did not take any other interest in its work. His portrait by Sir G. Kneller is in the rooms of the Society.

In September 1677 the Secretary, Henry Oldenburg, died after having served the Society in this position for fifteen years. Born in Bremen, he had entered as a student at Oxford in 1656 and later travelled in France until 1661 with Richard Jones, the son of Lord Ranelagh. He became a Secretary of the Society in 1662 under the First Charter, and until his death carried out his duties in the most careful and punctilious manner, though his lack of scientific training and experience sometimes involved him in difficulties. He disliked Hooke, with whom he found it difficult to work harmoniously, nor was it easy to conciliate the extreme sensitiveness of Newton, whom the suspicious and jealous criticisms of Hooke irritated profoundly, when questions of priority or accuracy were concerned. Nevertheless for fifteen years he served the Society devotedly and played an important part in making it known abroad and increasing its influence at home. His colleague, Dr J. Wilkins, had occupied himself with promoting the Society's general aims, and had left administrative matters to Oldenburg, to whom was entrusted in 1665 the regular publication of the *Philosophical Transactions* which he edited up to the time of his death. His portrait is in the possession of the Society. In 1717 the Society presented his son with a gift in acknowledgment of his father's long and valuable services.

In June 1678 the Duke of Norfolk began the demolition of Arundel House, so it became necessary to remove the library, which he had presented to the Society in 1667 at John Evelyn's suggestion, to the Long Gallery in Gresham College where space had been provided for it; regulations for its preservation and management had been drawn up by the Council which ordered that: 'An exact catalogue of all the books of the Bibliotheca Norfolciana be made apart, and also of all other books which shall accrue. No book shall be lent out of the Library to any person whatsoever. Any person desiring to use any book in the Library shall return it into the hands of the Library-keeper entire and unhurt. That the Library shall be surveyed once in the year by a Committee chosen by the Council to the number of six.'

This may have suggested to Sir George Ent, M.D., one of the original Fellows, the idea of presenting his library to the Society, which he did in the following year; the loss to the Society of Dr Jonathan Goddard's library four years before, because he had made no will, would have been well known to him. A catalogue of Sir G. Ent's library was compiled and published in 1680. The Council were now turning their attention to their library and its archives, and for their greater security resolved that the two Secretaries, and no one else, should have charge of the keys of the press in which the Society's books and papers were kept as these were becoming numerous and of considerable importance. M. Wicks, the Clerk, and H. Hunt, the Operator, were both at this time employed in cataloguing the Society's library. Hunt was made Keeper of the Library and Repository at the end of 1678, so he would seem to have shown himself to be exceedingly efficient in this class of work.

In the autumn of 1680 the Council recommended that the Hon. Robert Boyle should be elected President in succession to Sir Joseph Williamson, and this was approved by the Society at the Anniversary meeting. On 18 December however, Boyle wrote declining to accept his election, and explained his reasons for doing so in the following letter to Hooke, the First Secretary:

Though since I last saw you I met with a lawyer who has been a member of several Parliaments, and found him of the same opinion with my Council in reference to the obligation to take the test and oaths you and I discoursed of; yet not content with this, and hearing that an acquaintance of mine was come to town, whose eminent skill in the law had made him a judge, if he himself had not declined to be one, I desired his advice, and by it I found that he concurred in opinion with the two lawyers already mentioned, and would not have me venture upon the supposition of my being unconcerned in an act of parliament to whose breach such heavy penalties are annexed. His reasons I have not now time to tell you, but they are of such weight with me who have a great (and perhaps peculiar) tenderness in point of oaths, that I must humbly beg the Royal Society to proceed to a new election, and to do so easy a thing as, among so many worthy persons that compose that illustrious company, to choose a President that may be better qualified than I for so weighty an employment.

His health, moreover, was very indifferent and he passed much of his time at his sister's house in Pall Mall. Boyle being now resident in London had fitted up a laboratory in Southampton Street where he could carry on his researches, and he employed Hauckwitz, who had previously assisted him at Oxford, to continue the work for him on which he had been engaged. The Council very wisely desired to have an eminent man of science for their President, but when Boyle was unwilling to serve it was not easy to find a worthy substitute. Boyle had been the most obvious choice. Evelyn writes in his Diary on 30 November, 'the anniversary election at the Royall Society brought me to London where was chosen President that excellent person and greate philosopher Mr Robert Boyle, who indeed ought to have been the very first; but neither his infirmitie nor his modestie could now any longer excuse him'.

Having resolved that the President of the Society should be a man of science, the Council next approached Sir Christopher Wren who had been one of the philosophers since he was admitted as a fellow-commoner to Wadham College in 1649; Evelyn had met him in 1654 at All Souls College, Oxford, and described him as 'that miracle of a youth, Christopher Wren'. He accepted the offer and was elected President on

13 January 1681/2. He was a member of the Philosophers' Society of 1660, and became an Original Fellow of the Royal Society in 1663; later, as a Fellow of the Society, he had played an important part in advancing its reputation and helping to remove such difficulties as had arisen from time to time during the past seventeen years.

As early as 1657 he had been appointed to the Chair of Astronomy at Gresham College, a post which he held until 1661 when he was elected Savilian Professor of Astronomy at Oxford, succeeding Dr S. Ward. In 1661 he was appointed assistant to Sir John Denham, the Surveyor-General, and in 1663 was called upon to survey and report on the reconstruction of St Paul's Cathedral. From this time onwards he devoted himself to an architect's profession. Even for those times when men of exceptional ability were able to gain a mastery of all that was then known in various branches of learning, the scope and thoroughness of Wren's knowledge of whatever he undertook were quite exceptional. At Oxford he attracted the notice of Dr J. Wilkins, the Warden of Wadham College, and of Dr Seth Ward, the Savilian Professor of Astronomy; he assisted R. Hooke in his *Micrographia* and was highly praised for his attainments in anatomy by Dr T. Willis of Christ Church. He is said to have been the first in this country to carry out the experiment of injecting various liquids into the veins of living animals. Dr J. Wilkins in his *Essay towards a Real Character* credits him with having been the first to suggest that a standard measure of length might be derived from the length of a pendulum vibrating seconds at the place of observation.

At last the Society had for its president a man of great and varied scientific ability, who at this time was actively engaged in architectural work; the time which he could devote to the Society was limited, for he was very fully occupied with the designing and construction of a number of churches and other buildings in the city, and was also responsible for rebuilding St Paul's Cathedral after the Great Fire. But having undertaken the leadership of the Society, he threw himself heart and soul into the work, and during the two years of his presidency presided at thirty-one of the thirty-five meetings of the Council which were held, besides being present at most of the Ordinary meetings; he called upon the Treasurer for a report on the Society's financial position, which was considered by the Council, but the decisions which were reached are not recorded; the statutes relating to the payment of fees and subscriptions were also discussed. At his instigation the Council resolved that no Fellow should be recommended for election to the Council unless he had paid all his dues up to the previous Michaelmas. The list of Fellows who were in arrear with their payments was also considered by the Council and the names of twenty-three were struck off the register, nine of them being peers. No doubt while some of these had ceased to be interested in the work of the Society, there were others whose omission was due to negligence;

Sir John Bankes, who had taken a great deal of trouble to arrange the investment of Bishop Wilkins' bequest in the purchase of fee-farm rents at Lewes, Sussex, had only now completed the business and had handed over the formal deed of conveyance; yet his name occurred in the list of defaulters for the sum of £48. Another practical detail of administration was that the Treasurer and the two Secretaries were to meet the President on Mondays to deal with any financial matters which were outstanding. Unfortunately Wren's presidency of two years only was too short a time for his innovations to become part of the administrative routine, so several of them were discontinued after a while.

After two years Wren found himself obliged to decline renomination, but in those years he had done much to improve the efficiency of the Society. He had been a member of Council for five years before he became President and was therefore well acquainted with the questions which most urgently called for attention. One of these related to the buildings of Chelsea College which had been granted to the Society by a Royal Patent of April 1669 together with thirty acres of land. The building was in bad repair and the Society was not inclined to spend much on it. Several suggestions were made; Evelyn proposed that it might be let as a prison-house during the war for £100 per annum; Sir Robert Moray presented an offer of a noble member of the Society to plant the land with vegetables, etc., and to repair the house at his own cost, but nothing came of it. In 1681 it was brought to the notice of the Council that the king was desirous of resuming possession of Chelsea College, which he had granted to the Society, in order to build a military hospital on the site, and that he was willing to pay £1300 to the Society for the surrender of the estate. As great efforts had been made to find a tenant or to raise funds to build a house suited to the Society's needs on the site without success, the Council gratefully accepted the offer and in 1682 authorized the President, Sir Christopher Wren, to carry out the transaction; the money was invested in the stock of the East India Company, the bonds being placed in the Treasurer's chest where the Charters also were kept at this time.

The Secretaries in 1681 were Robert Hooke and Dr T. Gale, but the latter was replaced by Francis Aston at the end of the year; at the desire of the Council, however, Gale continued to deal with the Society's foreign correspondence, since there was as yet no Foreign Secretary. Hooke and Aston were regular in their attendance at the meetings of Council but Gale had rarely been present. The President prevailed on the Council to re-appoint the Anatomical and Georgical (Husbandry) Committees which had been appointed with five others in 1665, these being the most active in producing results; a Cosmographical Committee was also appointed. At this time only one-third of the members of Council were scientific men; the number of Council meetings was unusually large, there being fifteen and twenty-one in 1681 and 1682 respectively. R. Hooke and F. Aston, the Secretaries, were regular in their attendance but the average number of councillors, including the officers, who were present at the meetings did not reach nine. As a body, therefore, the Councils were not exceptionally active and left the President and the Secretaries to take the initiative.

Before the end of his tenure as President, Sir Christopher Wren drew the attention of the Council to the desirability of checking the indiscriminate admission of candidates who possessed inadequate qualifications, since the number of such cases was increasing. After discussing the matter fully the Council passed a special statute on 5 August 1682 in the following terms to deal with the matter:

Every person that would propose a Candidate shall first give in his name to some of the Council so that in the next Council it may be discoursed, *viva voce*, whether the person is known to be so qualified as in probability to be useful to the Society. And if the Council return no other answer but that they desire farther time to be acquainted with the gentleman proposed the proposer is to take that for an answer; and if they are well assured that the Candidate may be useful to the Society then the Candidate shall be proposed at the next meeting of the Society, and ballotted for, according to the Statute in that behalf; and shall immediately sign the usual bond, and pay his admission money as admission.

This may not have become the regular procedure at once, or was perhaps allowed to lapse after a few years, since it certainly was not in accordance with the Charter where the election of candidates to the Fellowship is declared to be the exclusive right of the President, Council and Fellows, and cannot be exercised by the Council alone. Some restriction on the admission of candidates who had insufficient qualifications was certainly needed, but it was not until 1730 that this matter was again reviewed.

At the end of 1682 the Council had to select a new candidate for the presidency, and invited John Evelyn to accept nomination. He writes in his Diary: 'I was exceedingly indangered and importun'd to stand the election having so many voices, but by the favour of my friends, and regard of my remote dwelling and now frequent infirmities, I desir'd that their suffrages might be transferred to Sir John Hoskins, one of the Masters in Chancery, a most learned virtuoso, as well as lawyer who accordingly was elected.' He only held the office for one year, during which he presided at eleven out of fourteen of the meetings of Council, but he continued to serve on the Council until his death in 1705; he very often presided at meetings of the Council as a Vice-President when the President of the day could not attend, and in many ways rendered most valuable service to the Society in whose well-being he was keenly interested.

In 1681 Dr Nehemiah Grew published his catalogue of the Society's Museum, and Daniel Colwall, who had from the first actively interested himself in its development, defrayed the expense of the illustrations. It is a folio volume of 435 pages with thirty-one sheets of plates, and appeared under the patronage of the Society. At this time the Museum or Repository contained several thousand specimens, mostly zoological, mineralogical or anthropological, which had been contributed by eighty-three donors, among whom were Boyle, Evelyn, Hooke, Pepys, Lord Brouncker and members of the Winthrop family in New England. In the following year Dr Grew was appointed Curator of the Museum.

At the Anniversary meeting of 1681 Robert Hooke, the First Secretary, was replaced by Dr Francis Aston, who took over the Minute-books, registers, etc., from him, as well as his secretarial duties. Francis Aston was born in 1645 and entered Trinity College, Cambridge, in 1661, being elected a Fellow in 1667. In 1669 he travelled on the Continent, and before starting he asked Isaac Newton, with whom he was on intimate terms, for advice on what he should observe; he received a long reply which would have perhaps been of greater use if the instructions had in some respects been more precise. He was a zealous and hard-working officer of the Society, taking his secretarial duties seriously; for instance, in 1680/1 when he was Second Secretary with Hooke as First Secretary, Aston was present at sixteen out of twenty-one Council meetings; during the next three years, when he was First Secretary, he attended forty-seven out of fifty-six meetings, and was prevented by illness from being at two or three others; the attendance of his colleagues, the Second Secretaries, during the same three years numbered only six. Later, Aston served on the Council in seven out of the sixteen years between 1694 and 1711, and in March 1712 was appointed a member of the committee which was to inspect and report upon the letters and papers relating to the dispute between the supporters of Newton and those of Leibniz over Newton's method of fluxions.

Aston never married. He died at Whitehall in June or July 1715 and bequeathed to the Royal Society the whole of his estate of forty-eight acres at Mablethorpe in Lincolnshire, which the Society still possesses, as well as a large number of books and other personal property. His portrait, by F. Kerseboom, is in the possession of the Society. The air-pump which is exhibited in the library of the Society was made for Aston in 1715 by Francis Hawksbee, junior. Aston had ordered it for himself just before his death, and when it was completed the Council decided to purchase it out of the funds left by him to the Society.

The other Secretary was Dr Robert Plot, an antiquary of wide interests and greatly interested in Natural History. He wrote a *Natural History of Oxfordshire* in which he describes many local customs and trades, and quotes many expressions and words in common use; he gives lists of plants and animals and describes the periodical stream at Assenden near Henley. Folklore figures largely, but he treats tales of haunted houses and supernatural manifestations with caution. He was interested in meteorology, and when discussing the barometric observations made at Oxford in 1684 in the *Philosophical Transactions*, made use of meteorological diagrams for the first time. He shows himself as an enthusiastic student of the new philosophy, but he had not yet wholly freed himself from the old legends and superstitions.

The quaint practice of the Fellows wearing a St Andrew's Cross of ribbon in their hats on the day of the Anniversary meeting is said by Aubrey to have come to an end while Plot was Secretary. In 1683 he was made the first 'Custos' of the Ashmolean Museum, and Professor of Chemistry at Oxford.

Hooke, being no longer a Secretary after 1681, then ceased to be responsible for publishing the *Philosophical Collections*, which he had undertaken in 1679; seven numbers appeared in 1679, 1681 and 1682. In their place the publication of the *Philosophical Transactions* was resumed in January 1683 and the 143rd number was published under the editorship of Dr R. Plot. Instructions were given in 1684 that duplicate copies of the Register and the Journal-books should be made in order to guard against their loss or damage by accident or fire, and this practice continued until the minutes were printed, which was introduced early in the nineteenth century. The Council also ordered that a particular (detailed) index to the registers, journals and accounts was to be kept regularly, but this does not appear to have been done.

During Sir J. Hoskins' year of office there was much activity in the Society and numerous experiments were carried out at the weekly meetings. Descriptions of what took place at these meetings are not numerous, and most of those which have been published are the work of visitors from foreign countries. Sir Thomas Molyneux, Bt., F.R.S., did however send to his brother, Mr William Molyneux, at Dublin a detailed description of one of these meetings in a letter dated 26 May 1683.[1] He writes:

The 23rd being Wednesday, I was at Gresham College; there I saw the Bibliotheca Norfolciana; afterwards I went to the Repository and viewed the rarities of that place, which do very much increase, there being new additions daily made. The Royal Society meeting together whilst I was in the house by the favour of Mr Hooke and Dr Green I was admitted to sit among them; the ceremony observed at their meeting is this: The President, one Sir John Hoskins, sits in a chair at the upper end of a table, with a cushion before him; the Secretary, Mr Aston, a very ingenious man, at the side on his left hand; he reads the heads one after another to be debated and discoursed of at the present meeting; as also whatever letters, experiments or informations, which have been sent in since their last meeting; of all which, as they are read, the Fellows which sit round the room, spake their sentiments and give their opinions if they think fitting; and of the chief matters discussed at this time was the cause of the inundation of the Nile. When this is over, if any of the Company have made experiments or have had particular information concerning anything worthy the notice of the Society, they then make it known.

[1] Published in Dublin, *Univ. Magazine*, 1841.

It will be noticed that only one secretary, F. Aston, attended the meeting, he being the First Secretary at the time, his colleague, Dr R. Plot, not being required to be present as well. In another letter Sir T. Molyneux, describing some of the Fellows whom he met while he was in London, writes:

Dr Green is a very civil obliging person [his name does not appear in the lists of the Society]; Hooke, the most ill-natured, conceited man in the world, hated and despised by most of the Royal Society, pretending to have all other inventions when once discovered by their authors to the world; Dr E. Tyson, a most understanding anatomist; Dr W. Croone and Dr F. Slare, both extraordinary civil and ingenious men; the first a very exact observer of the weather in whose study I saw several theremometers, hugroscopes and baruscopes.

As Sir John Hoskins was unwilling to serve as President for a second year, the Council recommended the Fellows at the Anniversary meeting of 1683 to elect Sir Cyril Wyche. He was a barrister and a Member of Parliament from 1661 to 1705; later he was secretary to the Lord Lieutenant of Ireland, and then one of the lords justices entrusted with the government of Ireland, 1693-5. He was sent as British Ambassador to Turkey in 1695. A statesman and lawyer of experience, he does not seem to have had any scientific knowledge or interests. He and his son were both elected Original Fellows of the Society in May 1663; he served fifteen times on the Council between 1681 and 1706, including 1683/4 when he was President, so it may be that he proved himself to be an experienced and useful member of Council even if his scientific knowledge was slight. In 1692, when he was Secretary of State for Ireland, he married the niece of John Evelyn, who describes him as being 'a man of perfect integrity, and a noble and learned gentleman'.

In 1684 a Society of about twenty members was founded in Dublin by William Molyneux; its aims were much the same as those of the Royal Society to which it regularly transmitted copies of its minutes and philosophical communications. A set of eleven advertisements, or rules for guidance, was drawn up for them by Sir William Petty, their president; in these he urged the members:

To apply themselves to the making of experiments rather than to reading discourses;
Not to despise simple experiments;
To weigh measure and compute in order to insure accuracy;
To provide instruments and apparatus;
To arrange for correspondents in numerous places from whom they may obtain from several places such observations as depend on the comparison of many experiments;
To calculate carefully their expenses and to procure the assistance of benefactors for extraordinary needs but not to pester the Society for useless members for the sake of their subscription.

The practical character of this advice is exactly what one would expect from a man whose outlook was so precise and businesslike as that of Petty, as was to be seen in his survey of the lands in Ireland, and in many of the proposals which he laid before the Society from time to time. The Dublin Society came to an end during the Jacobite revolt which took place in the last years of the century.

In November 1683 Michael Wicks, who had been appointed Clerk on the Society's staff in 1663, was dismissed. No reasons are given for this decision of the Council but probably his work had not given satisfaction to the officers. It was a year and a half before the balance of salary which was due to him was paid, a clear indication of the financial stringency prevailing at this time. The Council then ordered that a successor to Mr Wicks 'should be engaged to perform the laborious part of the Secretary's office', and on 5 December a Mr Cramer was appointed to be the new Clerk at £30 per annum. A week later however this decision had to be rescinded since it was pointed out to the Council that under the Charter the Clerk 'ought to be chosen by the whole Society in the same manner as the Treasurer and Secretarys'. Wicks therefore continued to act as Clerk until December 1684, after which he was employed from time to time as a copyist until 13 November 1695. He was then given a gratuity of £20 in addition to his payment as copyist. In this year Dr Denis Papin, who had frequently demonstrated experiments before the Society, was appointed to be a temporary Curator with a salary of £30 per annum. He had been elected a Fellow in 1680 and eight years later went to Marburg as Professor of Mathematics. His 'digester', which he demonstrated before the Society in 1684, was an apparatus for softening bones and extracting from them and other foods gravy and jelly, by treating them with steam under pressure. Evelyn describes a supper on 12 April 1682 at which bones of beef, mutton, birds and fish were treated in it and served up to the company which included several Fellows of the Society. The results were much approved by them.

In this year the death occurred of Dr William Croone or Croune, who was appointed on 28 November 1660 as the Society's first Register or Secretary. He was a Fellow of Emmanuel College, Cambridge, and from 1659 a Professor of Rhetoric at Gresham College, where he came into contact with other philosophers. In his will he planned the foundation of two lectureships, one at the Royal College of Physicians and the other at the Royal Society, but as he had not provided any endowment for them nothing came of his intention at the time; Lady Sadleir, his widow, repaired the omission by making provision for them in her will of September 1701.

In March 1683/4 Mr F. Aston, the Secretary, wrote to his colleague, Dr R. Plot, to say that he had received from Dr Martin Lister a paper on 'a new sort of Maps of Countrys, together with Tables of Sands and

Clays', the first geological paper which the Society had received. Mr Aston says:

I received from Mr Lister two schemes of the lands and clays found in England made by himself some twenty years since. He mentioned besides the great advantage of a map of the earths peculiar to some places and countries; he considers the sands and clays as two of the coats of the earth; the sand, probably the uppermost coat (for some reasons he gives) whence it comes to be washt to the body of rivers and the seashore. By this opinion I perceive may be given an account of sand beds too often attributed to the sea.

The paper was read before the Society and was published in volume 14 of the *Philosophical Transactions*. The author maintains that

we shall be better able to judge of the make of the earth, and of many phenomena belonging thereto when we have well and duely examined it, as far as human art can possibly reach, beginning from the outside downwards. As for the most inward and central parts thereof, I think we shall never be able to confute Gilbert's opinion, who will, not without reason, have it altogether iron.

Lister, who was born about 1638, was probably the Society's earliest geologist. About 1660 he began to note the occurrence of various kinds of soil and saw that this information could be conveniently shown on maps, the boundaries of different soils being indicated by appropriate colours. His paper, 'An Ingenious Proposal for a new sort of Maps of Countrys, together with Tables of Sands and Clays', was the earliest of its kind in this country. He was elected a Fellow of the Society in 1671 and served five or six times on the Council. On the other hand, some of Dr M. Lister's hypotheses lacked the support of reliable observations such as make his geology interesting, and some of his meteorological views are far from being convincing; in 1684, in a communication printed in the *Philosophical Transactions*, he ascribes the trade winds to the breathing of the plants in the Sargasso Sea, a fanciful suggestion which soon vanished before the correct explanation of them which was given by George Hadley in 1735, fifty years later.

Another pioneer geologist was Dr John Woodward who published his *Essay toward a Natural History of the Earth* in 1695. As a youth he was apprenticed to a linen-draper in London, but disliking the work he devoted all his spare time to science. Fortunately he came to the notice of Dr Peter Barwick, an eminent physician, who educated him with his own family. He then studied philosophy, anatomy and medicine with so much success that he was invited by Sir Ralph Dutton to go to his residence at Sherborne in Gloucestershire where he collected all the varieties of fossils and rocks that he could find, and extended his studies over as much of the surrounding country as he could reach. He was appointed Professor of Physic at Gresham College in 1692 and was elected a Fellow of the Society in 1693. He served on the Council for seven years between 1696 and 1710, but being very hot-tempered he at times came in conflict with his colleagues; in 1710 he grossly insulted Dr Hans Sloane. At this time geology was little studied in the Society, but a beginning had been made and much progress was recorded during the eighteenth century.

At the Anniversary meeting of 1684 Samuel Pepys, the diarist, was elected President and held the post for two years. Though not a man of science he was an exceptionally competent administrator, and was able to render the Society valuable service for the two years of his presidency. During that time he attended twelve out of the twenty-five meetings of Council which were held. He had been elected a Fellow of the Society in 1664, and in 1667 he and his friend Sir John Bankes had been very active in arranging the investment of the Bishop of Chester's legacy of £400 in fee-farm rents for the Society. The financial position still remained very unsatisfactory; forty-seven Fellows were ejected in July 1685 for the non-payment of the arrears of subscriptions which were due to the Society; at the Anniversary meeting no financial report was made, and only thirty-seven Fellows were present though some others came in later. The Fellows now totalled only 120 besides 20 foreign members, having fallen below the number reported at its first Anniversary meeting in 1663. At the Anniversary meeting of 1685 Francis Aston was re-elected Secretary, and Dr Tancred Robinson replaced Dr W. Musgrave who had served for a year only. At the next meeting of Council on 9 December both these officers unexpectedly submitted their resignations and declined to reconsider their action. That Aston, who had been a very efficient Secretary for four years and had been awarded an honorarium of £40 at the end of 1684 for his services to the Society besides having just been re-elected, should brusquely tender his resignation to the Council suggests that the causes of the trouble had started earlier. The minutes only record the fact of his resignation, but a letter from Edmond Halley to his friend William Molyneux of Dublin, dated 27 March 1686, gives fuller details. He writes:

On St Andrew's Day last, being our anniversary day of election, Mr Pepys was continued President, Mr Aston Secretary and Dr Tancred Robinson chosen in the room of Mr Musgrave. Everybody seemed satisfied, and no discontent appeared anywhere when on a sudden Mr Aston, willing, as I suppose, to gain better terms of reward from the Society than formerly, on December 9th in Council declared that he would not serve them as secretary, and therefore desired them to provide some other to supply that office; and that after such a passionate manner that I fear he has lost several of his friends by it. The Council resolved not to be so served for the future, thought it expedient to have only honorary secretaries, and a clerk or amanuensis upon whom the whole burden of the business should lie, and to give him a fixed salary, so as to make it worth his while, and he to be accountable to the Secretaries for the performance of his office. According to which resolutions Sir John Hoskins and Dr Gale were chosen Secretaries; and on January 27th last they chose me for their under-officer with a promise of a salary of fifty pounds per annum at least.

To this letter Mr Molyneux replied on 8 April 1686 saying he had heard of the late disturbances in the Society but was unwilling to trouble Mr Halley until matters were settled; he expresses himself well satisfied with the solution. He adds that to him it had always appeared unsatisfactory that the Secretaries should be annually elective, and that they had no established salary to encourage and recompense them for their labour. A Secretary had to spend his whole time receiving and answering letters and keeping the registers as well as other troublesome work; he expresses the hope that the Society will reorganize this part of their administration.

It is clear therefore that Dr F. Aston's action was due to the amount of work which his office entailed and for the performance of which he had very inadequate assistance; the brusqueness of his protest was probably due to his earlier complaints having produced no more effect than that of Oldenburg on the same subject in 1664. In 1683, after having been a Secretary for two years, Aston was excused paying the annual subscription of a Fellow since the Society had not been able to grant him any honorarium for those two years, a very inadequate recompense.

The Council were summoned to meet a week later when the President, Samuel Pepys, and sixteen members attended, and decided that a clerk should be appointed to assist the Secretaries in their work. The post was considered to be an attractive one in those days for the applicants included Dr Hans Sloane, Dr Denis Papin, the inventor of the Papin digester, who was at the time a demonstrator of the Society, a Mr Salusbury and Edmond Halley, who was then a Fellow; at each of two ballots at a meeting of the Society Halley gained the largest number of votes and was therefore appointed. He was instructed to ascertain what inventories of the Society's papers had been made when they were taken over by Mr Hooke and Dr Aston, to see what books and papers were missing and to enquire for them; he was also to attend all meetings of Council and to take down the minutes; while Dr Aston was requested to hand over the documents in his charge and to bring the Minute-book and Journal up to date. It was found that the books in the Secretary's keeping had been last checked on 27 February 1684 and a list of them is given in the Council minutes of that date. The whole matter seems to have been settled amicably, perhaps by the persuasive ability of the President, Samuel Pepys, whose long experience of men and affairs, and his knowledge of administration, were doubtless of great value to the Society. The Council granted an honorarium of £60 to Dr Aston for his past services, and he was excused from paying subscriptions until September 1687 since the honoraria, which the Council had been unable to pay to him when they were due, were still outstanding. They also granted a piece of silver weighing 60 oz. to Mr Musgrave, who had been Aston's colleague in 1684. We may conclude therefore that Aston had more grounds for his action than were known to Halley when he wrote to his friend Molyneux.

Charles II, the Patron of the Society, died on 6 February 1684/5. For some years previously the king had taken little interest in the affairs of the Society. No allusion to his death occurs in the Journal, and the Society held its usual meeting on the ninth of the same month. No effort was made to obtain the patronage of his successor, James II, who was well known to have little sympathy with the Society's work and aims. The name of Charles II will however always be honourably associated with the Royal Society for by giving his approval to the Society of Philosophers in 1660, and to the change of its name to that of the 'Royal Society' a few months later, he gave it assurance of his favour and support without which its beginnings would have been even more difficult than they were. With the grant to it of the First and Second Charters and by declaring himself its Patron he assured it of his protection against malicious attacks and calumnies which it was to need before long.

In January 1683/4 Halley met Sir Christopher Wren and Hooke in London and discussed with them the conclusion that the centripetal force of attraction was inversely proportional to the square of the distance since he found himself unable to prove it. Hooke maintained that upon that principle all the laws of the celestial motions were to be demonstrated, and that he had himself done this. Sir Christopher, to encourage the enquiry being undertaken, said that he would give either Halley or Hooke two months in which to bring to him a convincing demonstration of the principle; he would give a present of a book of the value of forty shillings to whichever of them did it. Hooke said that he had done it but would conceal his methods until others had tried and failed. Halley went home to struggle with the problem; but in August he went to Cambridge, having probably heard from Wren that Newton had discussed the subject with him some years before. Halley, says J. Conduitt who has left a record of the interview, asked Newton what the curve described by the planets would be on the supposition that gravity diminished as the square of the distance. Newton immediately answered, an ellipse; and on being asked by Halley how he knew this, replied 'I have calculated it'. He could not however find the calculation but promised to send it to Halley. Newton then set to work on recalculating the proof, and in a letter to Aston says that the work was requiring more time than he expected. He then undertook the greater task of elaborating a cosmic system developed according to rigorous mathematical formulation, and dependent only on his general law of attraction. By Easter 1685 the manuscript of the first book, except for some corrections and additions, was finished; and during the summer the second book was composed in a preliminary form. In the autumn he wrote to Flamsteed at Greenwich asking for certain astronomical data, and also for information about the tides, the major axes of Jupiter, Saturn and their satellites.

At the meeting of the Society which was held on 28 April 1686 Dr

Nathaniel Vincent, a Fellow of the Society, presented a manuscript entitled 'Philosophiae Naturalis Principia Mathematica' and dedicated to the Society by Mr Isaac Newton, wherein he deduces mathematically all the phenomena of the celestial motions assuming only a gravitation towards the centre of the sun decreasing as the squares of the distances therefrom reciprocally. It was 'ordered that a letter of thanks be written to Mr Newton, and that the printing of his book be referred to the consideration of the Council; and that in the meantime the book be put into the hands of Mr Halley to make a report thereof to the Council'. On 19 May the Society resolved that 'Mr Newton's "Philosophiae Naturalis Principia Mathematica" be printed forthwith in quarto, in a fair letter; and that a letter be written to him to signify the Society's resolution and to desire his opinion as to the volume, cuts', etc. Halley wrote to Newton conveying this decision of the Society and informing him that Hooke was contending that he had supplied some of the ideas and seemed to expect that he should be mentioned in the preface. When the Council met on 2 June they ordered that 'Mr Newton's book be printed and that Mr Halley undertake the business of looking after it, and printing it at his own charge; which he engaged to do'. The Council, who were much better informed of the state of the Society's finances than the Society generally, knew that the publication of Willoughby's De Historia Piscium had exhausted the funds at their disposal, and that even the salaries of their staff were in arrear. It was most fortunate, therefore, that there was one of the Society who was able and willing to come to its assistance in its dilemma.

Halley's father was a prosperous soap-boiler and merchant in the City of London, and from 1676 had given his son an allowance of £300 a year, which in those days was a large sum; so that Halley may have been able to set aside some of this allowance. In 1682 Halley married Mary Tooke, who succeeded to property in Norfolk and in Aldersgate Street, London. It seems clear, therefore, he was in fairly comfortable circumstances when he undertook to finance the printing and publication of the 'Principia' in 1686. Halley's father committed suicide in 1684 and Halley then became involved in a lawsuit with his stepmother on questions relating to his father's property and estate, but it is not certain what the outcome of it was. It would appear that it was not the salary of £50 attached to the clerkship of the Society which attracted him in 1686 so much as the interesting character of the work; and it was because he was fully convinced of the importance of the 'Principia' and of the necessity for its early publication that he readily accepted the risk of a possible financial loss; but the edition was quickly sold out and a second edition was soon exhausted, so there can be little doubt that he must have recovered most, if not the whole, of his outlay.[1] His action was the more helpful seeing how difficult

---

[1] I am indebted to Mr H. W. Robinson, Librarian of the Royal Society, for this information.

it was at that time to induce booksellers to print and publish mathematical works, which often lay on their hands for long periods without being sold. The 'Principia' was published about the middle of 1687, and the price of a copy of the first edition did not exceed twelve shillings for a volume consisting of sixty-four sheets and more than one hundred illustrations. It is indeed remarkable that Halley should have carried out his task in so short a time though Newton was not eager for speedy publication. A letter from him to Halley which is in the archives of the Society says: 'Pray take your own time; if you meet with anything which you think needs either correcting, or further explaining, be pleased to signify it to me'; and adds: 'I wish the printer be careful to mend all you note.'

In June 1686 the Council authorized the President, Samuel Pepys, to license the printing of Mr Newton's book entitled 'Philosophiae Naturalis Principia Mathematica', and dedicated to the Society, which he did on 5 July of this year.

At the Anniversary meeting of 1686 Pepys ceased to be President, and John, Earl of Carbery, who was elected as his successor, occupied the office of President for the three following years. He was a barrister and a Member of Parliament as well as having been the Governor of Jamaica from 1674 to 1678. He had no scientific interests and only attended five of the sixteen meetings of Council which were held during the three years that he was President; only two of these were held during 1689 as compared with fifteen in 1686, which does not suggest that administrative activity was being maintained. There were however some signs of improvement; after the two Secretaries, Sir John Hoskins and Dr T. Gale, had held office for two years, 1685–7, the former was replaced by Richard Waller, who was a zoologist, a good linguist and a keen man of business; he seems to have carried on his work in London as he had an address in Broad Street besides a country estate at Northaw in Hertfordshire. He had been elected a Fellow in 1681 and soon became interested in the administration of the Society. In 1687 he became at first the Second Secretary, but as Dr Gale rarely attended a Council meeting Waller virtually became the First Secretary and carried out all the duties of that office regularly until 1694, when Dr Hans Sloane replaced Dr Gale. During these seven years Waller attended eighteen out of the twenty-seven Council meetings which were held, while his colleague Dr Gale was present at only one.

At the Anniversary meeting of 1689 Thomas, Earl of Pembroke, was elected President in the place of the Earl of Carbery; his interests were centred in art and archaeology rather than in natural science so it was not to be expected that he would do much to promote activity in the Society. He held office for only one year and was not present at a single meeting of either the Council or the Society. After this unsatisfactory experience of electing a peer as their President the Council again endeavoured to prevail upon John Evelyn to accept nomination, but he was unwilling, and as he says in his Diary 'with great difficulty devolv'd the election on Sir Robert Southwell, Secretary of State to King William in Ireland'. His suggestion, which was a good one, was adopted and Sir Robert Southwell was re-elected annually as President for five years, 1690 to 1695. During his tenure of office the Council did not meet very frequently but out of the seventeen meetings which were held the President was present at sixteen. He communicated several papers to the Society, most of them on physiological and chemical subjects. Early in his career he had been employed in Dublin where he met Sir William Petty, who was then at work on the survey of the lands which were to be transferred to the army and to those of the so-called adventurers who had assisted in dispossessing the Irish. They became great friends and Southwell helped Petty on many occasions when the latter's impulsive actions were likely to get him into serious trouble. He was also a close friend of Evelyn who describes him as a sober, wise and virtuous gentleman.

In 1691 the Hon. Robert Boyle, one of the Society's most distinguished Fellows, died on 30 December at the age of sixty-five. He had been in failing health for some time and of late had taken but little active part in the administration of the Society, not having served on the Council since 1681. The sealed packets which he had deposited with the Society for safe custody in 1680, 1683 and 1684 were returned to his executors in February 1692.

One of the first acts of the new President was to impress on the Council the importance of resuming the publication of the *Philosophical Transactions* and of issuing parts of each volume as regularly as might be practicable. The Council therefore resolved 'that there shall be Transactions printed, and that the Society will consider of the means for effectually doing it. Dr E. Tyson, M.D., Dr F. Slare, M.D., Mr R. Waller (Secretary) and Mr R. Hooke, were desired to be assistants to E. Halley in compiling and drawing up the Transactions.' Steps were therefore taken to collect papers for publication and in February the 192nd number, forming part of the 17th volume, was published under the editorship of Halley. The remainder of the 17th and the whole of the 18th volumes appeared under the editorship of Richard Waller, the Secretary. Weld suggests that the improvement in the Society's invested funds in 1697, which were reported as now amounting to £250 of East India Company Stock and £800 of Africa Company Stock, was due to a considerable accession of Fellows in consequence of the reappearance of the *Transactions*; but this cannot have been the case, for the average yearly increase in admissions did not exceed seven, besides three foreign members for the years 1691-1700, or practically the same as for the whole thirty-seven years 1664-1700. The election of Dr Hans Sloane as Secretary at the Anniversary meeting of 1694 is much more likely to have produced such a result from a stricter treatment of those Fellows who were still in arrear with the payment of their subscriptions. The Society was just emerging from its most difficult period and the average number of Ordinary Fellows in the four five-year periods from 1691 to 1710 were 115, 125, 131 and 149. An improvement was indeed perceptible but no one could say yet if it would be maintained.

Towards the end of the century several foreign writers repeated a rumour then current to the effect that the Montmor Academy was the inspiration and source of the Royal Society. A recent investigation which has been made into the history of French scientific institutions of this time by Professor H. Brown of Baltimore University has proved the inaccuracy of this view very conclusively. Cassini, an eminent Italian astronomer and geodesist, in his *Recueil de Plusieurs Discours* of 1693, says that several foreigners were seen at the meetings of the Montmor Academy, and among others Mr Oldenburg who, having later gone into England and having inspired the English to form similar conferences, was the occasion of the foundation of the Royal Society. M. Thévenot, too, in his *Recueil de Voyages* of 1681, speaks of MM. de Rawneley (Ranelagh) and Oldenburg as 'having visited our Assemblies several times, and later established in England the one which still survives under the name of the Royal Society'. Fontenelle, in his *Histoire de l'Académie des Sciences*, writes that it is at least certain that the English gentlemen who laid the first foundations of the Royal Society of London had travelled in France, and had met in the houses of Montmor and of Thévenot. Neither he nor Cassini nor Thévenot can have verified their statements for Bishop Sprat's *History of the Royal Society* was translated into French and published in Geneva in 1669 and in Paris in 1670. Sprat certainly describes the beginning of the Society as having been laid at Wadham College, but Dr J. Wallis in his *Defence of the Royal Society* corrected this in 1678 and described the meetings which had been held in Gresham College as early as 1645. Robert Hooke, in a rather brusquely worded 'Answer to some Particular Claims of M. Cassini', says:

M. Cassini is in error concerning the beginning and original of the Royal Society....He makes Mr Oldenburg to have been the instrument who inspired the English with the desire to imitate the French in having philosophical clubs or meetings, and that this was the occasion of founding the Royal Society and making the French the first. I will not say that Mr Oldenburg did rather inspire the French to follow the English, or at least did help them and hinder us. But it is well known who were the principal men that began and promoted that design both in London and Oxford; and that a long while before Mr Oldenburg came into England; but the Society itself was begun before he came thither; and those who then knew Mr Oldenburg understood well enough how little he himself knew of philosophical matters.

Oldenburg became tutor to Richard Jones, the son of Lord Ranelagh, in 1654, and entered Oxford for study in June 1656. In 1657 he left Oxford with his pupil to travel abroad, and they were at Saumur until 1658, at Paris in 1659 and 1660, where he visited the Montmor Academy and the Cabinet Dupuy, and visited Leyden in 1661, in which year they returned to England. Richard Jones' mother was a sister of Robert Boyle, and in this way, no doubt, Oldenburg came into contact with some of those who formed the Philosophers' Society in 1660.

At the Anniversary meeting of 1694 Dr Hans Sloane had been elected a Secretary in place of Dr T. Gale and thereby became the colleague of Richard Waller who had done most of the work since he took up the secretaryship in 1687. Whether it was by mutual arrangement or by permission of Council that Sloane now became the First Secretary is not clear, but he attended nearly all the Council meetings, while Waller was seldom present. Sloane was living in London and was an energetic and zealous administrator, so he was able to commence his long career of service to the Society first as a Secretary, then as a councillor then as President, and lastly as councillor again until his death in 1753. It was not long before he made his influence felt, for in 1698 he arranged, probably with the support of Sir John Hoskins, a Vice-President, for the Council to meet no less than fifteen times to discuss the Society's financial position; he also revived the foreign correspondence of the Society which had been carried on actively until 1677, but after it had been placed in the charge of Dr T. Gale in 1680 little attention seems to have been paid to it. During the secretaryship of Dr Hans Sloane the letter-books contain copies of numerous letters which he addressed to persons at home and abroad inviting communications on subjects related to the aims of the Society; one of these reads:

The Royal Society are resolved to prosecute vigorously the whole design of their institution, and accordingly they desire you will be pleased to give them an account of what you meet with or hear of, that is curious in nature, or in any way tending to the advancement of natural knowledge, or useful arts. They in return will always be glad to serve you in anything in their power.

Thus he revived the practice which Oldenburg had initiated of maintaining close intercourse with the scientific men of other countries. He evidently appreciated the urgent need for the reorganization of the Society if it was to fulfil the aims of its founders, and with the opening of the eighteenth century the prospect seemed rather brighter, but he realized that for this to be maintained a capable and energetic president would be essential.

Dr Hans Sloane was held in great esteem both in London and in Paris during the first half of the eighteenth century. Such accounts of him as exist describe him as an eminent physician—he was the President of the Royal College of Physicians 1719-35—or as a collector of great zeal and discrimination. Not only did he bring together a large collection of minerals, botanical and zoological specimens, but he also formed a large library, a fine collection of manuscripts and of objects of art, all of which he bequeathed to the nation for the payment of £20,000 to his executors and which, with the Cotton Library and the Harleian Manuscripts, formed the nucleus of the British Museum. But this leaves out of account the great services which he rendered to science by his work in organizing the Royal Society, for this is less generally known, or at least is rarely referred to.

Hans Sloane was the seventh and youngest son of Alexander Sloane, a Scotsman who had married a daughter of Dr George Hickes, Prebendary of Winchester, and had settled in Ireland on receiving the appointment of receiver-general of the estates of Lord Claveboy, afterwards Earl of Clanricarde. He was born at Killeagh, County Down, on 16 April 1660; and from his youth he gave evidence of keen powers of observation and showed that he was mainly attracted by the study of the natural sciences; these interests suggested that he would be most suited for the medical profession. A severe illness interrupted his studies and delayed them for three years, so that it was not until his eighteenth year that he came to London to study chemistry and botany before taking up the profession of medicine. Here he developed his knowledge of botany with the help of Mr Watts, the Keeper of the Botanical Gardens at Chelsea, where he was a frequent visitor; he also made the acquaintance of William Courten. For part at any rate of the time which he spent in London he lived with Dr Thomas Sydenham, who was one of the most eminent physicians of his time and was greatly esteemed abroad. He encouraged Sloane in his medical studies, and probably recommended him to take his medical degree in France as he had himself done at Montpelier in 1659. While Sloane was in London he also made the acquaintance of the Hon. Robert Boyle and of John Ray, the naturalist, both of whom were later to be good friends to him. In due course he went to France, accompanied by Dr Tancred Robinson, a Fellow of the Royal Society, with the object of completing his medical education under more favourable conditions than he could find in England.

On landing at Dieppe he had the good fortune to make the acquaintance of Nicolas Lemery, an eminent French chemist, who assisted him in many ways during his stay in Paris. In July 1683 he took his degree as Doctor of Medicine at the University of Orange; after which he went to Montpelier with the introductions which Dr Sydenham had given to him, and remained there until the following May when he went back to Paris. At the end of 1684 he returned to London and in January 1685 he was elected a Fellow of the Society; a month later he was a candidate for the post of Clerk, or Assistant to the Secretaries, which was then vacant, but was not successful, E. Halley being selected. In 1687 he became a Fellow of the Royal College of Physicians. In the same year when Christopher Monk, second Duke of Albemarle, was appointed Governor-General of the West India Colonies, Dr Sydenham recommended that Sloane should accompany him as his medical adviser and as chief physician to the fleet. Sloane, on being offered the appointment, was greatly attracted by the opportunity which it would give him of scientific work in a little known region under exceptionally favourable conditions. He therefore accepted the post and asked that the ships' surgeons should be under his orders, that his salary should be six hundred pounds a year, together with three hundred pounds for the necessary preparations, and that if the fleet should be re-called he should have leave to stay in the West Indies. These requests were granted and he embarked in the Duke's frigate *Assistance* on 12 September 1687. Before sailing Sloane wrote to his friend Courten, whose acquaint-ance he had made in London a few years before, 'I design to send you what is curious from the several islands we land at,—which will be most of our plantations', adding 'I am extremely obliged to you beyond any in the world'. This was Sloane's introduction to collecting of which he afterwards made so striking a success. Courten was a member of a family of wealthy merchants trading in both East and West Indies where they had built up a great fortune. This was now much diminished, but Sloane's friend was still well off and devoted all his time and resources to forming a collection of rarities and curiosities of exceptional merit. Evelyn, who visited it in 1686, describes it in his Diary as 'such a collection as I have never seen in all my travels abroad—either of private gentlemen or of princes. It consists of miniatures, drawings, shells, insects, medals... minerals; all very perfect and rare of their kind; especially his books of birds, fishes, flowers and shells drawn and miniatured to the life.' Its value was estimated at that time to be about eight thousand pounds.

The first land visited by the expedition was Madeira, where Sloane made the most of his opportunities to collect everything that appeared to him to be of value or interest. Barbados was reached in November and here, as well as in other islands, he collected assiduously until the end of the following year (1688). Near the close of that year the Duke of Albemarle died suddenly, and Sloane accompanied the fleet which set sail for England on 16 March 1689. He reached Plymouth at the end of May with the whole of his collections; these aroused great interest among botanists on account of the large number of new species which they included—his friend John Ray being most enthusiastic. The Duchess of Albemarle was most generous in allowing him to occupy rooms in one or other of her houses while he was putting his collections into order.

Among those who were admitted to the Fellowship of the Society though they were not versed in experimental science was John Locke, the philosopher, who was elected in 1668. He was educated at Oxford and had published his first essay on Toleration in the previous year. As Newton had clarified the views of the late seventeenth century in physics so did Locke in philosophy, and their teachings formed the basis of eighteenth-century learning. His prose style is that at which the pioneers of the Royal

Society aimed and which Sprat so warmly commended, that is to say gravely reasoned and free from rhetorical ornament. He knew many Fellows of the Royal Society and later became one of Newton's most intimate friends. In 1693, Newton, who was then seriously ill with a nervous breakdown brought on by overwork, wrote in harsh terms to Locke and also to Pepys complaining of their attitude towards him. Fortunately both of them were reasonable men and realized that he must be in very bad health to have changed his attitude towards them so abruptly. Their enquiries confirmed this, and their replies, which assured him of their unbroken friendship towards him and the value they attached to his regard for them, cleared away all misunderstandings and re-established their mutual regard.[1]

For the next eight years the Council did not meet often but Sloane attended thirty-seven out of the forty-three meetings which were held. He lost no time in showing his industry and zeal and in 1699 fifteen Council meetings were held, an event which had not occurred since Sir Christopher Wren's presidency nearly twenty years before. The cause was, as usual, the desperate state of the Society's finances; for twenty years past no report on them had been made at the Anniversary meetings, and the last time that one had been presented the sum owing to the Society for unpaid fees and subscriptions had reached £2000. The number of Fellows was now at its lowest, being about 120. Sloane had lost no time in taking up his duties and it may be that the practice of advertising the publication of the *Philosophical Transactions* about this time was due to him; volume No. 18 was announced in *Houghton's Collections* for 15 February 1694/5 where the titles of the various articles with the names of their authors were given.

In 1694 Charles Montagu, who was then a Lord of the Treasury, arranged for the appointment of Lord Somers, Isaac Newton, J. Locke and E. Halley as a committee to consider the re-coinage of the currency and to draft a bill to this effect which would be laid before Parliament. This did not concern the Royal Society directly but all the members of the committee were Fellows of that body and the outcome of their labours had important results for it.

Early in this year Dr J. Wallis suggested to R. Waller, one of the Secretaries, that Newton should be asked to communicate any scientific work of his that was ready in order that it might be published by the Society and thus brought to the knowledge of scientific men in this country and elsewhere. On 4 July 1694, therefore, the Society 'ordered that a letter be written to Mr Isaac Newton praying that he will please to communicate to the Society in order to be published his treatise of light and colours and what other mathematical or physical treatises he has ready by him'. A few years later when conditions had changed Newton did communicate to the Society his treatise on 'Optics' and it was published in 1704.

[1] Cf. L. T. More, *Isaac Newton, a Biography*, pp. 382-7.

In March 1695/6 Mr Montagu wrote to Newton to inform him that King William had approved of his appointment as Warden of the Mint, thus providing him with an official post and an assured salary. He therefore moved to London and was brought into much closer contact with the Royal Society than had been possible so long as he lived at Cambridge. He was a member of its Council in 1697–8 and again in 1699–1700, as also was Robert Hooke, but they were never present at the same meeting; in 1697–8 there were five meetings and in 1699–1700 there were four only; of these Hooke was present at five in 1698, and two in 1700, but Newton was not present at any of them.

At the end of 1695 Sir Robert Southwell ceased to hold the office of President and Mr Charles Montagu was elected to replace him. He was a man of great ability and early became one of the Lords of the Treasury, being raised to the office of Chancellor of the Exchequer in 1694. He was President for only three years and during this time his public duties prevented him from occupying himself with the affairs of the Society, so that he was not present at any of the ten meetings of Council which were held.

In 1698 Halley resigned his post as Clerk to the Society in order to take charge of a scientific expedition in the ship *Paramour* which had been fitted out by the Government for determining the latitudes and longitudes of a number of places in the English settlements in America; as well as to make observations in terrestrial magnetism in continuation of the work which he had carried out in this subject some years earlier. In this ship, on the first sea journey to be undertaken for a purely scientific object, Halley sailed for two years, October 1698 to September 1700, and reached 52° 24′ south latitude in the Atlantic. The account of his voyage was not published until 1775[1] though his journal of it was preserved in the records of the Board of Longitude in London. Nevertheless immediately on his return to England, Halley set to work on his material and produced his famous magnetic sea charts, in which the distribution of the magnetic declination over the oceans was shown graphically. His Atlantic chart was published in 1702 and was dedicated to Queen Anne's consort, Prince George of Denmark.[2] During his absence his place on the staff of the Society was taken by Jezreel Jones who was then a traveller and collector to the Society.

At the Anniversary meeting of 1698 the Fellows elected Lord Somers to succeed Lord Halifax as President. He had had a brilliant legal career and became Attorney-General in 1692, being appointed Lord Chancellor in 1697. His duties as one of the great officers of State left him little time for those of the President of the Society; he consequently attended none of

[1] *A Collection of voyages chiefly in the Southern Atlantic Ocean.* Dalrymple. London, 1775.
[2] 'Edmond Halley and the Story of his Charts.' Professor S. Chapman, *Royal Astronomical Society, Occasional Notes.* June, 1941.

the meetings of Council, his place being taken by Sir John Hoskins or Sir R. Southwell, the Vice-Presidents, who had Dr Hans Sloane, the First Secretary, to assist them. The President did however attend the Ordinary meetings from time to time.

At the end of the seventeenth century the position of the Society could not be described as satisfactory. The number of Fellows had been falling steadily for twenty years, and though there had been a slight improvement recently no one could say whether this marked the turn of the tide, or was but a temporary recovery which was unlikely to be maintained. Of the financial position little can be said, for no reports had been made to the Anniversary meetings since 1677, and these were not resumed, even partially, until 1716.

The selection of peers and statesmen for the position of President had not been a success, for none of them had held the office long enough to effect any real improvements in the organization. Hans Sloane had done all that he could as First Secretary and was doing his best to introduce several improvements, but he lacked the support which a capable and energetic President could have given him. Newton was available, and was the only one among the Fellows with the prestige, ability and mental powers to grasp the situation and to remedy it; but he would not accept nomination so long as Hooke's bitter criticisms would, he knew, be directed against him, and might involve him in controversies which it was his constant aim to avoid.

It was known that Gresham College would not be able to provide accommodation for the Society indefinitely, and so far all attempts to make other arrangements had failed. With the abdication of James II and the acceptance of the throne by King William the disturbed conditions which had prevailed in former years had come to an end and the times appeared to be exceptionally favourable for the Society's return to its earlier prosperity if only a new start could be made on better lines, and if any improvement which was achieved were not merely maintained, but developed. Unless this was practicable it seemed very doubtful whether continuance of the Society's activities could be guaranteed; and yet was there anything in those days which was so well worth preserving? The faults and difficulties which had brought it so low were evident and fully recognized. Newton, who had resigned his Lucasian professorship, had settled in London, where as Master of the Mint he was in possession of a salary adequate for his requirements, and the demands of his post on his time and energy would not prevent him from undertaking the Presidency of the Society if he was prepared to accept it.

The standard of learning among the more eminent of the philosophers of the seventeenth century, to whose activities we owe the formation of the Royal Society, was exceptionally high, and the period has been fittingly described as the century of genius. Not only had they mastered what was then known of their special subjects such as mathematics or astronomy but had acquired much of what the education of those days included, namely a thorough classical training and also philosophy as it was then understood. The Universities required that candidates for a degree should satisfy the examiners in theology and in the philosophy of Aristotle. Mathematicians often extended their studies to some branches of medicine, and students of medicine sometimes acquired a wide knowledge of mathematics and other subjects. Sir Robert Moray studied mathematics, classics, music and chemistry, though he was a soldier and diplomat rather than a learned student; Sir William Petty not only received a thorough classical education from his Jesuit teachers at Caen but also studied anatomy and medicine in French and Dutch Universities as well as mathematics and other subjects; and as much may be said of others. What the scholarship of such men included has been summarized already for several of them when their activities both before and during the early days of the Society were described; but it has to be remembered that they were but few especially at first. In the latter half of the seventeenth century men of science in the Society, even when this description was freely applied, did not exceed thirty at first and at no time were there more than fifty. Even in the eighteenth century when the membership of the Society was growing steadily there were at first about fifty and at no time did they exceed one hundred and fifty; of these the greater number had not attained any considerable distinction in their studies, and never added materially to the reputation of the Society by their work. Throughout the history of the Society we find that its scientific reputation was created, maintained and advanced by quite a small number of its Fellows who were endowed with exceptional talent and scientific foresight. It is to Sir Isaac Newton and those of his successors who were similarly gifted that the Society owes its scientific reputation which has been gradually built up during the past three centuries. The outstanding achievements of the leaders of science occupy an enduring place in the Society's record at all periods of its history, but there were also a much larger number of men who devoted their energies to the promotion of natural science on a more modest scale, without ever rising to the brilliant successes achieved by the few. They played however an important part as members of Councils, and of committees in dealing with the Society's business both scientific and administrative. It is from the contributions of these men as they have been recorded in the minutes of the Council and the journals of the Society's meetings that its history can be reconstructed.

Unfortunately the Society's policy was not to make as full a use of such men as it might have done, for both Councils and advisory committees contained more Fellows whose interests were literary than those who could by any extension of the term be classed as scientific. It might be sometimes that those who had a special aptitude for administrative work rendered useful service on a Council, but the inclusion of from ten to fifteen non-scientific Fellows among the twenty-one councillors undoubtedly weakened its scientific influence on the Council's policy (cf. Appendix II C).

The account which has been given of the formation and development of the Society during the second half of the seventeenth century has recorded the names of those who were the most active, and we have seen that in these early days mathematics and medicine attracted the most active students of the new philosophy. Newton, James Gregory of St Andrews, who was described by his colleagues of those days as second only to Newton, and a few others, set an example to many: and the work of men like W. Harvey, J. Goddard, G. Ent, R. Lower did the same in medicine and anatomy. W. Gilbert, R. Boyle and R. Hooke showed the way to chemists and physicists, though these were few in number at the time, and J. Evelyn, Ray, Hans Sloane and others showed what botanists and zoologists could do on similar lines.

The communications which were offered to or asked for by the Philosophers' Society and later by the Royal Society show that practical applications of science, and the methods of industrial technology were much more appreciated in the earlier years by the Fellows than more academic studies of mathematical and physical subjects. This is not surprising since the latter were to some extent encouraged at the Universities, especially at Cambridge, though their activities were mostly devoted to classical studies. The same experience was met with in Paris when the influence of the State was mainly exerted in promoting industrial progress, and technical efficiency rather than scientific education. Though the number of those who were devoting themselves to promoting the new philosophy was small, and communication between them was slow and difficult, the close contact which they maintained with skilled opticians, instrument makers, clock- and watchmakers and other skilled artisans, acted powerfully to the advantage of both parties. The acquaintanceship with such men was greatly valued by R. Boyle, J. Gregory, J. Flamsteed and others, while R. Hooke, himself a highly skilled physicist and mechanist, gave much help and advice to technical industry and profited much by their co-operation in return.

The Royal Society and academies in Paris and elsewhere collected a large amount of information about industrial processes, but it was not until the next century that much of it began to appear in the educational schemes. Neither the Universities nor the professional institutions appreciated the need for this widening of their outlook, and those who sought for something more than a stereotyped classical and literary training were obliged to seek it in the Universities of Edinburgh, Glasgow, Utrecht, Leyden, Paris and Montpelier and elsewhere.

In the next century the growing communities of the dissenting creeds encouraged a more enlightened spirit of enquiry in their academies, which was welcomed by the scientific Fellows of the Society, many of whom had joined the presbyterian or puritan fellowship. Sir John Finch and Anthony Lowther, a brother-in-law of William Penn, were both Original Fellows of the Society, while Dr Richard Lower, also of the Society of Friends, was elected a Fellow in 1667.

In several ways the existence of a majority of non-scientific Fellows was already showing itself to be a serious hindrance, but many years were to pass before it was effectively dealt with.

During the last years of the seventeenth century the proportion of men of science in the Society varied between twenty and thirty per cent of the Fellowship and there was no sign as yet that it was increasing. In the Councils the scientific members were from seven to eight in number out of the total of twenty-one councillors, so in neither sphere were they numerous enough to carry much weight. At the time this was not of very much importance, for the minority included several of the pioneer philosophers who had been active in bringing about the foundation of the Society, and their names have already been recorded here together with the branch of knowledge in which they were specially active. Even though the Councils and the Ordinary meetings of the Society might not always maintain the standard which was desired, there were always several of these men carrying on their own researches in the fields of investigation which they had made their own. It was the original work of these men at this time, and of others like them in future years, which built up and continually advanced the prestige of the Society despite the fact that an unscientific majority controlled the Councils and the Society's meetings. The situation was a strange one; the early pioneers probably thought that they and their successors would be able to maintain the control of the Society's scientific activity, and underestimated the weight of the dead hand of the majority. On the other hand, there is no doubt that the existence of an organized and recognized Society as a meeting point for its members, where fellow-workers could meet and discuss matters that concerned them, and which would publish such original work by its Fellows as was considered to be new and important, rendered most valuable service to science. These more eminent Fellows of the Society, who devoted themselves untiringly to such research as they considered to be important, did much for many years to counteract the influence of the unscientific majority.

REFERENCES

BIRCH, T. *History of the Royal Society.* 1660–1687.

BROWN, Professor H. *Scientific Organizations in Seventeenth-century France.* Baltimore, 1934.

BRYANT, A. *Samuel Pepys: The Years of Peril.* Cambridge, 1935.

CLARK, G. N. *Science and Social Welfare in the age of Newton.* Oxford, 1937.

FITZMAURICE, Lord EDWARD. *Life of Sir William Petty.* London, 1895.

HOOKE, R. *Diary.* Edited by W. H. ROBINSON and W. ADAMS.

MASSON, F. *Robert Boyle.* London, 1914.

MORE, L. T. *Isaac Newton, a biography.* 1934.

PEPYS, S. *Diary.* 1660–1670.

ROBERTSON, A. J. *Life of Sir Robert Moray.* 1922.

Royal Society, Notes and Records of. 1940.

Royal Society, Record of. 4th ed. 1940.

SCOTT, J. J. *The Mathematical Work of John Wallis.* 1938.

SPRAT, T. *History of the Royal Society.* London, 1667.

TURNBULL, Professor H. W. Tercentenary Memorial Volume on James Gregory. Edinburgh.

WELD, J. C. *History of the Royal Society.* Vol. 1. London, 1848.

CHAPTER IV

SIR ISAAC NEWTON AND SIR HANS SLOANE:
1701–1740

T HE FIRST of the three periods into which the Society's history is
conveniently divisible ended with the presidency of Lord Somers,
by which time many of its early difficulties had been overcome and
less troublous years seemed to lie ahead.

The election of Sir Isaac Newton as President at the Anniversary meeting of 1703 introduced the second period which lasted until the death of
Sir Joseph Banks in 1820. It differed in many respects from the one which
preceded it and which has been described in the earlier chapters; order was
introduced into the administration, the financial control was improved
and the payment of fees and subscriptions was made more punctually. The
average number of candidates elected in each year was only twelve in the
first decade, but by the middle of the period it had risen to double that
number and gave promise of a further increase. It was a period of progress towards more efficient administration, but shows less evidence of
intellectual brilliance than the preceding one. There were certainly a
number of eminent scientific men among the Fellows who communicated
the results of their work to the *Philosophical Transactions*, but the Society
as a whole seemed to have lost much of the enthusiasm which characterized
the founders of the Society, and to have relapsed into the self-satisfied
attitude which was characteristic of much of eighteenth-century life. This
may be due in part to the growing number of those who took no interest
in science; it is not easy as yet to estimate their influence, but it was certainly beginning to be noticeable in the Councils. With the end of the
seventeenth century forty years of work by those who had built up the
Royal Society led to a century in which the aims of the educated classes of
the country and with them of those who had formed the Society developed on new lines. The knowledge which had been acquired was now
to be shared by wider circles than hitherto, and was to be utilized in order
to improve the conditions of human life. In this, the age of Reason, an
active interest was taken in this world, the life upon it and its phenomena
rather than on what was to be expected in a future life; there was a greatly
increased confidence in human understanding and what it could achieve,
the judgments of authorities of earlier times being weighed against the
evidence that could be brought forward in support of them. The ordered
procedure of nature was accepted as indicating the uniform relation of
cause and effect without any supernatural or magical interference. John
Locke's advocacy of toleration contributed largely to provide a more

humane conception of life for the people generally, though it was late in the new century before this had affected the lives of the labouring classes to whom better education was but slowly made available. In the Society the scientific Fellows were active in promoting the branches of science in which they were individually interested, and, if their example was followed in a lesser degree by their colleagues whose interests lay in literary or philosophical subjects, this is fairly attributable to the fact that scientific matter occupied most of the meetings of the Society, and formed the subject of the majority of the communications which were offered. These literary Fellows and others of similar interests formed more than half the members of the Council and to their feeble interest may be attributed the scanty attendance at the Council meetings to which allusion will be made later. When Dr Hans Sloane arranged in 1699 that the Council should meet specially for a thorough discussion of the financial position of the Society, he evidently realized that unless something could be done and done quickly the end of the Society could not be long postponed. The arrears due from Fellows for their unpaid subscriptions and fees had been accumulating for forty years and only comparatively small amounts had been paid off from time to time; certain unavoidable expenses had to be met and the number of Fellows who were contributing regularly had been falling off for several years past. The average number of Fellows for each year of the five-year period 1671-5 had been 215, the highest so far reached; by 1691-5 this had fallen to 115, and might well fall even lower. To deal with the situation the Council held fifteen meetings in 1699 at which about half its members were present; many proposals were put forward, including one to enforce the regular quarterly payments of sub-scriptions, but no definite scheme was adopted. The average number of Fellows for this five-year period 1696-1700 was 125 or ten higher than in the one before, but even this could not be taken as indicating the beginning of a permanent improvement, since it might well be no more than a temporary increase.

The last twenty years of the seventeenth century had been a period of great difficulty for the Royal Society; the number of Fellows had fallen off; financial resources were very low and but for the £1300 which the Society had received from King Charles for releasing at his desire the Chelsea Hospital estate which he had given to them in 1669, the position would have been desperate; the frequent changes of the Society's officers at short intervals had also prevented the adoption of any systematic reform of the administration. With the opening of the eighteenth century pro-spects were becoming brighter. The theory of the divine right of kings which had prevailed for so long had been done away with by the Common-wealth, and the writings of the philosopher John Locke, who had been elected a Fellow of the Society in 1668, provided a new political theory for the nation which had been eagerly adopted by its leaders. Those who had accepted it as their guide in organizing the revolution which brought about the abdication of James II in 1688 had decided upon the introduction of a constitutional monarchy and it was in accord with this that the group of Whig leaders invited William of Orange to take the throne under the conditions laid down in the Declaration of Rights which Parliament had adopted in 1689. Some years earlier Lord Somers had written a tract in which he maintained the absolute authority of Parliament to limit, restrain or to qualify the right to the succession; and six years before the revolution he had laid it down: 'If they mean by those *lovers of commonwealth principles* men passionately devoted to the public good and to the common service of their country—who believe that kings were instituted for the good of the people and the government ordained for the sake of those who are to be governed, and therefor complain or grieve when it is used for contrary ends, every humane and honest man will be proud to be ranked in that number.'[1] Similar views were also held by a good many of the leading men in the Society, and notably by Newton, as may be seen from his letter on allegiance to King William which he wrote to Dr John Covel, the Vice-Chancellor of Cambridge University in February 1688/9, and which is quoted by L. T. More.[2] The new principles of government for which the Whig party now stood offered a good expectation of a period of comparative quiet during which the Society could reorganize its administration and plan one which might be expected to develop successfully even with resources as slender as those of the Society still were; at any rate several of the older and more distinguished Fellows were of this opinion and might be trusted to do all in their power to promote it. But this only became possible when Robert Hooke died on 3 March 1703 after forty years of close association with the Society, at first as demonstrator and Curator, and then as Fellow, councillor and Secretary. His biographer, Richard Waller, describes him as an active, restless, indefatigable genius, but of a melancholy, mistrustful and jealous temperament which became more strongly marked as the years passed by. He adds that Hooke had formed a project for the better housing of the Royal Society, and to this end 'proposed building a handsome fabric for the Society's use, with a library, repository, laboratory and other conveniences for making experiments'; but he would never put his plans in writing nor make a will, so that when he died no record of his intentions was in existence. As one of the Surveyors of the city of London, Hooke had acquired considerable wealth and at the time of his death must have been a rich man.

From 1696 Newton had been living in London where he held an official appointment, first as Warden and from 1699 as Master of the Mint, so he might have been elected to succeed Charles Montagu (Lord Halifax) as President of the Society at the end of 1698 had it not been well known that

[1] Campbell, *Lives of the Lord Chancellors*, IV, p. 76.
[2] *Isaac Newton, a biography*. New York, 1934, pp. 348.

he would not accept nomination so long as he was liable to be attacked by Robert Hooke and to have his views criticized with the malicious jealousy which was characteristic of Hooke's temperament. Hooke, after ceasing to act as a Secretary, was elected a member of Council in November 1684 and also of fourteen other Councils between 1684–1702, so that he was on the Council in almost every year and frequently attended their meetings. As already mentioned, Newton was a member of the Council in 1697–8 and again in that of 1699–1700, but not in any other year until he was elected President at the Anniversary meeting of 1703 (cf. p. 111).

At the Anniversary meeting of 1703 Lord Somers retired from the presidency and Sir Isaac Newton was elected in his place; he was then sixty years of age. The Treasurer, A. Pitfeild, had replaced A. Hill in 1700, and the Secretaries were R. Waller and Dr Hans Sloane, who had been acting in this capacity since 1687 and 1694 respectively. Most of those who had held the position of President in the Society since the retirement of Lord Brouncker had done so for brief periods only and were too fully occupied with their own affairs or their public duties to give adequate service to the Society; several of the Secretaries had held their posts for too short a period to initiate and carry out improvements in the administration even if they had recognized the need for them. However, during the last decade of the seventeenth century the Secretaries, Richard Waller and Hans Sloane, who had been ably supported by Sir John Hoskins, a Vice-President, had done much to stay the decline in efficiency which had set in, and to take steps towards establishing a better and more orderly state of things. Sir John Hoskins had been President in 1683 for one year, and for twenty years afterwards as a Vice-President had frequently presided in place of Presidents who were often absent. But more than this was needed, and until there was a strong and capable president who appreciated what the true aims of the Society were, and was determined to adapt its policy so as to ensure their advancement, the Secretaries, however zealous, were working under great difficulties.

THE OFFICERS

In December 1701 Newton resigned his professorship at Cambridge, and his fellowship shortly after. He had been connected with Trinity College for forty-one years, and the change to an official post in London and to the presidency of the Royal Society brought him into quite a different field of activity. In 1705 he was knighted by Queen Anne during her visit to Cambridge University, and was the first man of science to be so honoured in this country. Having been accustomed to form his conclusions after careful enquiry and full consideration he resented being asked to give his reasons for them, and was exceedingly sensitive to criticism however well meant, since he sometimes saw ill-natured interference where none was intended. He was neither a good public speaker nor ready in debate, which made it difficult for him at times to conduct the business of the Society's meetings as efficiently and as quietly as he would have wished, so it is not surprising if he was considered by some to be imperious and somewhat intolerant. His senior Secretary, Dr Hans Sloane, who was an administrator of great ability, assisted him loyally from his ten years' experience of the Society's business, but he too was considered by some to be impatient of criticism and correction.

Hitherto the Society had met on Wednesdays, but as on this day the President was always occupied with his duties at the Mint he arranged for the meetings of the Council and the Society to be held on Thursdays so that he might be free to attend them regularly and to devote himself to the Society's business. It is on this day of the week that the Council and the Society still hold their meetings.

For the next four and twenty years the Society was to be directed by a man of great scientific eminence, who both understood and appreciated the Society's aims, and whose time was not now so fully occupied by his official duties or his own researches as to hinder the efficient performance of his presidential duties, which he carried out with exceptional zeal and regularity. He attended 161 Council meetings out of the 175 which were held between the date of his election and the end of 1726, shortly before his death, besides being present at most of the Ordinary meetings. This was a new conception of a president's duties, for during the preceding twenty-five years three presidents had attended no Council meetings and three others had been present at only about one-third of those which had been held. It would seem that Newton, besides setting an example of regular attendance, impressed on his colleagues the great administrative importance of the officers being present as often as possible for, as we shall see, most of the presidents who followed him maintained a high standard of regular attendance at the meetings of the Council. His efforts were ably seconded by his Secretaries, especially by Dr Hans Sloane, who was present at 116 out of 132 Council meetings between 1694 and his resignation in 1713, when he was replaced by Edmond Halley. This painstaking devotion to duty on the part of the officers resulted in the Society being provided with an administration which, with occasional amendments and additions to meet the needs of changing times, served the Society well until 1823 when the rapid growth of its scientific activities necessitated a revision of its statutes, and closer attention being given by its Council to its administrative procedure. The stress which was laid on the regular attendance of the officers raised no difficulties, since no rights or liberties of the Fellows generally were affected, but it might have been otherwise with an urgently needed reform which the President can hardly have overlooked. Had the President found it possible at this time to strengthen the Councils by increasing the number of scientific men elected as councillors until they formed a well-established majority in each Council, it is reasonably certain that the attendance at them would likewise have increased, and the scientific activity of the councillors would have been accelerated many years earlier than it actually was. But any such attempt to increase the scientific representation on the Councils would probably have been objected to by the body of Fellows since the election of the councillors was left to the vote of the Fellows at the Anniversary meeting. But it was by some such change as this, had it been possible, that increased scientific activity could have been attained for, as will be seen in Chapter VII, it was precisely by such action as this that the policy of the Society was completely changed between the years 1820 and 1840, and its character as a scientific institution was rapidly altered for the better. The scientific councillors increased from eight in 1820 to twelve in 1821 and sixteen in 1823 out of twenty-one, but to attempt any such drastic change in the early part of the eighteenth century would doubtless have given rise to strong opposition on the part of the non-scientific majority and would have led to ill-tempered opposition similar to that which the purchase of Crane Court in 1710 caused, and with more justification.

Sir Isaac Newton's sensitive temperament disposed him to avoid such contentious debates as far as possible, and it is not at all surprising that he should have allowed the constitution of the Councils to remain unchanged in view of the almost certain opposition that any action would have aroused. A century later a reconstruction on these lines was carried out without any serious opposition.

In March 1702 Sloane's friend, William Courten, died, having made him his residuary legatee and sole executor, so that Courten's library and collections were now added to those which Sloane had already made for himself, the whole forming the finest and most varied collection of its kind then existing in this country. He added to it constantly throughout his life and at his death in 1753 it included about 40,000 printed books, 3500 manuscripts and nearly 80,000 objects of various kinds. The work entailed by the addition of the Courten collections to his own must have been very great, but for the moment matters of even greater importance claimed his attention; Sir Isaac Newton having been elected President at the Anniversary meeting of 1703 the task of reorganizing the Society's administration lay before him and his First Secretary.

Dr Hans Sloane, who for ten years was Newton's colleague as Secretary, for fourteen more as a member of his Council, and who served for fourteen and a half more as his successor as President, ably assisted him, and later carried on the policy which Newton initiated. Ten years' work as Secretary between 1693 and 1703 had shown him how much there was to be done, what faults had to be remedied, and in what direction improvements could best be made. He was now forty-three years of age, full of energy, a well-known physician and in receipt of a considerable income from his profession. His zeal and force of character now and then brought him into conflict with some of the Fellows who could not see the necessity for the innovations which he advocated; but this is the price which all reformers have to pay for the attainment of their aims, though it may be long before their justification is realized and acknowledged. It was indeed most fortunate for the Society that two such men as Newton and Sloane, together with R. Waller, who became a Secretary in 1687 and who had been Sloane's colleague from 1693 to 1713, should have taken up and carried on for some forty years the administration of the Society at this stage. Dr Hans Sloane was held in great esteem in both London and Paris during the first half of the eighteenth century, and he is usually described as being an eminent physician and a collector of great zeal and discrimination; but the services which he rendered to the Society as an organizer and an administrator have not usually been appreciated at their full worth.

In 1713 Dr Hans Sloane resigned the post of Secretary which he had held since 1694, and E. Halley was elected in his place. By this time Sloane, who had built up a large and lucrative practice in London, had decided to move from his house in Bloomsbury to Chelsea where he purchased a large house, the Manor House, in which he could keep and display his collections which were by now the most important in the country. He continued to be an active member of the Council and attended its meetings regularly until 1727 when he was elected President on the death of Sir Isaac Newton. Only a year later, in 1714, Richard Waller, who had been Sloane's colleague as Secretary for twenty years, died and was succeeded by Dr Brook Taylor, a distinguished mathematician; he only held office for four years when Professor John Machin, Professor of Astronomy at Gresham College, replaced him, and carried out the duties of Secretary very conscientiously for twenty-eight years until 1747 when he resigned. He only failed to attend 32 out of 180 Council meetings. Edmond Halley, who replaced Dr Hans Sloane in 1713, was succeeded by Dr J. Jurin in 1721; he too was most regular in performing all the duties of his post which he held until 1727. Dr Rutty replaced him for three years and then Dr Cromwell Mortimer became Secretary retaining the post for the next twenty years. The importance of the strict performance of their duties by the officers had by now been fully realized.

COUNCILS

The powers which the Charter gave to the Councils to make statutes and regulations and to carry on the administration of the Society as well as to execute any resolutions that the Society might adopt have already been mentioned. Councils were not required to obtain the Society's approval for these acts since only the selection and election of the members of Council and the election of candidates for the Fellowship were reserved to the Fellows assembled at an Anniversary or a General meeting. The decisions taken by the Council were recorded in the register of their minutes and it is from these that we can form an idea of each Council's initiative and activity. The increase in membership during the period which is now being considered, and the improvement in the financial position which accompanied it, might reasonably be expected to influence the work of the Council and to be reflected in their minutes; but as this was not the case, it is worth while examining the matter more closely. The average number of scientific and non-scientific men who served on the Councils in each twenty-year period is given in Appendix II C, as well as the average number of meetings held annually and the average number of members who attended a meeting. It will be seen there that more than half of the twenty-one members of Council were selected from those Fellows who were not interested in, or had but a slight knowledge of, science.

The average number of meetings held annually was at first only seven, but this increased as the administration was improved; only about half the full number of councillors were present, but difficulty of transport in those days may have prevented some from attending as often as they might have done, unless they lived in London. No sign of increasing activity on the part of the Councils was as yet apparent.

With so small an attendance of councillors, the four officers, who were almost always present and were presumably well acquainted with the business which was to be brought before the meetings, would often be able to direct the Council's decisions in accordance with their views although they might not always have an actual majority. From time to time complaints were heard that the Council's proceedings were controlled by the officers, but if the members did not avail themselves of their right to attend the meetings and to take part in the discussions they must accept their share of the blame. It was to the advantage of the Society, as we have seen, that officers' tenure of their posts should not be too short, as in that case they had not the time to become familiar with the details of their work and their responsibilities. The system in fact was not working too well and one fault at any rate lay in the number of non-scientific Fellows who were being admitted to the Fellowship; they greatly outnumbered their scientific colleagues since they comprised from 65 to 70 per cent of the total; the Fellows could elect as many candidates as they pleased, and as no standard of the scientific efficiency to be possessed by candidates is laid down in the Charters the Council could not introduce one.

In the following chapters we shall see that these same difficulties continued throughout the remainder of the eighteenth century in spite of Sir Hans Sloane's attempt in 1730 to overcome them, and it was not until after 1830 that a solution was found which was eventually brought into full operation in 1848. Now that the administration of the Society was beginning to operate on more settled lines, it is worth while seeing how the members of Council were being selected from the Fellows who were available. A similar investigation has been made for the period of Sir Joseph Banks' presidency, 1778-1820 (p. 199), and a comparison of the two statements is of considerable interest. Out of the 157 councillors who were elected between 1701 and 1740, 60 per cent served for one to three years only, 29 per cent for from four to eight years, and 11 per cent for from nine to eighteen years. Eighty years later 88 per cent of the councillors served for one or two years only (see p. 199). There were only from about 150 Fellows in 1701 to 300 in 1740 available from whom the councillors could be selected at this time and ten had to retire each year, so that those who had proved themselves to be efficient were often re-elected for considerable periods; Dr R. Mead, for example, served on the Council for forty-five years and Edmond Halley, the astronomer, for thirty-four years, periods which were certainly unreasonably long.

THE FELLOWSHIP

Under the statutes of 1663 any of His Majesty's subjects who held the rank and title of Baron or of any higher rank were privileged to be proposed at one meeting and to be elected at the next; these were usually but few in number, rarely exceeding 11 per cent of the Fellows, and of them only a very small proportion took any active part in the work of the Society or in promoting its aims.

Of much greater importance was the number of scientific Fellows in the Society at any time as compared with those who had had no training in scientific work and who for the most part took little interest in it. At the time of the formation of the Society these latter had outnumbered the scientific Fellows in the proportion of about two to one and at some periods they formed an even larger majority. The increase in the number of Fellows which began during the last years of the seventeenth century had given hope of improvement; the average yearly number of them in the five-year period 1691–5 had been 115, and reached 125 in the next one; but from this time onwards the number of Fellows increased continuously until the year 1847 when the new statutes restricting the election of candidates came into force (see Appendix II A). In the period which we are considering in this chapter the average annual number rose from 131 in the five years 1701–5 to 290 in that of 1736–40, the increase being greater in the second half of the forty-year period when the effects of better administration were producing their results. In 1700 the number of Ordinary Fellows had been 125, and in the course of the next forty years 530 were added by election, an average of thirteen annually; by the end of 1740 the number of Fellows was 301, so that 354, or about nine annually had died, resigned or had been removed from the list for non-payment of subscriptions. It is interesting to examine these figures more closely; in the first decade 97 Ordinary Fellows were elected, but in the second the number was nearly doubled, reaching 182, but this fell to 132 and 119 in the last two periods. It is highly probable that the increased confidence in the Society's future, which was due to Sir Isaac Newton's tenure of the presidency together with the Society now being in possession of a house of its own, is reflected in the large number of candidates who were elected between 1711 and 1720. The growth of the Society, though steady, was still slow; on the other hand, the increase in the number of foreign members at this time was remarkable, for the number admitted in each successive decade rose from twenty-four in the first to forty-eight in the next, then to sixty-three, and in the last one to eighty-three. It is clear that the reputation of the Society as an important scientific institution was growing, but that the numbers of the foreign members elected should rise in so short a period from 24 to 49 per cent of the total Fellowship suggests that some of them were being accepted without sufficient regard being paid to the value of their scientific qualifications; some years later the Council took steps to restrict their numbers and to introduce a more critical selection of the candidates whose names were put forward for election as foreign members. The increase of the scientific Fellows to one-third of the total Fellowship was a good sign, and may be due to Sir Isaac Newton's acceptance of the presidency and to confidence that greater attention would be paid to the advancement of the Society's scientific interests, but this rate of increase was not maintained.

With the increase in the number of Fellows the Society was now being more widely represented in various professions than it had been. Mathematicians and astronomers still formed one of the larger groups, but they were outnumbered by the physicians and surgeons who together accounted for about 60 per cent of the total, and did not fall much below that figure until the early part of the nineteenth century (see Appendix II B).

It has been stated that at the time of George I's accession (1714) at least eleven members of the medical profession were Fellows of the Royal Society;[1] but this must refer to the most eminent and widely known of its members only, for in 1698 there were nineteen physicians and surgeons out of 119 Ordinary Fellows, and by 1740 there were sixty-three out of 301, corresponding to 16 and 21 per cent respectively. The medical profession was already a large one and a considerable number of its members holding the Fellowship of the Royal College of Physicians, or an M.D. degree at one of the Universities, were proposed as candidates and were readily elected to the Fellowship of the Society. Up to 1860 the members of this profession always formed the largest group of the scientific Fellows, there being in that year eighty physicians and thirty-seven surgeons out of 330 scientific Fellows.

FINANCE

The financial position of the Society at this time is not easy to summarize since neither the Council minutes nor the Journal-book contains any account of each year's receipts and expenditure between 1677 and 1716; investments of £250 in East India Company Stock, and of £800 in Africa

[1] B. Williams, *The Whig Supremacy*, p. 365.

Company Stock were made in 1697, but no details are given. As has already been mentioned the financial position at the end of the seventeenth century had given rise to grave misgivings, but somehow or other that crisis was weathered, though no details are available. When the Crane Court houses were purchased in 1710 payment could only be made with the aid of contributions made by eight of the Fellows which totalled £650,[1] and by a mortgage on the property of £900 on which 6 per cent was paid; also in 1713, when £300 had to be paid to the Secretary, R. Waller, for building the Repository Gallery, the Treasurer was glad to accept the offer of H. Hunt, the Society's Operator and Librarian, to lend this sum to the Society. At the end of 1716 the Treasurer's balance is stated to have been £562, and £600 of Bank Annuities were bought in 1717, so the more efficient and careful administration which had been introduced was improving the situation. Still it does not look as though financial difficulties had disappeared, and it is not surprising that Sir Hans Sloane, with his long experience of the Society's business, should on taking up the presidency in 1727 have examined the financial position closely. He obtained from the Attorney-General an opinion that the Society would be within its rights in suing any defaulting Fellow for his arrears provided that he had signed the Obligation, and this greatly strengthened his hand by making the Society's position clear for the guidance of Councils in future. They could now deal effectively with the defaulters, though they were at times dilatory in using their powers. Fellows who had not paid their fees or subscriptions by 1730 were at once warned and legal action was taken. No record exists of the amount which was recovered by this action, but it must have been considerable, for the receipts of the year 1730 are reported as having amounted to £1700, though the Treasurer's balance in hand at the end of 1727 had been £564, but had fallen to £9 two years later; £900 of South Sea Bonds had been purchased in 1723, in addition to £600 Bank Annuities bought in 1717. The position of Fellows' subscriptions was gone into carefully in 1729. R. Gale, an antiquary, had succeeded A. Pitfeild as Treasurer at the end of 1728, but no report was made to the Anniversary meeting. The sums which various Fellows owed to the Society at this time were checked during the next few years and some of them made offers which were accepted by the Council; others excused their shortcomings on various grounds, some of which were admitted. Some disputed their liability and others were excused for the whole or a part of their debt. Others again were 'put in suit' and their debts recovered from them by legal action. It would seem that the Treasurer's supervision had been far too lax for some years past and that it was high time that such an enquiry should be held. At the end of 1731 Sir Hans Sloane heard that a small estate of forty-eight acres of arable land in the village of Acton near Fulham was for sale and negotiations for the

[1] Cf. *Record of Royal Society*, 4th ed., 1940, p. 141.

purchase of it were begun; by the following summer this freehold estate was bought for £1600 from a Mr Pannet of Kensington together with a dwelling-house and outhouses; it was let at the time to one, Roger Life for £65 per annum; there was also a quit rent of £1. 4s. 0d. to be paid to the Duke of Kingston. At a Council meeting in July the Treasurer was author-ized to sell South Sea Bonds to meet the cost of this estate which had been paid for out of the Treasurer's cash. This property remained in the posses-sion of the Society until 1882 when the thirty-five acres of it which still remained were sold on very advantageous terms whereby the finances of the Society were placed on a satisfactory footing (see p. 279).

Under these improved conditions the Council decided to fix the re-muneration of each Secretary at £60 per annum.

In December 1731 the Council decided that the payment of subscrip-tions might be compounded for by a payment of £25 over and above the admission fee, but it would seem that this was not allowed generally; the question of authorizing a composition fee to cover future annual sub-scriptions came up again in 1753, and it would appear that until then such payments of this nature as were accepted were only allowed by the special permission of the Council who took each case into consideration before giving a decision.

For the years 1718 to 1727 the average balance which remained in the hands of the Treasurer as reported at the Anniversary meetings was £390, so that the Society's position was now improving. A resolution which was adopted by Council on 30 November 1717 says that any Fellow who presents the Society with the sum of £50 or upwards shall have his bond delivered up to him; he could not then be sued for non-payment of his subscriptions in future. It is not known to what extent advantage was taken of this, but as it was not customary until many years later to circulate such information to the Fellows this form of compounding for subscrip-tions was probably but little known and seldom applied for.

In 1720 the Treasurer was instructed to sell £600 Lottery Annuities and to reinvest the proceeds in South Sea Bonds; three years later the invest-ment of the whole of the money then in the hands of the President and Treasurer was by order of the Council to be in East India Bonds, and at the Anniversary meeting of 1723 it was reported that £900 of South Sea Bonds were in the Treasurer's chest. Such isolated pieces of information as these are all that we have from which to form an opinion on the financial position. It might be expected that after what had been done in 1729 financial matters would not have troubled the. Council for several years to come, but within ten years the same difficulty recurred; at the Anni-versary meeting of 1739 the Treasurer had to report a deficiency in the year's working of £95, and at the following Anniversary meeting this had grown to £240. The Council at once appointed a committee of enquiry which met at the beginning of the following January and reported on 14 January 1741, so the matter which they had to investigate cannot have presented any great difficulty or complexity. The committee who were entrusted with the enquiry did their work thoroughly, and reported to the effect that 'the total number of Fellows was then 293 of whom 152 had compounded for their subscriptions [presumably in 1730] and 139 who had signed the Obligation to pay annually owed £1844. 16s. od. in arrears', the equivalent of about three years' subscriptions. Two Fellows had been exempted from paying subscriptions but no reason for this special consideration is given; both were medical men of distinction. The power to sue those who were in default had evidently not been utilized. The committee further reported that the Society's revenue, exclusive of subscriptions and fees, was £232, from which taxes and other expenses to the amount of £92 had to be deducted; the annual expenses of the Society were estimated at £380. Taking the number of Ordinary Fellows at 301 and of those who had compounded or had been exempted at 141, there were 160 who should have been paying an annual subscription of £2. 12s. od., which would provide £416. Admission fees varied from year to year and might be as much as £20 on the average. Other receipts were:

Fee-farm rents (Wilkins bequest)	£24	0	0
Mablethorpe rents (Aston bequest)	27	0	0
One house in Crane Court	24	0	0
Two houses in Coleman Street (Paget bequest)	97	10	0
Estate at Acton	65	0	0
	£237	10	0

The total receipts should therefore have been about £650, from which £92 for taxes, etc. had to be deducted, so that there should have remained £558 to meet the annual estimated expenses of £380. That this was quite feasible is shown by the fact that in 1734 the Society had been able to invest £500 in East India Bonds. James West had taken over the treasurership from Roger Gale in November 1736, so he may not have had time to gain much experience, and Sir Hans Sloane's health had deteriorated considerably of late so that in 1741 he was obliged to retire; these may have been contributive factors in bringing about so unfavourable a situation which certainly should not have been allowed to arise.

Although the Council's power to control the administration of the Society and to modify its policy was almost unrestricted, changes in its membership had but little influence on the policy which it followed; occasionally some decision, some line of action, or the advent of a president of strong personality or decided views was followed by well-marked changes, but for the most part variations in the membership indicate influence which operated for a year or two only. The purchase of the houses in Crane Court necessitated extra meetings of the Council and led to an increase in attendance of councillors in 1710, and the adoption of new statutes in 1728-30 had similar results at that time. Presidents like Sir Isaac Newton and Sir Hans Sloane, who were not only men of great ability but devoted themselves to the improvement of the Society, naturally influenced the type of man selected to be a member of the Council and the proportion of scientific members showed a definite increase in consequence.

During the early part of the eighteenth century the slight improvement, which was noticeable at the end of the seventeenth, had been maintained, the number of Fellows increased, and by 1740 the Society's financial position was such that, with reasonable care and foresight, the Treasurers need have had no anxiety about it. In the country generally there was a feeling of relief that the turbulent days of Stuart times were at an end and that more peaceful and prosperous days lay ahead. A similar optimism affected the Society and its Councils.

The account of the general advances which took place in the Society's administrative arrangements during the years when Sir Isaac Newton and Sir Hans Sloane were Presidents of the Society makes a suitable background against which the history of these forty years may be studied. Both of these men introduced notable improvements which continued to form a part of the Society's procedure for many years to come. That this was the case bears out the suggestion which has already been made that at this moment in its history the Society urgently required the guidance for some years of a president who understood its needs and had the will and force of character to carry out the improvements which were called for. Fortunately two such men were available and they guided the Society for forty years until its direction passed into other less competent hands with disappointing results.

There was one side of the Society's activity which had been very strongly developed in the earlier years but which now suffered a temporary decline. The application of science to industrial practice and technical professions was then eagerly fostered, but in the eighteenth century the communications which were received by the Society were of a more academic character than had been customary in the past.

When Dr W. Croone died in 1684 he left a scheme for two lectureships which he intended to found, one of which was for the Royal Society. In his will however he made no provision for endowing them, but his widow, who subsequently became Lady Sadleir, remedied the omission, and in her will dated 25 September 1701 bequeathed to the Society one-fifth of the clear rent of the King's Head Tavern, in or near Old Fish Street, London, at the corner of Lambeth Hill, 'for the support of a lecture and illustrative experiment for the advancement of natural knowledge on local motion or (conditionally) on such other subjects as in the opinion of the President for the time being, should be most useful in promoting the objects for which the Royal Society was instituted', the remainder being paid to the Royal College of Physicians, also for the support of a lecture to be delivered before them. A decree in Chancery of 1728 empowered the Society to devote the whole of the net annual profits of the legacy to the payment for a single lecture and its attendant expenses but in those days this amounted to £3 only. It is now about £49. In 1739 the College of Physicians proposed to the Society to 'get rid of Lady Sadleir's donation'; but the Council were of opinion that this could not be legally done and so refused their consent. The first lecture was read in 1738 by Dr Stuart on the 'Motion of the Heart'.

The Society was not the only institution at the beginning of the century which was endeavouring to extricate itself from its difficulties. One of the two trustee bodies which were responsible for the administration of the estate of the late Sir Thomas Gresham, the Mercers' Company, was at this time financially embarrassed; the property which they administered was a very valuable one but the buildings of Gresham College were in a bad state of repair and extensive renewals were needed to make them suitable for the purposes of an important educational institution. It was therefore proposed that some of the land should be sold and the proceeds used to meet the cost of rebuilding the College on a more convenient and suitable plan. If this could be arranged the Trustees of the College would then be able to provide the Society with the accommodation which it needed; in return the College would continue to gain largely from a close association with the Society, since its fine library and the collections in its museum would be accessible at all times to the professors and others of the College who might wish to make use of them; the scientific meetings which would be held there and the communications which were read and discussed at them would also increase the prestige of the College as the centre of learning and scientific education in London. An Act of Parliament which would give the trustees the power to deal in this way with the objects of the trust was however necessary, and in 1701 the trustees of Gresham College, with the consent of all the professors of the College except Hooke, arranged that a bill should be introduced into Parliament for rebuilding the College on the grounds 'that it had become old and ruinous, and the repairs thereof very expensive; but the said College standing on a considerable area of ground, and great part of it lying waste, good improvement might be made by rebuilding it'. The bill was passed by the House of Commons but was thrown out in the Upper House on the petition of Hooke, who had from the first strongly opposed the project. When Hooke died in 1703 the trustees again brought forward their bill, at the same time informing the Society that they were desirous of 'accommodating them with conveniences for their meetings, repository and library'. The Council returned their thanks to the trustees and desired Sir Christopher Wren to view the design and to consider what accommodation the

Society needed to carry on its work and for its normal expansion. This he did in a report entitled 'Proposals for building a House for the Royal Society' which is preserved in its archives; in it the dimensions of the various meeting-rooms, etc., are set out. He recommends that its essential requirements would be met by providing

a place at the College so seated in the said ground that the coaches of the members (some of which are of very great quality) may have easy access and that the building consist of these necessary parts:

1. A good cellar underground, so high above it as to have good lights for the use of an elaboratory and housekeeper.
2. The storey above may have a fair room and a large closet.
3. A place for a repository over them.
4. A place for the Library over the Repository.
5. A place covered with lead for observing the heavens.
6. A good staircase from bottom to top.
7. A reasonable area behind it to give light to the back rooms.

All which may be comprised in a space of ground 40 foot in front and 60 foot deep.

A clause to include these requirements, as well as a hall and almshouses, was inserted in the new bill, and also another requiring that all this additional accommodation should be built within five years of the passing of the Act under a penalty of two thousand pounds; but exception was taken to this and the bill was rejected by the House of Commons at the first reading.

This ruled out all prospect of obtaining suitable accommodation at Gresham College within any reasonable period of time, and as its trustees had previously expressed the desire that the Society should find other quarters the Council now set about the matter in earnest. On 21 April 1703 it was resolved that the Society should purchase a place of abode for itself; and a committee of the Council, consisting of Mr T. Isted, a councillor of ten years' experience, Mr A. Hill, the late Treasurer, Dr F. Tyson, who had already served on the Council for six years, Sir John Hoskins, an ex-President—he died two years later before the committee had completed its work—Dr H. Sloane (Secretary), Mr A. Pitfeild (Treasurer), was appointed with instructions to 'consider of a place to build on or buy, and to lay their thoughts before the Society'. They met frequently and inspected various houses and sites in several parts of the city but none appeared to be suitable. The Duke of Bedford offered 'an estate of inheritance, or a lease of ground for sixty-one years', but these were not accepted by the Council. In 1705 the Council received a communication from the Mercers' Company to say that they had resolved 'not to grant the Society *any room at all*' at Gresham College. This decision was not wholly unexpected, and a committee had already been appointed by the Council to superintend the removal of 'the goods of the Society, in case any warning comes from the committee of Gresham College'; at any rate it caused the Council to increase their efforts to obtain accommodation elsewhere.

The Council next sought the assistance of the Crown in their difficulty and addressed a petition to Queen Anne for a grant of land at Westminster, but this produced no result. They then applied to the trustees of the Cotton Library, which was at that time established in Cotton House, Westminster, requesting permission to meet in their rooms, but here again they were disappointed; and no better success was met with in many other directions which were explored by the committee in the seven years during which they continued their enquiries, nor could suitable accommodation be found at any price which the Society was able to afford. At last, on 8 September 1710, the President informed the Council that a house belonging to the late Dr Brown in Crane Court, Strand, was to be sold, 'and being in the middle of the town and out of noise, might be a proper place to be purchased by the Society for their meetings'. Accordingly 'the President, Sir Christopher Wren, as being specially conversant with the Society's requirements, Mr A. Hill an ex-treasurer, Mr A. Pitfeild (Treasurer), Dr H. Sloane (Secretary), Dr R. Arbuthnot, Mr R. Waller (lately Secretary), Mr Ch. Wren, Mr T. Isted, Dr R. Mead, Sir John Percival and Dr W. Cockburn, four of them to be a quorum, were ordered a committee of the Council to take care of this matter'. On 20 September the President reported to the Council 'that he and several members of the Committee had been to view Dr Brown's house and found it very convenient for the Society'. The Council then resolved to buy the interest of the house for £1450, the votes of twelve of its members being in favour and only one against it, such action being within their powers under the Charter. When it became known that the Council had decided to acquire a house for the Society and discontinue the use of the rooms in Gresham College which it had occupied for so many years, considerable dissatisfaction was expressed by certain of the Fellows. A pamphlet entitled 'An account of the late Proceedings in the Council of the Royal Society, in order to remove from Gresham College into Crane Court in Fleet Street' was published anonymously at this time criticizing the proceedings of the President and the Council, enlarging on the inconveniences of Crane Court and complaining of the expense which had been incurred. There was at this time no ready means of informing the Fellows of any action that the Council might decide to take in the interest of the Society, and therefore the anonymous writer of the pamphlet may not have known that the change was the result of the demand of the College trustees themselves, the Council having no choice in the matter. This explanation does not however hold good, for Dr John Harris, the Second Secretary who, together with another member, Mr W. Clavell, left the Council meeting of 26 October without voting either for or against the completion of the purchase. Dr Harris had been elected Second Secretary on the previous 30 November and had attended six out of the fifteen Council meetings which had been held between that date and October 1710, so that he must have been fully cognizant of all the circumstances, and must have known that the Council could not obtain any accommodation at the College; he did not stand for re-election at the following Anniversary meeting and Richard Waller, the former Secretary, replaced him. No copy of this pamphlet exists in the Society's library but there is one at the British Museum, and from this Mr Weld in his *History* quotes as follows:

The President gave orders at night to summon as many Fellows as were in town, or could be found, to meet at Gresham College on 1st September previous to a meeting of the Council which was held on the same day. At this Extra-ordinary meeting he told them that they were without any being of their own; that their continuing in Gresham College was very precarious; that Dr Brown's house had been proposed to them and a committee had viewed it, and that he thought it very convenient for the uses of the Society. He added that he had called them thither that he might hear what objections they had to offer against the proposal, that the Council might consider of them and take their final resolutions accordingly. The profound silence that followed sufficiently exprest a general surprise; till the President after a little while began the debate, and addressing himself to some particular members asked their objections. They told him that the very *embryo* of the Society had been formed in Gresham College, and that they kept their weekly meetings in that place some time before they obtained the Royal Charter of Incorporation; that the Society had continued there almost ever since, even in their most flourishing condition; that they en-joyed the same freedom and convenience as formerly without the least dis-turbance or impediment, and therefore they hoped to hear the reasons that induced him, and a few others who appeared as zealous and earnest, to remove from thence. Till that question was debated and determined, it was out of season to inquire into the inconveniences of the house he had recommended.

The President was not prepared (or perhaps not instructed) to enter upon that debate; but freely (though methinks not very civilly) reply'd, that he 'had good reasons for their removing, which he did not think proper to be given there'. The Acting (First) Secretary (Dr Hans Sloane) who has engrossed the whole management of the Society's affairs into his own hands, and despotically directs the President, as well as every other member, took upon him to relate a fact which he thought would determine every vote. He told them that one of the Gresham Committee ask'd him (not long ago) why the Royal Society did not remove from Gresham College since the City had several times sent them warn-ing to that purpose? This does not appear to have satisfied the members, who were of opinion that there were no grounds for removing; they likewise re-monstrated 'that that season of the year, and the short notice he had given of this meeting, made it very improper to determine an affair of so great importance to the Society at that time, and therefore they moved that the debate should be adjourned to St Andrew's Day, or at least to some other extraordinary Meeting'.

This the President would not hear of; they then offered to give him their opinion either by ballotting or voting viva voce—but in vain; his scruples were immoveable; so that some of the gentlemen with warmth enough ask'd him, To what purpose then he had call'd them thither? Upon which the meeting broke up somewhat abruptly, and not only the members of the Society, but most of those of the Council also, left the President with Dr Hans Sloane (Secretary) and Mr R. Waller (the late Secretary) and one or two more to take such measures at the Council as they best lik'd.

In these days disputants were in the habit of expressing their views in debate with a violence of phrase and a vigorous criticism which would be thought intolerable at the present day. Dr John Woodward, a physician, even better remembered as an early pioneer in geology, quarrelled more than once with his colleagues at the Society of which he became a Fellow in 1693. When a Professor of Physic at Gresham College, he had a difference of opinion with Dr R. Mead on a medical matter which led to his fighting a duel with Mead at the gate of the College. He was elected a member of Council in November 1696 and served on it for seven years between then and 1710. In that year he expressed himself to Dr Hans Sloane at a meeting of Council in such insulting terms that the other members demanded that he should apologize; this he refused to do and was in consequence formally expelled from the Council. He brought an action against the Council demanding his reinstatement but was unsuccessful. It is advisable therefore to hesitate before accepting as true the accusations and criticisms which were freely made in discussions without any evidence of their reliability being produced.

The meeting which was called to discuss the purchase of the houses in Crane Court does not seem to have been very tactfully handled though the situation was certainly a difficult one. The Society had been notified in 1705 by the Mercers' Company that no accommodation would be available for the Society at Gresham College, and even before this it was known that it might be called upon to remove its property from the College at any time. A committee of the Council had been seeking suitable accommodation elsewhere ever since 1701 without meeting with any success, and the President and officers may well have despaired of finding such a house as they wanted. The Society could not afford to pay much for one since its funds were very low; no financial reports of this period exist, but the general impression obtained from all that we know is that bankruptcy was threatening, if not imminent. The President was not at his best in handling a meeting where there was a conflict of views, especially when he had made up his own mind very definitely; the First Secretary, Dr Hans Sloane, was not popular, and though he must have had full knowledge of the whole question he does not seem to have used it to the best advantage. It would have been better perhaps if a full statement of the position and all that had led up to it had been made to the meeting to begin with, but this may not have been in accordance with the custom of that day when those entrusted with authority did not as a rule explain the reasons for their decisions.

The meeting realized that the power of final decision lay with the Council and that though any views they might express would be considered by it these could only be of an advisory character. A week later the Council resolved that 'the Committee appointed on 8 September, four of them to be a quorum, be appointed to contract for and purchase Dr Brown's house and another small one adjoining it for a sum not exceeding £1450, not including the repairs; and for seeing the deeds and writings prepared to be perfected, and the house fitted up for the reception of the Society with all necessary expedition'. On 26 October the President reported to the Council that he had, with the committee appointed for that purpose, agreed to the purchase of the two houses in Crane Court for £1450. The Council then ordered that the Treasurer should pay £550 to Mr Brigstock and Dr Brown's trustees, and that the offer of Mr Collier, who was ready to advance £900, the remainder of the cost, on a mortgage on the two houses with interest at 6 per cent, should be accepted. This meeting of the Council was attended by thirteen members, but two of them, Dr J. Harris, one of the Secretaries, and Mr W. Clavell, withdrew without voting.

At last, after fifty years during which it had been dependent on the hospitality of benefactors or other institutions, the Society had now a house of its own. By the time that the necessary repairs had been carried out at Crane Court, and the Society's property had been transferred from Gresham College to the Society's new quarters, opposition to the change had disappeared and no more complaints were heard. The Society carried on its work there for seventy years, during which its Fellows added greatly to its scientific reputation by worthily discharging the obligation laid upon it in its Charters, namely 'the improvement of Natural Knowledge'. By that time its importance had been widely recognized and the Government of the day offered to provide the Society with accommodation in Somerset House, where the Anniversary meeting of 1780 was held.

Ward, in his preface to his *Lives of the Gresham Professors*, speaks of the year 1710 as having been a very unfortunate one for the College by reason of the removal of the Royal Society to Crane Court. While the Society held their meetings at the College, such of the professors who were members of it were treated as Honorary Fellows being excused from annual payments, and felt little inconvenience from the lack of a College library; but after the books of the Society were removed they became very sensible of that disadvantage.

Those who have described the history of the Society while it was passing from the seventeenth century to the eighteenth have laid stress upon the importance of Sir Isaac Newton's acceptance of the presidency, on the dispute with Flamsteed over the publication of the latter's astronomical observations which embittered their relations for several years, and also on the Society's acquisition of a house of its own; but these ten years, 1701–10, merit a fuller consideration, for they constitute a period of exceptional importance in the Society's history since they were a turning-point in its fortunes and its efficiency. With the presidency of Sir Isaac Newton and the Society's acquisition of a house of its own the years of difficulty and steady decline were left behind, and a period of steady progress was setting in; by the year of Newton's death the number of Fellows was double what it had been when he became President, and the Treasurer was able to report that he had a balance in hand of over £600 besides certificates for £600 Bank Annuities, and £900 in South Sea Bonds in the name of the Society. From this time onwards both the number of Fellows and the financial resources of the Society continued to increase, and in spite of the temporary difficulties the administration, which was steadily improving in efficiency, showed itself to be quite capable of dealing with the problems which arose. The Society was now fairly launched on the career of high distinction which it has since followed successfully; and Newton's acceptance of the presidency and the possession of a house of its own were two important factors which convinced many that the Society had survived its years of difficulty, and had an assured career before it.

An entry in the Journal-book dated 16 February 1703/4 records that 'Mr Newton presented his Opticks to the Society. Mr Halley was desired to peruse it, and to give an abstract of it.' In the preface, which was written a short time after Hooke's death, Newton says: 'To avoid being engaged in disputes about these matters I have hitherto delayed the printing.' The researches which are discussed in this work had been made many years before, but publication had been postponed until the bitter and jealous criticisms which might have been expected from Hooke could no longer be made.

On the Anniversary day of 1704 Prince George of Denmark was elected to the Fellowship of the Society and the Journal-book records that 'The Society were extremely pleased with the honour the Prince did them in suffering them to choose him a member' and the Council desired the President and Secretary to wait on the prince with the Statute-book to have the honour of his subscription. This they did on 7 December 1704. In the same year Flamsteed had drawn up a summary of his Catalogue of Stars on which he had been at work for some years past. A copy of it, which he had given to a friend, was by chance shown to some Fellows of the Royal Society at one of its meetings and was then delivered to the Secretary to read publicly. Much interest was aroused, and the President and the Secretary were desired to enlist the aid of Prince George to meet the cost of printing the Catalogue. In this way the Society was inadvertently brought into an affair which Flamsteed had intended to keep in his own hands and to carry out with the assistance of his own friends. The prince agreed to Sir Isaac Newton's suggestion that he should support the work and a committee approved by him was appointed to supervise the publication, but Flamsteed was not included. As this committee was not appointed by the Council of the Society the difficulties and misunderstandings which arose subsequently do not properly belong to this history, but they will be found set out in a masterly manner by L. T. More in his biography of Newton.[1] The outcome was that Flamsteed, being greatly dissatisfied with the result of the committee's work when his Catalogue was published, bought up and burned all the copies of that issue which he could acquire, and printed a new Catalogue under arrangements made by himself; this was published in 1726, six years after his death.

With a committee actively engaged in seeking for new accommodation for the Society the President was able to take up its reorganization starting from the action which Sloane had taken in Council some years earlier. In 1706 he proposed that 'Every person newly elected a member of the Royal Society do, before his admission, pay his admission-money and give a bond to pay his weekly contribution, except foreigners'; and 'that no person be capable of being a member of the Council, who hath not given a bond to pay his weekly contributions, or who hath not paid them till the quarter-day then last past'. These were approved, but at a subsequent Council meeting an exception was made in the case of the professors of Gresham College who might be admitted to the Fellowship of the Society without paying admission fees, or giving bonds to pay their weekly contributions; and at a meeting held a few months later the officers of the Society were excused from paying annual subscriptions while they were acting in their official capacity.

Sir Godfrey Copley, Bt., of Sprotborough, Yorkshire, who was Controller of Army Accounts in 1704, had been elected a Fellow of the Society in 1691, and though his interests were not scientific he served for fourteen years as a member of the Council, between 1692 and the time of his death in 1709. In his will, which was proved in the Prerogative Court on 11 April 1709, he bequeathed to Sir Hans Sloane and Abraham Hill, Esq. (a former Treasurer of the Society) the sum of 'One hundred pounds in trust for the Royal Society in London for improving Natural Knowledge, to be laid out in experiments, or otherwise for the benefit thereof, as they shall direct and appoint'. The income of the trust was at first paid to Dr J. T. Desaguliers, who was elected a Fellow of the Society in 1714 and had been excused paying the annual subscription on account of the number of experiments which he showed at the meetings. Later he was appointed Curator of the Society's collections, a post which he held until 1743. He does not appear to have received any fixed salary but was

[1] *Loc. cit.* Chapter XIII.

remunerated by donations of from ten to twenty pounds, and occasionally of even larger sums.

In 1710 a volume of the *Philosophical Transactions* was published which contained an article by John Keill stating that Leibniz had published Newton's Fluxions under another name and with a different method of notation. Gottfried von Leibniz, who had been elected a Fellow in 1673, sent a letter to the Society in 1711 asking that the difference of opinion regarding his claim and that of Newton for priority in the discovery of the Calculus might be submitted to the judgment of the Fellows. The Society therefore on 6 March 1711/12 appointed the following to be a committee 'to inspect the letters and papers relating to the matters in dispute': Dr R. Arbuthnot, Mr A. Hill, Dr E. Halley, Mr W. Jones, Professor J. Machin and Mr Barnet. On 20 March of this year Mr F. Robartes, a musician and a writer on the theory of sound, was added to the committee; on 27 March Bonet, the Prussian Minister, and A. de Moivre and Dr Brook Taylor, both mathematicians, with F. Aston, a former Secretary of the Society, were included on 17 April. On some previous occasions the Society had declined to intervene in or to adjudicate upon claims or complaints brought by one Fellow against another; but this was evidently considered as being a special case since the documents in question were in the Society's possession. The committee presented their report to the Society on 24 April 1712 and it is preserved in its archives. In it they express the opinion that Mr Newton was the first inventor, and recommend that the letters and papers forwarded by the committee be made public. The Society agreed unanimously with the report and 'ordered that the whole of the matter from the beginning, with the extracts of all letters relating thereto, with Mr Keill's and Mr Leibniz's letters, be published with all convenient speed that may be, together with the Report of the said Committee' (Journal-book, vol. xi). This publication appeared in 1712 under the title of the *Commercium Epistolicum*. On 28 April 1714 Leibniz wrote to Mr J. Chamberlayne, chamberlain to Prince George of Denmark, saying that he entirely disapproved of the Report of the Committee, and of the *Commercium*, which he had not seen, and asked him to lay his letter before the Society. It was brought before the Society on 20 May 1714 when a resolution was adopted to the effect that 'it was not judged proper (since this letter was not directed to them) for the Society to concern themselves therewith, nor were they desired to do so. But if any person had any material objection against the *Commercium*, or the Report of the Committee, it might be reconsidered at any time.' The subsequent discussions and arguments which were put forward by supporters of the two philosophers who were primarily concerned are ably discussed by L. T. More in his biography of Newton.[1] Leibniz died in 1716.

[1] *Loc. cit.* Chapter XIII.

For some years past there had been many delays in the payments which were due to the Society for the fee-farm rents at Lewes (Wilkins bequest) and numerous applications which had been made by the Treasurer in 1679 and subsequently had been of no effect; by 1703 the sum due to the Society amounted to £450. As there appeared to be no other means of reaching a settlement, the Council instituted an action in the Chancery Court in 1704 to obtain payment of the arrears due to the Society, and also to establish who was liable for the payment of the rent in the future. Nine years later, in 1713, the Court gave its decision in favour of the Society's claim for the arrears, and ordered that in future a payment of £24 per annum should be made by Lord Abergavenny and his heirs. In 1939 this annual payment was redeemed by the Abergavenny Estate by a payment to the Society of £570.

At the end of 1710 Queen Anne appointed the Society to be Visitors and Directors of the Royal Observatory at Greenwich, the warrant making the order being dated 12 December 1710. The President and such others of the Council as he should think fit were to be constant Visitors of the Observatory; and they were authorized to demand of the Astronomer Royal within six months after the end of each year a copy of the observations which had been made. The Visitors could also require the Astronomer Royal to make such astronomical observations as they might think proper, and if any of the instruments were defective they were to notify the Board of Ordnance so that they might be exchanged or repaired. The issue of this warrant annoyed Flamsteed, the Astronomer Royal, extremely, and he attributed it to Newton's desire to hinder the work at the observatory, but of this there seems to be no evidence. The Visitors were: The President, Sir Isaac Newton, Mr F. Robartes, Dr John Arbuthnot, physician and mathematician, Dr Edmond Halley, astronomer, Dr R. Mead, M.D., Mr A. Hill (ex-Treasurer), Sir Christopher Wren, Mr Ch. Wren, and Dr Hans Sloane, Secretary. This was not an advisory committee to which exception could fairly be taken; besides the President four at least of the members were mathematicians of some distinction, and the others were men of varied administrative experience. The Astronomer Royal's objections were not supported by Mr Secretary St John, afterwards Lord Bolingbroke. The Visitors directed Flamsteed to send copies of his observations to the Royal Society and they represented to the Board of Ordnance the inefficiency of several of the instruments, only to be told that the Board had no fund from which such expenses could be met. Faced with this reply and the Astronomer Royal's obstructive attitude the Visitors could accomplish little until Flamsteed's death which took place in 1719.

Richard Waller's reappointment to the post of Secretary when Dr G. Harris resigned in 1710 coincided with a change in the allocation of the Secretaries' duties. It had hitherto been customary for the First Secretary, who was responsible for recording the business transacted at meetings of the Society and of the Council, to attend them, the Second Secretary replacing his colleague if he should be ill or unavoidably absent. The Council had never adopted any special resolution to this effect, but it had been the custom ever since 1662; after 1710, however, when the Society first met in its own house in Crane Court, both Secretaries attended every meeting except when prevented by illness or other urgent matter; and from the end of this year Waller, who had again become Dr Hans Sloane's colleague, was present with him at nearly every meeting, whereas during the preceding fifteen years he had seldom been present more than once a year. This was a very wise innovation since it relieved the First Secretary of some of his work which had long been recognized as being too heavy for one officer to carry out efficiently, and it also kept both officers in touch with all the Society's business. No reasons are given for the change but the need for it must have been fully realized by the President, and it was probably one of his reforms. Both Oldenburg and Aston had complained of the excessive amount of work which a Secretary was required to do, but that it should be shared between the Secretaries does not seem to have occurred to anyone though it was a simple and natural solution. The selection of Oldenburg to be a Secretary in the First Charter was an excellent one for he was well educated, had considerable literary experience and was also a good linguist. The editing of the *Philosophical Transactions* was naturally entrusted to him; it interested him and fitted in well with the extensive foreign correspondence which he had built up, although it added greatly to his work. Such clerical work would not have appealed so much to Dr J. Wilkins, and in 1668 he became Bishop of Chester. For the next ten years Oldenburg's colleague was changed every two or three years while Oldenburg himself became almost indispensable. In 1717 the Council made a gift to his son Rupert in consideration of his father's valuable services to the Society. After his death these duties, and with them the regular attendance at the meetings of the Council and the Society, were performed by the more enterprising of the two Secretaries, for example by Hooke, Aston and Sloane, while their colleagues played quite a secondary part. From 1694 when Hans Sloane became Waller's colleague he attended almost every Council meeting out of eighty-six, while Waller was only present at ten in as many years, so firmly had the practice been established that one Secretary alone should concern himself with administration. It was high time that this was changed.

When the property in Crane Court was taken over a room or rooms had to be built in which the Society's collections could be housed and Richard Waller, one of the Secretaries, undertook to carry out this for £200; but in 1712 he had to report to the Council that his estimate had been greatly exceeded since the cost of the work amounted to £400. After considerable discussion Waller agreed to accept £300 as payment in full on condition that he should be registered as a benefactor who had presented the balance of £100 to the Society as a gift; this was agreed to and his name appears among the benefactors of the Society. On 21 February 1711 he had made a will by which he left £1000 to the Society but this bequest he revoked by a codicil dated 19 June 1714 'for several good and weighty reasons'. There were still the repairs and renewals to be done at Crane Court, and though their cost did not amount to anything like the £1800 estimated by the author of the anonymous pamphlet of 1710 the sum of £800, which included the cost of the repository, was one which the Society at this time found it very difficult to raise. The President gave £120, Lord Halifax £100, Mr R. Waller, a Secretary, £100, as explained already, and some smaller sums came from other Fellows. At this juncture a member of the staff of the Society most generously came to its assistance.

From the early days of the Society to the end of the seventeenth century various Fellows had taken charge of the collections of curiosities in the repository and of the books forming the library, but in 1696 the Council decided to appoint a permanent official to carry out this work. Henry Hunt was therefore appointed on 25 November 1696 to be the Keeper of the Library, and subsequently he became also Keeper of the Repository and Housekeeper. Hunt had first entered the employment of the Society on 9 January 1672/3 as a boy to assist Robert Hooke with his experiments. From then onwards for forty years, until his death in 1713, he always lived with the Society and was one of its most loyal and best workers. Hooke records in his Diary on 4 June 1672/3 the arrival of Hunt from the country, and again on 9 January that he first came to stay; subsequent entries throughout the Diary tell of the work he did in his early years. In 1676 Hunt was appointed Operator to the Society in succession to Richard Shortgrave, who had held the post from 1663. It is recorded in the Council minutes of 2 November 1676 that 'Mr Henry Hunt having been proposed to succeed in Mr Shortgrave's place, the Council having heard the several good testimonies given him of his ability and honesty, received him to be Operator to the Royal Society'. His work in this capacity and in that of Keeper of the Library and Repository was highly valued and he earned the respect and esteem of all the officers and Fellows with whom he came in contact. Hunt was paid at first twenty pounds a year as Operator and then forty pounds a year as Keeper of the Library and Repository. In addition, he engraved plates for the *Philosophical Transactions* and other works. Some of his drawings, which are excellent, are in the possession of the Society and others are in the British Museum. He was careful in the management of his own affairs, and when the Society was moving from the rooms which it had occupied at Gresham College to its new quarters in Crane Court he was able to lend the Society the sum of £462 to assist in paying off the mortgage on its new property. Soon afterwards gifts of money were received from Fellows for the Society's house and this enabled the Council to repay to Hunt £262 together with the interest due. By the following January, however, the Treasurer was again in difficulties and Hunt advanced another £200 to pay bills which had fallen due. Two years later when the Council decided to repay Richard Waller £300 of the amount which he had expended in building the Repository at Crane Court, Hunt again came to the Society's assistance by lending an additional £250 on 26 July 1712. Altogether the Society then owed Hunt a sum of £650 and interest, but it is doubtful whether this was ever repaid, as no mention of it occurs in the Council minutes; the Society was not then, nor was it for some years to come, in a position to repay that amount.

Hunt, whose death was reported to the Council on 29 June 1713, had lived his whole life from boyhood onwards with the Society and in its service. He was probably an orphan and he never married. Hooke, until his own death in 1703, had befriended him constantly and always treated him more like a son than a subordinate or an assistant.[1]

Vacancies in the administrative staff were now becoming inconveniently frequent and that caused by Jezreel Jones' departure on an expedition to the coast of Barbary had to be filled up. At the Council meeting on 5 November 1701 'Dr Hans Sloane, the Secretary, told the Councell that he had imployed Mr Humfrey Wanley in doing the business of a Clerk since Mr Jones's departure but withall he had told him that he must not expect any certain salary'. However, we find that the Council did vote each year a sum of £15 to be paid to Humfrey Wanley, who continued to serve the Society until the end of 1706. Wanley was a man of wide learning; he went to Oxford in 1695, and became an assistant in the Bodleian Library in the following year. He prepared the index to E. Bernard's Catalogue of Manuscripts in 1697 and one of Anglo-Saxon Manuscripts in 1700. In this year he became Assistant Secretary of the Society for the Promotion of Christian Knowledge, and from 1702 to 1708 was the Secretary of that Society. He was an enthusiastic archaeologist and became a Fellow of the Society of Antiquaries when it was formed in 1717. Wanley resigned upon receiving another appointment in 1706, perhaps a more permanent post on the staff of the Society for Promoting Christian Knowledge, and he was immediately elected a Fellow of the Royal Society; he was subsequently employed to catalogue the Harleian Manuscripts and later became Librarian to the first and second Earls of Oxford. He died in 1726.

Although Dr John Thorpe's appointment as Clerk was not recorded until 4 February 1707/8, he had carried out the duties of the office since November 1706 when Wanley left. He had been elected in 1705 but resigned his Fellowship on taking up the post of Clerk; in 1713 when he gave up the Clerkship he was reinstated in his Fellowship. From June 1713,

[1] I am indebted to Mr W. H. Robinson, Librarian of the Royal Society, for details of Hunt's career, and of other members of the Society's staff.

when Hunt died, until the following November, Thorpe was appointed to perform the additional duties of Housekeeper and Keeper of the Repository and Library. For this extra work his salary of £40 a year was increased to £60, but he resigned before the end of the year. He then went to live in Rochester where he soon had a very successful practice as a physician. He assisted Dr Hans Sloane in some of his work.

At the Council meeting on 7 December 1713 it was resolved 'that the same person should officiate as Clerk who should now be Housekeeper, Keeper of the Repository and Library'. The salary of the person chosen was to be £30 per annum and 6d. for every page he copied of minutes, papers and letters. At the same meeting Mr Alban Thomas, who was then twenty-seven years of age, was appointed. He was the son of the Rev. Alban Thomas, Rector of Blaenporth, in Cardiganshire, and he matriculated at Jesus College, Oxford. He was an intimate friend of Edward Lloyd the antiquary, and in 1708 became Librarian of the Ashmolean Museum. He is said to have had a mastery of French and Latin as well as a good knowledge of mathematics and experimental philosophy. Thomas was also, for a time, a reader at the University Press at Oxford. While still in the service of the Society, he received, on 9 June 1719, the degree of M.D. from Aberdeen University. Early in 1723 he was implicated in some way with the Jacobite plots of that time and had to go into concealment; the enforced absence from his work at the Society which this entailed led to his dismissal in March 1723; as he saw no hope of advancement if he remained in London, Thomas returned to his birthplace in Cardiganshire where he settled down to the life of a country doctor. He died in 1771.

From the many candidates who applied for the vacancy caused by Thomas' dismissal Francis Hauksbee, junior, was chosen. His uncle had for many years past been a curator of experiments to the Society and also possessed a shop in Crane Court where he made air-pumps and other scientific instruments. The nephew had worked with his uncle for many years and had, no doubt, assisted him in experiments which were shown to the Society. A specimen of his work is preserved in the air-pump which F. Aston had ordered from him but which had not been completed when Aston died in 1715; the Society however acquired the instrument, paying for it out of Aston's estate which he had bequeathed to them. F. Hauksbee, junior, carried out his duties as Clerk to the Society with great care and diligence throughout the forty years which he spent in its service until his death in 1763. This long period of service was greatly to the advantage of the Society, for frequent changes among the members of the administrative staff were nearly as undesirable as those among the officers.

In 1714 an Act of Parliament was passed offering a reward to anyone who could produce a satisfactory method for determining the longitude at sea, a matter of the highest importance to seamen; it constituted also a Board of Longitude and named the President of the Royal Society as one of the Commissioners. Other Commissioners were added from time to time and the Astronomer Royal was one of these. Many applications for grants were considered by the Board, one of the earliest of which was for a survey of the coasts and headlands of Great Britain which was suggested and supported by William Whiston, a mathematician who was a pupil of Newton. The Board in 1714 offered the sum of £20,000 for any practicable method of determining longitude within half a degree, and in 1728 John Harrison submitted plans of his first chronometer which was completed in 1735 by George Graham; he designed a second and a third type in 1739 and 1757, finally completing his fourth and best type in 1759. These are now exhibited in the Royal Marine Museum at Greenwich. The payment of the award was delayed by a dispute between Harrison and the Board of Longitude in which King George III supported Harrison, and also, it is said, by Dr N. Maskelyne, the astronomer, who favoured stellar observations in preference to chronometric methods. Harrison was awarded the Copley Medal by the Society in 1749 for his horological work. He eventually perfected his chronometer to the satisfaction of the Commissioners, and the Royal Society was officially consulted about his final type of the instrument. It was upon a recommendation of the Council that Harrison was awarded sums amounting in all to £15,000 during the period 1737–73, thus ultimately receiving the whole amount of the award.

In 1717 Newton, in his official capacity as Master of the Mint, laid down the rule that one troy pound of standard gold should be coined into $44\frac{1}{2}$ guineas, thus one troy ounce of standard gold would be equal to £3. 17s. $10\frac{1}{2}d$. Standard gold consisted of eleven-twelfths of fine gold and one-twelfth of alloy so that 85 shillings was the parity figure for an ounce of fine gold. This decision remained unaltered until 1939, when the Chancellor of the Exchequer, in his speech on the Finance Act for that year, gave the reasons for modifying the parity definition, and reminded the House of its past history. This furnishes one of numerous instances where the State even as early as this was making use of the Royal Society and its most eminent Fellows to advise in scientific matters of difficulty and technical importance; this practice continued to increase throughout the eighteenth and nineteenth centuries and has been greatly developed in recent years as is recorded later in Chapters VIII and IX.

The Fellows were now beginning to take more interest in the Society, and as they became more confident that it would increase in prosperity and efficiency in the years to come they were more ready to contribute to it. Bishop John Wilkins in 1672 had left £400 to the Society; and it will be remembered that Robert Hooke was said to have purposed devoting the greater part of his estate to providing a house worthy of the Society, but in the end he made no bequest to carry out his plan. Francis

Aston in 1715 bequeathed the whole of his estate to it without laying down any conditions; in 1717 Dr E. Paget, who had been elected in 1682 and had served on the Council in 1694 but did not take any other active part in its administration, bequeathed two houses in Coleman Street which brought in a rental of about £100 per annum. They remained in the Society's possession until 1835 when they were required by the city of London in connection with the adoption of a scheme for improving the approach to London Bridge; an agreed sum of £3150 was paid to the Society for them, and this was invested, the income being paid over to the Society regularly by the Charity Commissioners.

The honoraria to be paid to the Officers of the Society had frequently been considered by the Council in past years; Oldenburg had complained of lack of assistance in 1664 and had finally been granted £50 a year; Aston had made a similar complaint in 1685 but no regular rate of payment had then been decided upon. The matter seems to have been left in abeyance in later years, perhaps because both Richard Waller and Sir Hans Sloane were rich men, but in 1720 the President came to the conclusion that it was time that something should be done. He therefore proposed to the Council in October 1720 that £50 should be paid to each of the Secretaries annually and this was agreed to; in 1741, when the financial position had improved, the Council increased the amount to £60. No honorarium was proposed for the Treasurer whose work was neither so heavy nor so continuous as that of the Secretaries.

For many years past considerable interest had been taken in antiquarian studies, and in 1572 a Society of Antiquaries had been founded by Archbishop Parker principally with the object of preserving the ancient documents of historical value which the recent dissolution of religious houses had gravely imperilled; its members also desired to keep alive the knowledge of the past history of this country, its people, its customs and its laws. The Society had intended to petition Queen Elizabeth for a Charter of Incorporation but no effect was given to this. James I was not disposed to support the Society lest its activities should become political or subversive, and dissolved it in 1604.

Historical studies of this character hardly fell within the scope of the Royal Society, and though communications of an antiquarian or archaeological character sometimes appeared in the *Philosophical Transactions* they were not in favour. The present Society of Antiquaries was established in 1717 and was granted a Charter of Incorporation in 1751. During the eighteenth century a considerable number of its members were proposed for election into the Royal Society and were admitted; they usually formed part of the non-scientific group of the Society and in a few cases were elected to one or other of the officers' posts; most of the Treasurers before the nineteenth century had archaeological interests but these do not seem to have been accompanied by any marked financial ability. After the revision of the statutes in 1847 the number of those who belonged to both the Royal Society and to the Society of Antiquaries fell off very considerably.

John Flamsteed, the Astronomer Royal, died in 1719 at Greenwich and Edmond Halley was appointed to succeed him through the good offices of Dr R. Mead, who wrote, 'I have been so happy as to get Flamsteed's Place for Dr Halley by means of my Lord Sunderland'. Dr R. Mead, who was a very eminent physician, had been elected a Fellow in 1703 and was a man of considerable influence; he served on the Council in 1705 and was a member of it continuously from 1707 until 1752, a period of forty-five years in all.

In August 1701 A. von Leuwenhoek of Delft had written to the Society promising that a cabinet containing nineteen of the microscopes which had been made by him should be sent to the Society at his death, which took place in 1723. He bequeathed them 'as a mark of my gratitude and an acknowledgment of the great honour which I have received from the Royal Society'. In the following year the Council presented his daughter with a handsome silver bowl bearing the arms of the Society in testimony of their esteem for her father. These instruments remained for many years in the Society's possession and are mentioned by Adams in his *History of the Microscope* as being there in 1798. Fifty years later Sir James South drew the attention of the Council to their absence from the Society's rooms; enquiries were then made in all likely quarters but the microscopes have never been recovered; it seems to be fairly certain that they passed out of the Society's possession at some time between 1800 and 1830. It was not until much later that store ledgers, in which every item of the Society's property was registered as a matter of routine, were introduced, so that the whereabouts of each object whether on loan or in store could be checked periodically. This very necessary control was instituted in 1935.

In October 1719 Robert Keck, a Fellow of the Society, bequeathed the sum of £500 to form a fund the income from which was to be paid to the Fellow who should be charged by the Council with the duty of carrying on the Society's correspondence with men of science in other countries. He had served on the Council in the years 1716–18, and no doubt had been impressed by the very considerable amount of work with which the two Secretaries had to deal. In 1720 the Council resolved that this bequest should be invested in 5 per cent Bank Bonds or in East India Bonds, but it was May 1723 before the Society received the money which by then amounted to £547. 10s. 0d. for interest and principal after the legal costs had been deducted.

At a meeting on 11 April 1723 the President informed the Council that Mr P. H. Zollman, who had been recommended by Mr Walpole for appointment as a Clerk to the Society, was not qualified to hold that post, since he had not been naturalized in this country; he therefore recommended to the consideration of the Council 'whether the said Person might not be of good service to the Society through his skill in many languages to serve them in the Capacity of an Assistant to the Secretarys in managing a foreign Correspondence. Whereupon Mr Zollman being put up by Ballot for the said Place he was accordingly chosen, his salary being £20 per annum.' He was elected a Fellow of the Society in June 1727, and resigned his post as Assistant for Foreign Correspondence in 1728.

The post of 'Foreign Secretary' was not created until much later, and for a century the income of the Keck Fund was paid to the Fellow who was charged by Council with the task of dealing with the foreign correspondence for the time being. Such men were not, strictly speaking, officers of the Society and were not usually members of the Council, though Dr James Parsons was a councillor in 1753 while he was a recipient of the grant, and Dr M. Maty had been on the Council in 1759 for two years before he took charge of the foreign correspondence. The custom by which the Fellow in charge of the foreign correspondence did not attend meetings of Council, unless specially summoned, was not satisfactory. He received his instructions from the Secretaries or from them through the Clerk, but he did not hear any discussion of them which might take place at the Council. Had Dr Charles Hutton been present at the Council meetings in 1779 and 1780 the disagreement between him and the members of Council which led to his resignation in December 1780, and caused much ill-feeling at the time, might not have arisen. It was not until 1823 when the statutes of 1776 were being revised during the presidency of Sir Humphry Davy that the Council resolved that the Foreign Secretary should in future rank with the two principal Secretaries, and that a new office, that of the 'Secretary for Foreign Correspondence' should be instituted.

From its earliest days the Society had been in the habit of receiving many communications on the subject of the weather of this and other countries; these were in some cases published and in others were passed on to those of its members who, like Halley and some others, were specially interested in meteorology. Among these was Dr James Jurin, M.D., who was elected to the office of a Secretary at the end of 1721 in the place of Halley and held it until the end of 1727. He was a very useful and active member of the Society, attending every Council meeting of the thirty-nine which were held during his six years as a Secretary. He was both a physician and a good mathematician who communicated many papers to the *Philosophical Transactions* up to the time of his death in 1750; he was an ardent supporter of Newton's views. On becoming Secretary he proposed that the Society should provide competent observers who were willing to take observations with the necessary instruments; and undertook to transmit to such correspondents the instruments. Notice of this new departure was given in the *Philosophical Transactions* where it was announced that 'Ingenious travellers are now furnished with extraordinary accommodations that were not known to former ages; as thermometers, barometers, hygrometers, microscopes, telescopes, micrometers, exact scales and weights promptly to weigh liquors, and with other circumstances to examine the intrinsic value of all coins, medals or metals; pendulum watches, instruments and indexes for magnetical variations, and inclinatory needles, and other helps to ascertain longitudes, and other manifold contrivances for manifold uses'.

At this time several Fellows of the Society were active in promoting inoculation as a precaution against smallpox; among these were Sir Hans Sloane, Dr Jurin, Dr Williams, Mr Wright and Mr Gale, who published several papers on the subject in the *Philosophical Transactions*. The subject aroused much interest at this time and there are in the Society's library two large folio volumes of manuscripts dealing with the statistics of inoculation.

In 1726 the President's health did not allow him to attend the Council and Ordinary meetings as regularly as had been his habit; he was not present at the Anniversary meeting when he was re-elected for another year, but attended two Council meetings in February and March. After the second of these he was taken ill and died on 20 March in his eighty-fifth year. He was buried in Westminster Abbey on 28 March. He was the most illustrious of the Presidents of the Royal Society, and though he may not have added much to his scientific publications during the twenty-four years of his presidency, the services which he rendered to the Society were of the utmost value to it. He had enabled the Fellows to acquire a house of their own; he had initiated an administrative system which, as developed by his colleague and successor Sir Hans Sloane, served the Society well for a century; by the regularity of his attendance at both Council meetings and Ordinary meetings he set a standard which his successors strove to maintain; and from him the Society learned to expect from its presidents and its other officers that close attention to its affairs and prudent guidance of its policy without which no continued advance would have been possible.

Of the three portraits of him which the Society possesses one, painted in 1717 by Charles Jervas and presented by Newton in 1717, hangs in the Meeting Room.

On Sir Isaac Newton's death in March 1726/7, the Council elected Sir Hans Sloane, who had been associated with him during the whole of his presidency, had acted as a Secretary for nine years before then, and had been a member of Council ever since. In 1728 shortly after becoming President Sir Hans Sloane raised the question of adopting an address to be presented to the king, George II, praying him that he would become the Patron of the Society as his father had been, and would inscribe his name in the Charter-book. There had been a suggestion some ten years earlier that this should be done, and a page of the Charter-book was prepared to receive his signature, but no further steps were then taken; on the present occasion however the proposal was agreed to. The king received the President and other members of the Council graciously, and signed his name in the Statute-book as Patron.

In these years the Society's museum was receiving considerable additions from time to time from its Fellows and other quarters. The Winthrop family, who had settled in New England in 1630, had kept up the connection which John Winthrop and his son had formed in the seventeenth century, and these two had contributed to the Society collections of zoological, botanical and mineralogical objects from time to time. Their descendants of the next generation did not inherit their interest in observing and making experiments to develop the natural wealth of the colony, but a grandson who bore their name, the son of Walt Winthrop, was a worthy successor to his grandfather. This John Winthrop was born in Boston on 25 August 1681, took his degree at Harvard University in 1700 and became Hollisian Professor of Mathematics at Cambridge, New England. He considered himself aggrieved by the action taken by the Courts and Legislature of Connecticut in certain legal proceedings arising out of the settlement of the estates of his father and uncle, and therefore went to England to seek redress from the Privy Council, and to assert some kind of hereditary claim to the governorship of Massachusetts or Connecticut, to which he believed that he was entitled. After his arrival in this country he continued to live in London; in 1734 he was elected a Fellow of the Royal Society. Dr Cromwell Mortimer, who at this time was the Secretary responsible for the editing and publication of the *Philosophical Transactions*, describes him in Vol. 40 as a generous benefactor to the Society's repository to which he contributed 'above six hundred curious specimens chiefly from the Mineral Kingdom, accompanied by a list containing an accurate account of each specimen. Since Mr Colwall, the Founder of the Museum of the Royal Society, you have been the benefactor who has given the most numerous Collection; and it is hoped your generous Example will be followed by some of the present Members, by which means our Repository may soon become one of the most conspicuous in Europe.'

Of another consignment Dr Mortimer says: 'You have sent to England many rare curiosities for the Royal Society, which, although by the disingenuity of the Pilot they missed their port and were not laid up in the intended Repository, are some of them to be seen in a recent Museum at Cambridge.'

When Sir Hans Sloane purchased the manor of Chelsea in 1716 he presented the freehold of the botanical garden there to the Company of Apothecaries on condition that it should be preserved as a physic garden, and that the company should present yearly to the Royal Society at one of its meetings fifty specimens of plants which had been grown in the garden during the preceding years, and which were all to be specifically distinct from each other until two thousand had been presented; this condition was fulfilled by 1761. This collection of plants was frequently consulted and was a notable addition to the museum. Though the Society was in no way responsible for Chelsea Physic Garden, its Council always took a serious interest in it and its welfare. Some years later when the Company of the Apothecaries proposed to close it, the Society's Council strongly urged a reconsideration of the matter, which was conceded and the garden was maintained.

Immediately after his election as President in March 1727 Sir Hans Sloane introduced a change in the Society's procedure which was recorded in the Journal as follows:

The President took occasion to mention an alteration he judged ought to be made in a matter of form hitherto used in the Ordinary meetings, which is to make a difference of solemnity in laying the Mace when the President is in the Chair, and not laying it when the Chair is supplyed by a Vice-president. He observed that there was no foundation for making any such difference, and that as the Vice-president is invested with the powers of a President in every respect in his absence, he ought also to be attended with the same solemnity in the Ordinary meetings.

His contention was agreed to and the Mace has since then been laid on the table whether President or Vice-President is in the Chair.

He also called attention to the fact that copies of the Journal-books, etc. had not been made regularly in accordance with a resolution of the Council adopted many years previously (January 1683/4). The Council thereupon resolved that the work should be put in hand at once and Thomas Stack was appointed as copyist. He worked for the Society for a number of years, making copies of all the books, arranging the letters, papers, etc. He eventually became a Doctor of Medicine, and was elected to a Fellowship of the Society in 1737. The President was sixty-seven years of age when he took up his new post, and at once introduced several changes in the Society's constitution which, in his opinion, were urgently needed. Of these the most important were: (a) to exempt the foreign members from paying subscriptions since their position was honorary, and they should have no claim to take part in the administration of the Society; (b) to sue such Fellows as were in arrear with their payments which they should have made to the Society; and (c) to make it compulsory for every candidate to be approved by the Council on the recommendation of three Fellows, one of whom at least was to be a member of Council, before the candidate could be balloted for. The last of these propositions was passed as a statute and acted upon from 1728 to 1730, all candidates in those years being approved by the Council before being balloted for at a meeting of the Society. In the latter year the desirability of limiting the number of Fellows was taken into consideration by the Council, but before any statute was adopted a case was drawn up and submitted to the Attorney-General (afterwards Lord Chancellor Hardwicke) for his opinion. The points on which guidance was desired were firstly,

Whether it is any infringement of the rights and privileges of the Fellows that a candidate should be approved by the Council before being balloted for by the Fellows generally: considering that the rejection of a candidate by the Council does not disqualify him from being put up again?

and secondly,

Whether the Council cannot by virtue of their general power of regulating the body, limit the number of the members thereof; or at least make such laws as may check the too great increase of the body with new members unfit for answering the end of the Institution?

The opinion of the Attorney-General on the first point was that

The Charter having joined the President, Council and Fellows together in the election of Fellows, as members of the entire body, and having directed such elections to be made by a major part of them all, without giving any preference in those acts to the Council, I think the Council should not make a Statute whereby to assume a negative to themselves, which seems to me to be the effect of this Statute. Therefore I apprehend this Statute not to be warranted by the Charter.

His opinion on the second point was

Considering that the Charter hath left the body at large without limiting the number of Fellows, and considering also the nature of this foundation, I think the Council cannot make a Statute to limit the Fellows to a certain number. But they may make reasonable Statutes or bye-laws to describe and ascertain proper qualifications of persons to be elected Fellows in such manner as may best answer and promote the ends of an Institution so useful to the learned world.

On the receipt of this legal opinion the Council rescinded the statute which they had adopted in 1728 and substituted for it the following:

Every person to be elected Fellow of the Royal Society shall be propounded and recommended at a meeting of the Society, by three or more members, who shall then deliver to one of the Secretaries a paper signed by themselves, signifying the name, addition, profession, occupation and chief qualifications of the Candidate for election, as also notifying the usual place of his habitation; a fair copy of which paper with the date of the day when delivered, shall be fixed up in the Meeting Room of the Society at Ten several Ordinary Meetings, before the said Candidate shall be put to the Ballot.

Saving and excepting that it shall be free for everyone of His Majesty's subjects who is a Peer, or the son of a Peer of Great Britain or Ireland, and for every one of His Majesty's Privy Council of either of the said Kingdoms, and for every Foreign Prince or Ambassador, to be propounded by any single person, and to be put to the Ballot for election on the same day; there being a competent number for making elections.

This statute was passed on 10 December 1730 and the first 'paper' or certificate was lodged with the Secretaries on 25 February 1730/1, since when such a certificate has been required for every candidate whether subsequently elected or not. These certificates have been bound in a series of folio volumes which are preserved at the rooms of the Society forming a most valuable series of documents relating to the candidates who have been proposed as Fellows of the Society. This legal opinion was of great importance in laying down precisely the course to be followed in the admission of candidates.

It is interesting in the first place to note that the Council had realized the need for some kind of control over the number of Fellows to be elected and, what was even more important, the nature and scope of their qualifications. At the end of 1729 the number of Fellows was 254 and this was rising steadily, reaching 301 by 1740. So far the prospect of an excessively large membership was remote; but the fact that out of this total 202 had but little knowledge of natural philosophy, or any interest in it, was disturbing. Herein lay the real danger to the Society and its future, and it was many years before an effective and practical remedy was found. Sir Hans Sloane, and Sir Isaac Newton before him, had both foreseen how serious a hindrance an unscientific majority might be to the aims of a scientific institution, but it was difficult to see how this was to be remedied under the Charter. Another assurance which the Attorney-General had given that Fellows in arrear with their subscriptions who had signed the Obligation, that had been required from every Fellow since 1674, might be proceeded against in a court of law for the recovery of their unpaid fees and subscriptions was most important; had it been regularly put into operation in all cases where sums due to the Society had not been paid after a reasonable interval many of the financial difficulties which were experienced might have been avoided. But these arrears were too often allowed to accumulate in the hope that some payment would soon be made, and Treasurers often preferred to send reminders of overdue arrears instead of asking the authority of the Council to prosecute the defaulters. However, in the present case all defaulters were informed that legal action would be taken forthwith and a considerable sum due to the Society was recovered.

According to Dr Kippis the Council had been for some time in the habit of approving the election of candidates before their names were submitted to a meeting of the Society; but this henceforth was discontinued, at any rate in principle. From time to time in later years complaints were made that the officers had means of assuring the election of candidates whom they favoured, but even if this did occur occasionally to describe it as the usual practice was exaggeration.

In 1736 James Hodgson, F.R.S., the nephew of Flamsteed, presented to the Society the clock which Thomas Tompion had made in 1676 to the order of Sir Jonas Moore, in order that it might be given to Flamsteed, together with some other instruments and books, for his work at the Royal Observatory at Greenwich for which no equipment had been provided by the State. During Flamsteed's life it stood in the great room of the observatory; and on his death in 1719 it passed with other property to his executors; seventeen years later it came into the Society's possession. It was in the Society's house at Crane Court but it seems to have passed out of the Society's keeping at some time in the latter half of the eighteenth century. Early in the nineteenth century many attempts were made to trace it but these were of no avail until 1925, when it appeared in a London sale room with some other property of the Lowther family in whose possession it was said to have been for many years. It seems likely that it was borrowed from the Society by Sir James Lowther, Bt., who was elected a Fellow in 1736, and after his death in 1802 it had remained in the house of the family until the history of its acquisition had been forgotten. At the sale it was sold and went to America, but two years later it again appeared in this country for sale; it was then bought by some who were interested in acquiring it for the British Museum and presented to the National Collections there. It is now exhibited in the King Edward VII Gallery, and has been described and illustrated in the *British Museum Quarterly* (vol. III, Plate XXX).

The attention which the Society had devoted to industrial and technical schemes, and the contributions to them which natural philosophy could make, had occupied much of the time at its meetings in the seventeenth century, but recent achievements in mathematics and physics had partially diverted attention from practical applications of science. There were probably many who looked forward to a closer relationship between the theoretical and practical branches of knowledge. In 1721 a proposal was made in an anonymous pamphlet that a Society should be formed to promote Arts and Manufactures; and a copy of it was sent to the Royal Society. The Council discussed the suggestion but decided that the Society was not then in a position to embark on so large and costly a scheme. Its resources were increasing slowly but they were not sufficient for any considerable undertaking, and only eight years later serious steps had to be taken for dealing with the recovery of sums which were due to it.

The President was now beginning to feel the effects of a long and very busy life; he still attended all the meetings of Council, but these had become fewer of late though a recurrence of financial stringency shows that all was not going too well; for two years, 1738 and 1739, the Treasurer had reported a deficit on the year's Income and Expenditure account. The position was investigated and steps were taken to amend the faults which then came to light, and which should have been dealt with sooner by the Council who had ample powers to do so.

In 1741 the President, who was now eighty-one years of age, represented to the Council that he would not stand for re-election, and desired them to select another candidate. This the Council with much regret proceeded to do after having, at his suggestion, authorized the annual allowance to be paid to each Secretary being increased from £50 to £60.

Sir Hans Sloane had served the Society continuously for fifty-one years (from November 1690 to 1741) and as councillor or as an officer of the Society had been actively associated with every improvement in its administration which had been introduced. He was to live for another twelve years and at each Anniversary meeting he was re-elected a member of the Council. Though he could not attend its meetings his advice was sought on many occasions. The sixty-three years of continuous executive assistance and wise counsel which he gave to the Society and his activity in promoting its well-being have not been equalled by any other Fellow of the Society in its long history.

The Society could look back with considerable satisfaction on what had been accomplished in the past forty years. The officers were attending the meetings of Council and of the Society regularly; the payment of fees and subscriptions was being made more punctually, and when this was neglected legal means of enforcing it were available. The Society was suitably settled in a house of its own and a satisfactory procedure of administration was working reasonably well. There seemed to be no grounds for anticipating serious difficulties. The financial position was improving and, if the Treasurers dealt firmly with those whose payments were overdue, there should certainly have been no difficulty in building up a reserve of invested funds. Some embarrassment was sure to occur before long, since the number of Fellows who were neither scientific men by training nor interested in science or technology was increasing more rapidly than the men of science; also there was as yet no standard of qualifications which candidates had to possess, and under the conditions which then existed it was impossible to impose one. Men of science were still in the minority both on the Councils and at the meetings of the Society. Nevertheless the orderly and effective administration which had been evolved under the last two presidents gave good hopes that a way would be found of overcoming even these restrictions whereby the Society might attain the aim of its founders of being an institution wholly devoted to the promotion of Natural Knowledge.

As the property of the Society had of late been increased by several bequests, the Council recommended that the Society should memorialize the Crown for a licence to purchase or hold lands, etc., in mortmain. No copy of this petition exists in either the Journal- or Minute-book of the Society though it is frequently alluded to. It was stated to have been lost or mislaid in the office of the Secretary of State and this led to another being drawn up in the same terms and submitted by the Council. Mr J. C. Weld however ascertained that the documents in question were preserved in the State Papers Office and has printed a copy of the Petition in his *History*.[1] It bears the date 23 June 1724 and is signed by the President, Sir Isaac Newton, and eleven members of the Council including the two Secretaries, J. Machin and J. Jurin. The petition was submitted to the king, George I, on 30 June 1724 and was referred to the Attorney-General or the Solicitor-General for their 'opinion what His Majesty may fitly do therein'. On 21 July 1724 the Attorney-General reported that in his opinion a royal licence to acquire and hold landed property to a yearly value not exceeding one thousand pounds might be allowed; a licence to this effect was granted 17 December 1724.

By the end of the period which has just been described the President's health was beginning to fail since he was seventy-nine years of age in 1739, and the activity of the other officers was unequal to making up for his failing initiative. The number of Council meetings fell to three in the year and the Treasurer had to report that his receipts were £95. 16s. 6d. in 1739 and £240 in 1740 short of the expenditure. Seeing that the President had dealt satisfactorily with a similar position only ten years before, a recurrence of a deficit so soon was not expected. Recorded history offers no adequate reason for this, but the general slackening which always crept into the administration when the President was no longer able to exercise the influence that he had hitherto given to it naturally suggests itself as a contributory cause. The activity of the Councils fell off markedly during the last years of Sir Isaac Newton's life, and a similar cause was now in operation again. Except for this disadvantage the Society's present condition and its outlook had improved markedly during the forty years which were now ending, so that it may be of interest to examine one of the last years of Sir Hans Sloane's presidency, the year 1739, in some detail. The Fellows numbered three hundred and of these one hundred were scientific men.

The officers were James West, the Treasurer, who had replaced A. Pitfeild in 1732 about the time that the Acton estate had been bought with the payments made by Fellows who had been in arrears with them for some years. He should not therefore have found himself in similar difficulties only eight years later. The Secretaries were Professor John Machin, who had held the office since 1718, and Dr Cromwell Mortimer, M.D., who had replaced Dr W. Rutty in 1730. The Council of the year 1739–40 were well chosen, ten of the twenty-one councillors being scientific

[1] Vol. I, p. 431; *Record of the Royal Society*, 4th ed. p. 285.

men, and of these several were of considerable repute; Dr Edmond Halley, John Hadley and Dr John Machin, were astronomers; William Jones was a mathematician; George Graham was celebrated as a horologist and instrument-maker; Sir Hans Sloane, Dr Richard Mead and Dr Cromwell Mortimer were physicians; Peter Collinson and Isaac Rand were botanists as was also Sir Hans Sloane. Besides the President, Dr E. Halley, Isaac Rand and Dr R. Mead were elderly men, Halley being eighty-three at this time. Only three Council meetings were held in the year, and of the councillors eight attended none of them. A considerable number of communications were received during the year and were read by the Secretaries at the Ordinary meetings; new instruments and improvements to those already in use were also described. The Ordinary meetings were evidently well supported, and it was only the administrative work which for the time being showed weakness. Applications were occasionally received for grants in order to carry out experiments and researches but the Society's funds did not admit of these being made, and a century had to pass before the first bequest to the Society in support of scientific research was received.

In the early part of this chapter it was remarked that though Sir Isaac Newton had greatly improved the regularity of the officers' attendance at the meetings of the Council and the Society he did not attempt to increase the number of scientific men on the Council; it is suggested that had it been possible to have fourteen instead of only seven or eight scientific councillors on each Council the attendance at each meeting might well have been doubled, and the Society would have anticipated its action a century later when Councils in which there was a majority of scientific councillors impressed the same character on the whole of the Society's policy. But apparently so drastic a change was considered impracticable then, and so favourable an opportunity did not occur again until a century later, as will be described in Chapter VII. For the time being therefore scientific activity in the Society was left mainly to about eight or nine councillors. There were however some men of scientific distinction who, though they served from time to time on the Council, were doing the Society and science a far greater service by the additions which they were making to Natural Knowledge by their work and by their communications which were published in the *Philosophical Transactions*. Fortunately there were several of such men at this time, some were nearing the end of their labours but others might be expected to have a life's work before them.

Sir Isaac Newton was the principal of these, but after communicating his paper on 'Optics' to the Society in 1703 he turned his attention mainly to the reform of the administration. Other mathematicians of note were A. de Moivre [1697][1] who wrote on fluxions, on the theory of chances and

[1] The dates in square brackets indicate the year of election to the Royal Society.

life annuities; the Scottish mathematician, C. Maclaurin [1719], who died when comparatively young, distinguished himself at Edinburgh; he published a number of mathematical works. Brook Taylor [1711], who was for a few years a Secretary of the Society, was a distinguished mathematician who published the theorem which bears his name; it has recently been found that the same theorem had been discovered some forty years earlier by Professor James Gregory of St Andrews.

John Flamsteed [1676], the first Astronomer Royal, lived until 1720, six years before his authorized Catalogue of Stars was published. He was succeeded by Edmond Halley who was already sixty-four years of age. On his death in 1742 James Bradley [1718], who had for twenty years been Savilian Professor of Astronomy at Oxford, succeeded him on the recommendation of Lord Macclesfield. Bradley had discovered the aberration of light in 1729, and the nutation of the earth in 1748, both of which discoveries he communicated to the Society. John Hadley [1717] was a mathematician and scientist who greatly improved the construction of astronomical instruments.

In biological science Sir Hans Sloane [1684] was distinguished in botany and in medicine; Stephen Hales [1717] studied Newtonian physics and astronomy at Cambridge and applied the methods of research which he then learned to his researches in animal and vegetable physiology, in which he carried out much highly important work on blood pressure and on the sap pressure in growing plants. Later in life he devoted himself to the application of science to hygiene, ventilation, the purification of water and the preservation of foods. Other botanists were James and William Sherard.

Besides Sir Hans Sloane, Dr R. Mead and Dr Arbuthnot as well as the surgeon and anatomist William Cheselden had a high reputation; James Lind, who served in the West Indies in 1739, published his treatise on scurvy in 1754.

These men and others associated with them kept alive the Society's reputation for scientific research which the Councils of that time were doing but little to advance.

REFERENCES

CLARK, G. N. *Science and Social Welfare in the age of Newton.* Oxford, 1937.
DAMPIER, Sir WILLIAM. *A History of Science.* 3rd ed. Cambridge, 1942.
KIPPIS, Dr A. *Biographica Britannica,* 1778–1795.
MORE, L. T. *Isaac Newton, a biography.* 1934.
Royal Society, Notes and Records of. 1939–41.
Royal Society, Record of. 4th ed. 1940.
WELD, J. C. *History of the Royal Society.* Vol. I. London, 1848.
WILLEY, B. *The Background of the Eighteenth Century.* London, 1939.
WILLIAMS, B. *The Whig Supremacy 1714-1760,* vol. XI of Oxford History of England. 1939.

CHAPTER V

A GROWING ADMINISTRATION: 1741–1778

AT THE end of the first forty years of the eighteenth century during more than half of which the Society had had Sir Isaac Newton for its President, its administration had been planned on sound lines and was developing satisfactorily. His genius was not limited to his scientific researches but showed itself in many other fields in which he was interested, not the least important of these being the reorganization of the Society's administration; in this he was ably seconded by Sir Hans Sloane, at first as a Secretary, then as a councillor and after 1726 as his successor as President. The result of their able guidance was that the Society found itself in 1741 in a much more satisfactory position than it had known hitherto; the number of Fellows had increased considerably, and there seemed to be every prospect that it would continue to grow so long as Councils carried out their duties with zeal and energy; the financial position had been put on a sounder basis after the unfortunate experiences of 1728–30 and 1739–40 had been rectified. Only the watchful supervision of the councillors, to whom the Treasurer's reports were rendered annually, was needed, so that steps might be taken as soon as the sign of any decrease in the efficiency of control appeared without waiting until a deficit had to be reported and arrears of unpaid subscriptions necessitated drastic action. The Fellows had every reason to congratulate themselves on the Society's position which promised well for the future but for one disturbing symptom.

Although on the whole there had been a steady improvement, one branch of the Society's organization, and that an important one, had not shared in this: the Councils had not been meeting on the average as frequently as they did formerly, and when they did meet the average number of members who attended had not shown any tendency to increase, being rarely more than nine or ten out of twenty-one. The number of meetings would depend on the amount of work which there was for the Council to do, and when we call to mind all that had happened in the forty years that had just ended there was certainly sufficient business to keep the Councils of several years to come fully employed. It must however be remembered that both Presidents had reached a considerable age when the tenure of their office came to an end, and the Council meetings held in their later years were fewer than there had been previously. But this does not explain why the attendance of councillors at their meetings did not improve when there certainly was plenty to be done and the financial difficulties of earlier years had to some extent passed away. If the Councils, to whom the Society looked to see that it continued to contribute its full share to the 'improving of Natural Knowledge', were meeting less frequently and were being poorly attended, the scientific results would certainly deteriorate, and it would be of importance to trace the causes and to remove them. There is no contemporary record to show that this question was ever taken up by any Council, and as neither the number of the meetings nor the attendance of the members would be known to the general body of the Fellows, it is not surprising that nothing was done. To-day, such an investigation is much more difficult to carry out in the absence of many of the original documents, and a single forty-year period is insufficient to provide any reliable solution of the problem. It will therefore be preferable to return to this matter in the next chapter when the Councils of the years 1701 to 1740, 1741 to 1780 and 1781 to 1820 can be discussed together, and their activities during the eighteenth century can be better estimated (see p. 203).

When Sir Hans Sloane laid down the presidency in 1741 he left to his successors an administration which was fully adequate for the Society's needs at that time, and which could be readily developed to meet any new conditions which might arise. He had shown the Council how financial difficulties could be avoided by the simple means of making prompt use of the power which they already had to remove defaulting Fellows from the lists of the Society and to take legal action to recover whatever sums such Fellows owed to the Society, instead of allowing the arrears to accumulate over a period of years as had happened on several occasions. The excessive number of foreign members was another matter needing correction, but this presented no difficulty. It was also high time that the Society should accept publicly its responsibility for the *Philosophical Transactions* which had hitherto been published by one of the Secretaries nominally on his own responsibility, but in foreign countries the *Transactions* had always been regarded as the official publication of the Society, whatever the view of the Council might be.

There was however one part of its organization which from the first had presented a problem of considerable difficulty. Although the aims of the Society were defined in the Charters as the 'improving of Natural Knowledge' the non-scientific Fellows, who had neither the training nor the interest to advance these aims, had hitherto constituted the majority of the Society, and still provided the greater number of its councillors. The majority of them were not interested in scientific matters nor were they competent to discuss them; a few probably had had experience of administration and could render useful service on the Council, but in most cases their attendance would be infrequent.

THE OFFICERS

At the Anniversary meeting on 30 November 1741 Martin Folkes, a Vice-President, was in the Chair, and at the ballot for electing the officers for the ensuing twelve months he was elected President. He was at this time fifty-one years of age, and had been a Fellow since December 1713; in 1717, and again from 1719 to 1741, he had been a member of Council and had often been appointed to act as a Vice-President by Sir Isaac Newton, and also by Sir Hans Sloane when he was President. He had been educated at Saumur University, and when that institution was suppressed in 1695 he returned to England and entered Clare Hall, Cambridge. There he studied classics, philosophy and mathematics, but later his interests were mainly devoted to archaeological and literary subjects, and in 1750 he was elected President of the Society of Antiquaries, which had been established in 1717 though it was not granted a charter of incorporation until 1754. In 1742 he presented the Royal Society with a portrait of himself which had been painted by William Hogarth.

As President he took his duties seriously and presided at forty-two out of the fifty-two Council meetings which were held during his term of office, thus following the example set by his two predecessors; but only five of these meetings were held annually. This was notably fewer than under either of the preceding presidents, which suggests that neither the scientific nor the administrative activity of the Society was being adequately maintained, since Sir Hans Sloane had left several important matters which demanded further investigation at the earliest opportunity. Dr Thomson, in his *History of the Society*, criticizes the character of the communications printed in the *Philosophical Transactions* during Folkes' presidency as being in many cases 'trifling and puerile'. Pamphlets by one John Hill, a quack doctor, as well as others by an anonymous author, also comment with considerable severity on the lowering of the standard of the Society's meetings and publications which was noticeable at this time. Hill cannot be considered as an impartial critic for he is said to have been a disappointed candidate for the Fellowship, but competent opinion in those days seems to have been convinced that the Society's standard of learning was in fact deteriorating. There was no sufficient reason for this falling off, for the membership was increasing and financial difficulties had for the most part been overcome. The officers and the members of Council must therefore be held largely responsible for this unfortunate state of things. That it should have occurred within a few years of the accession of a President to whom scientific research made but a slight appeal shows that the influence of the President at this time on the Society's progress was very real, even though at this time both Secretaries were scientific men of ability and experience.

The other officers included James West as Treasurer, an office which he had held only for five years; he was a politician and also an antiquary of wide interests, being a discriminating collector of manuscripts, rare books, prints, coins and pictures. He was a joint secretary to the Treasury from 1741 to 1762. He does not seem to have been an energetic administrator and the financial difficulties which arose in 1740, as well as a clerk's defalcations in 1765, indicate a lack of effective control. The Secretaries were John Machin, an astronomer, who had been elected a Fellow in 1710 and was Secretary from 1719 to 1747; he held the professorship of astronomy at Gresham College until his death in 1751. During the last few years he was in poor health and could only attend the Council meetings infrequently. The other Secretary was Dr Cromwell Mortimer, M.D., who had occupied the post since 1730 and undertook the editing of the *Philosophical Transactions* from that time until the Council entrusted the work to a committee of the Society in 1751. He also acted as an assistant to Sir Hans Sloane, which may have added unduly to his work. At this time therefore two of the officers were men of science even though the other two were not, which was not a reasonable proportion for a scientific institution of high standing; but even this ratio was not always maintained in the eighteenth century, for later it fell even lower. The part that the officers of the Society have normally played in keeping the Society's work at a high level has been so important that their qualifications during the years 1741–1820 may profitably be examined, since this period was not notable for a consistently high standard of learning. In education the well-to-do classes of these years had fallen below the standard of earlier times; in the larger schools history and science were neglected, and a knowledge of external nature, as Bacon calls it, was left to the dissenting academies to provide, where a higher standard of education was attained. The number of graduates admitted to Oxford and Cambridge during this century decreased very considerably. Those who joined the non-scientific group of the Society during this period showed no enthusiasm for the advancement of the Society's aims, and left this to the officers and to such members of Council as might be interested. When Lord Macclesfield succeeded Martin Folkes as President he appealed to the Fellows to assist him in raising the standard of the Society's work and Lord Morton followed his example, but such improvement as resulted was not maintained.

From 1662 when the First Charter nominated the members of the first Council the Treasurer had not been a man of science; no reason is given for this but he may have been considered as representing the non-scientific group of Fellows who were from the beginning in a majority. As the years went by this same practice was followed, and Abraham Hill, Daniel Colwall, Alexander Pitfeild, Roger Gale, James West, Samuel Wegg, William Marsden and Samuel Lysons successively held the office of Treasurer until 1819. They were historians, antiquaries, archaeologists, literary men, etc., but none of them was a man of science. With the appointment of Captain Henry Kater (1827–30), a geodesist, the old practice came to an end. Throughout the seventeenth and eighteenth centuries therefore there were never more than three of the officers who could be men of science, and even this proportion was not always attained.

This was most prejudicial to the prestige of the Society, for the other members of the Council did not attend the meetings regularly and were content to leave the management of the business to the officers. Complaints of communications being inefficiently reported on to the 'Committee of Papers' were frequently made, and these formed one of the principal grounds of criticism in the early part of the nineteenth century. The officers were said to act too much on their own authority, but this would be the inevitable result of laxity on the part of the other members of Council.

The latest year in which there were three scientific officers was the last one of Sir Hans Sloane's presidency; after this the proportion fell off. In the first period, 1741–80, there were two scientific officers in twenty-two of the years, only one in eleven years and none at all in the other six years. In the second period, 1781–1820, the position was definitely worse: there were two scientific officers in eight of the years and only one in twenty-eight years; in four years there were none. The man of science who had formerly held these posts was being replaced by those who were antiquaries, historians or librarians. That a scientific society of the distinction to which the founders of the Royal Society had aspired should not have had a single man of science among its officers for ten years out of eighty shows clearly that drastic changes in its administration were already overdue before 1820.

COUNCILS

In the last chapter when the period 1701–40 was being discussed it was pointed out that the Society had surmounted many of its more serious difficulties and that a time of steady development had arrived; it was reasonable to expect therefore that the executive body of the Society, its Councils, would have seized the opportunity to be more active than they had hitherto been. But this had not been the case; they had not met more frequently nor had the attendance of members been noticeably larger at the end of Sir Hans Sloane's presidency than at the time of Sir Isaac Newton's election. Habits and customs which have grown up during half a century are not to be changed in a day, so it has seemed preferable to leave any discussion of this point until more information could be utilized. The period of thirty-eight years with which this chapter deals is especially suitable for there was a steady increase in the number of Fellows; the Society's income was adequate to its needs and was growing; and on three or four occasions the Society was invited to draft instructions for the scientific work of exploratory expeditions, which gave ample opportunity to the Council to put forward plans for research of many kinds and to support them with vigour. Besides this the Society undertook the comparison of standard weights and measures, as well as a piece of geophysical work—the determination of the deflection of the plumb line from the vertical by the mass of Mount Schiehallion in Scotland. In the latter part of the period the Council undertook the revision of the statutes of

1663 which were no longer suited to the changed conditions. There was therefore ample opportunity for the members of Council to show their interest in both the scientific and administrative work of the Society by their presence at its meetings, but there is little evidence of this (cf. Appendix II C).

From 1741 to 1752 Martin Folkes, who was primarily an antiquary, though he had studied philosophy and mathematics, was President; from 1752 to 1763 his place was taken by Lord Macclesfield, a man of science and a well-known amateur astronomer, who was most active; he was followed by Lord Morton, a mathematician and an amateur astronomer, who held the position for five years. James West, a former Treasurer, and Sir John Pringle, an eminent physician, complete the list. This period should therefore be a typical one. The first point that comes out clearly (see Appendix II C), is that the proportion of scientific councillors hardly varies whether the President and officers are men of science or not; that science should be always represented by a minority in the Council of an institution which was established for the improvement of Natural Knowledge was manifestly unsatisfactory. This represents a similar division in the Fellows of the Society where the scientific Fellows constituted about one-third only of the whole Fellowship. The influence of Lords Macclesfield and Morton and of Sir John Pringle is clearly shown in the increased number of Council meetings which were held during the last half of the period, but the average number of councillors who attended any meeting showed a very small increase. It seems that the majority both of the Fellows and of the councillors were content to leave the work of the Society to be carried on by the officers and a few others who were interested in promoting its aims, and these were probably drawn for the most part from the scientific body of the Fellows. There is no sign of any attempt to increase the scientific representation on the Council.

The members of the Council had before them at the end of each year a detailed report from the Treasurer on the financial position of the Society, and they then nominated five of their number, who together with the same number nominated by the Society, audited the Treasurer's accounts for the past year and reported on them to the Anniversary meeting. This gave the councillors ample opportunity, indeed imposed it on them as a duty, to examine the accounts critically and make themselves thoroughly acquainted with them. Nevertheless, as we have seen, the total amount of unpaid subscriptions and fees was often allowed to increase until the Treasurer's funds were insufficient to meet the claims which he received. A committee was then appointed which advised that legal action should be taken against those in default. But it is strange that none of those who were specially charged with the control of the finances, that is to say the Treasurer, the other councillors, or the auditing committee of each year, should have called attention to the financial position before it had become

serious. This habit of accepting whatever was laid before them and neg-
lecting to exercise their rights of criticism was not only harmful to the
Society but perpetuated one of the greatest evils with which reformers of
later years had to contend.

In the early days of the Society when its resources were meagre and
when such questions as came before the Council were decided without
establishing precedents which might become embarrassing later on, the
councillors' duties were not so important as they came to be later; but it
seems to have become customary by the eighteenth century to accept
returns and reports with but little criticism. This was specially marked in
the case of the Treasurer's accounts which were presented to the Council
but not to the Anniversary meeting until a special order was made in 1830
that they should be circulated to every Fellow. If the accounts were not
to be made public the habit of accepting them without careful examination
would be quickly formed, and would soon become an established custom.

THE FELLOWSHIP

The satisfactory growth of the membership of the Society, which had
begun during the years when Sir Isaac Newton and Sir Hans Sloane were
guiding its policy, continued during the period with which this chapter
deals. From 301 in 1740, the number of Ordinary Fellows had risen to
460 by 1780 without including the foreign members; these had risen to a
maximum of 160 between the years 1760 and 1770, but were beginning
to decrease by 1780. Taking the analyses which have been made of the
Fellowship of the years 1740, 1770 and 1800, the increase is well shown in
both the scientific and the non-scientific groups but unfortunately to a
much greater extent in the latter category. Seventy-one per cent of the
Fellows are classed as non-scientific and about the same in 1800 (see
Appendix II A), but after that, as will be shown later, a marked change
took place.

The original intention of the Society had been that the class of foreign
members should be reserved for scientific men of other countries whose
work had been of exceptional distinction; such definitions are never easy
to apply, and there can be no doubt that admission to this group had been
and still was being much too readily accorded. This would be facilitated
by the non-scientific members being always in the majority on the Council
for they could not be expected to know anything of the respective merits
of candidates. As will be seen from Appendix II A, the foreign members
numbered forty more than the scientific Fellows in 1770; by 1780 their
numbers had begun to fall off in consequence of the Council having resolved in 1766 that not more than two should be elected in any year until
the total number of foreign members had been reduced to eighty. There
was no falling off in the number of candidates who were proposed for
admission to the Fellowship and were elected by the Society; nine hundred

and fifty of them were so elected between 1740 and 1780, and the number in any one decade differed but little from that in the others. The loss by deaths and resignations was therefore very considerable, and this is not surprising since two-thirds of the candidates had no knowledge of scientific subjects, nor did they take any interest in the aims of the Society. The social distinction, which they may have hoped to gain from its Fellowship, probably proved to be much less than they had expected, so that what they gained for their subscription was not in their opinion worth the outlay to them.

The increase in the number of scientific Fellows between 1740 and 1770 was lamentably small, being only fourteen, but the next period of thirty years shows a change for the better since the increase rose to thirty-six; the proportion between the two groups remained about the same, the number of the scientific members being less than half that of their colleagues.

FINANCE

The deficit of £95. 16s. 6d. reported by the Treasurer in 1739 and which was followed by one of £240 in the next year provides a useful starting point for a review of the situation during the period 1741-80. Out of a total of 293 Fellows 152 had been allowed to compound for the payment of their annual subscriptions, presumably under the Council's resolution of 1729; the two distinguished surgeons, James Douglas and William Cheselden, had been exempted from paying their annual subscriptions but the reasons for this concession are not stated. The rest, who numbered 139, had signed the obligation to pay their subscriptions annually, but the total payments which they had made fell short of what they should have amounted to by £1845. The committee further reported that the whole revenues of the Society, exclusive of fees and subscriptions, was only £232 per annum, from which taxes and other payments amounting to £140 had to be met. They also estimated that the annual expenses could not be less than £380, so that the punctual payment of subscriptions was of urgent importance. This was nothing new, and the same had been said of the financial position in 1728 which Sir Hans Sloane took prompt and energetic measures to redress. It is difficult therefore to understand how another crisis of precisely similar character could have been allowed to occur only ten years later if the members of Council had taken any interest in the finances of the Society when they had the Treasurer's accounts before them. Weld only says that the Council made great exertions to disengage the Society from its difficulties; but no explanation is forthcoming to tell us why they should have arisen again so soon. That more than half the Fellows should have been allowed to compound for their subscriptions by paying one more year's subscription if they had already paid £25 or about ten years' subscriptions was most imprudent.

The Committee presented their report to the Council on 14 January 1741; in this the names of all the Fellows were shown in three lists, the first of which contained all those who had discharged their obligations under the arrangements which had been approved in 1730; the second included those who had not yet been admitted nor had given their bond; and lastly those who had given their bond but were in default. This last category numbered 139 out of the total Fellowship, and they owed to the Society the sum of £1845 or about £13 each. The Council ordered that letters of reminder should be sent at once to all defaulting Fellows, and if the sum due from each was not remitted within twenty-one days legal proceedings were to be taken. No explanation is forthcoming why similar action should not have been taken at any time during the previous ten years. At the end of 1742 the Council reported that the receipts during the preceding twelvemonth had amounted to £1422, and the sum of £500 was invested in East India Company's Bonds.

The practice of accepting a payment of £25 as covering all future subscriptions to the Society had been allowed by the Council in 1730 as a temporary expedient, but this, which was practically a compounding fee, was regarded as being so favourable that by 1739 half the Fellows had availed themselves of it, thereby landing the Society in fresh financial difficulties, but the Council did not introduce any additional conditions or restriction; each case was to be decided by them as it arose and sometimes very leniently. In February 1753 two Fellows, Robert Hoblyn, elected in 1745, and Henry Baker, elected in 1740, applied to be granted the remission of their unpaid arrears and to be allowed to compound for their subscriptions in future. The Council were at first prepared to agree, and fixed the payments to be made by them at £20 and £10 respectively; but the Treasurer, who had learned by his experience in 1739, objected, pointing out that the subscriptions of thirty-three other Fellows were in arrears at the time, and that the sums proposed in no way represented the loss that the Society might incur under the resolution. The Council thereupon rescinded their earlier resolution and adopted a new one which laid down: firstly, that no Fellow should be allowed to compound for arrears already due, and, secondly, that any Fellow who had paid £21 *before admission* should be excused his annual subscriptions in the future. The question of composition fees came before the Council on several subsequent occasions, in 1774, 1834 and in 1921, when the statute dealing with them was revised. It is surprising that a Society should have dealt with this problem of composition fees so unscientifically when the work of three of their earliest Fellows, Sir William Petty, John Graunt and Edmond Halley, had founded the study of Vital Statistics about a century before. From this time until 1834 the sums received from composition fees were credited to the general funds of the Society, and were treated as income, but in that year, on the proposal of the Treasurer, J. W. Lubbock, the Council ordered that in future they should be invested, and this has been the regular practice ever since.

When discussing this vexed question of the payment of subscriptions, to which we shall return later, it should not be forgotten that throughout the seventeenth and much of the eighteenth centuries postal rates were high, the service was slow, and when letters from London reached a country town they had to be passed on by local arrangements to the village or house where the person lived to whom the letter was addressed, and this was often at some distance. The remittance of small sums may also have been difficult for those who had no agent in London to make the quarterly payment of 13s. 4d. on their behalf to the Society. It was not until 1866 that the Council resolved that the year's subscription should be paid in one sum, but as late as 1868 two Fellows were still paying the old weekly rate. At the outset most of the members of the Society were living in or near London, and would attend most of its meetings, in which case the payment of a weekly subscription would present no difficulty, but later when they were more widely distributed the inconvenience of remitting small sums was a real one as may be seen from diaries of the eighteenth century.

At the present time it is very difficult to furnish accurate information about the Society's income and expenditure in former days; the minutes of the Anniversary meeting only record the total sums received and expended during the past year, but investments were not reported; in the minutes of Council they are sometimes mentioned. As an example, the total receipts were reported to the Anniversary meetings of 1735, 1737 and 1739, but not to those of 1736 or 1738, the average of the five years being £780. Weld says that when James West was elected to the office of Treasurer in 1736 the average income of the Society was £760; the committee however found that only £232 were received in 1739 besides whatever was received from the 113 Fellows, who should have provided £295 annually but apparently did not. With no detailed statements of income and expenditure, the matter cannot now be carried further, but as the receipts in 1741 exceeded the expenditure by £297 improvements could be made and were then introduced. After this there was an improvement and it seems likely that other sums were invested in 1749 and in 1752, but details are not available. When Martin Folkes retired from the presidency at the end of 1752 the invested capital of the Society, according to Weld, amounted to £3000. During the presidency of Lord Macclesfield, 1752-64, this part of the administration seems to have been more effectively controlled, for £600 of 3½ per cent Consols and £1300 Bank Annuities were bought during those years.

Early in 1763 Francis Hauksbee died having been Clerk to the Society for forty years, and a successor had to be appointed. On this occasion the choice made was an unfortunate one which strongly emphasized the need for careful and continuous supervision of the Society's office. On

3 February 1763 the Society selected E. M. da Costa, who had been a Fellow of the Society since 1747, as his father M. da Costa (elected 1736) had been before him; a relation, perhaps the father of the latter, is said to have been a physician in the suite of Queen Charlotte when she came to England from Portugal to be the wife of Charles II. Emanuel Mendes da Costa is described as a naturalist who studied conchology and collected fossils on which he published several papers. He appears to have been an able man with various interests, and among other things his opinion on old silver and jewellery was held by experts to be of value. A memoir on the Seymour family recently published mentions that one Aaron Lazarus, a jeweller who specialized in old trinkets and valuable silver, was in the habit of dealing with Lord Francis Seymour about 1761; Lazarus consulted E. M. da Costa on the age of the Bodleian bowl in the Ashmolean Museum, and the probable use to which it was put. Able though he may have been in other fields, his duties as Clerk to the Society were very negligently performed. It appears that at this time the arrears of subscriptions due from Fellows had again been mounting up though this should not have been allowed; the Council therefore ordered that defaulters should be more strictly treated and sued if payment was not promptly and regularly made. They instructed da Costa to this effect and agreed to pay him a commission of one shilling in the pound on all the arrears that he recovered. Investigations which were made a year or two later showed that almost immediately after his appointment as Clerk he had commenced to appropriate part of the funds entrusted to him; considerable deficiencies in his accounts having come to light they were reported to the Council on 10 December 1767; he was therefore suspended from duty and on 17 December 1767 was dismissed from his employment by the Society. The total deficiencies due to his frauds were found to amount to more than £1492, but of this sum one thousand pounds were repaid to the Society by his guarantors, who sued him for the amount. His collections and library were seized and sold, and he was sentenced to five years' imprisonment. After his release he eked out a living by lecturing, but died in poverty in 1791.

Da Costa's successor was another Fellow of the Society and on this occasion the choice was a more fortunate one. John Robertson, who was appointed Clerk, Housekeeper and Librarian on 7 January 1768, had previously been Master of the Mathematical School of Christ's Hospital and First Master of the Royal Naval Academy at Portsmouth. He gave of his best in the service of the Society and was respected by all, officers and Fellows alike. Unfortunately he only lived to work for the Society for nine years, since he died on 11 December 1776 while still in its employment. The Council, who held Robertson in much esteem, then appointed his eldest son, John, as his successor, and his widow was made Housekeeper.

The average income of the ten years 1761-70, as reported to the Society, was £1100, but no investments are recorded until 1774, when some of the land belonging to the Acton estate was sold and the proceeds were presumably invested. Weld says that the average annual income of the Society in 1768 was £1450; the discrepancy between the two figures cannot now be explained but it may be that the sum given by Weld includes amounts which were invested later on.

Having reviewed the position of the Society on Martin Folkes' accession to the presidency, and described the officers with whom he was associated in the administration of it, we have discussed the activities of the Councils during the period 1740-80 as well as the growth of the Fellowship which was showing considerable improvement. The financial position in spite of some unfortunate crises was on the whole becoming more stable, and with careful supervision need have caused little anxiety either to the Treasurer or to the councillors. We may now pass on to the various occurrences which make up the history of the institution; some are of considerable importance in its future development; others may seem to be insignificant in themselves but supply information in matters of greater interest.

In March 1741/2 Edmond Halley died; he had been elected a Fellow of the Society in 1678 and had served the Society as a councillor or a Secretary for the greater part of these sixty-three years. Lord Macclesfield, a distinguished amateur astronomer, and a Fellow of the Society, at once wrote to the Lord Chancellor, Lord Hardwicke, to enlist his support for Dr James Bradley, urging that he should be appointed to succeed Halley as Astronomer Royal on the grounds that, besides his other abilities, Bradley took the greatest interest in astronomical observing, so that there was every reason to expect that science would reap much valuable information from the opportunities which he would enjoy at Greenwich Observatory. Lord Macclesfield also represented that such a man as Bradley, whose genius was already widely recognized, would undoubtedly greatly enhance the reputation of this country for the promotion of astronomical science. Bradley had been elected Savilian Professor of Astronomy at Oxford twenty years before, and had made the acquaintance of Lord Macclesfield at Oxford, where they had worked together on astronomical problems; he had been invited to carry on his observations at Shirburn Castle in Oxfordshire, where Lord Macclesfield had built and equipped an observatory in 1739. Bradley had been elected a Fellow of the Society in 1718 and had communicated to it his paper on the Aberration of Light in 1729; later, in 1748, he contributed another of great importance on the Nutation of the Earth's Axis which was also published in the *Philosophical Transactions*.

In October 1743 the Royal Society Club, or the Club of the Royal Philosophers as it was called until 1795, was definitely established and held its first meeting. As early as 1645, or perhaps even earlier, some of the philosophers were in the habit of meeting informally in order to discuss the new philosophy and to exchange views, either at the lodgings of one of them or at a tavern where they dined together. It was at such meetings that John Winthrop, during his visit to this country in 1641–3, may have discussed with the philosophers of that day his and his father's suggestion that, if the political conditions in this country did not improve, some of their number might wish to emigrate to the Massachusetts colony in New England where liberty of action and freedom of speech would await them. From this time until 1660 the philosophers who were resident in London continued to hold such informal discussions at Gresham College, at Dr Goddard's house or at the Mitre Tavern in Wood Street. The same practice continued after the Royal Society had been formed in 1660, and the custom of dining together at a tavern after the weekly meeting continued as it was appreciated by many, besides providing a convenient occasion for discussions and for the exchange of views.

Pepys, in February 1664, after his admission as a Fellow, records adjourning 'to a club supper' at the Crown Tavern where 'excellent discourse till ten at night, and then home'. Again, after being with Lord Brouncker and Hooke at Gresham College at a meeting of the Society, he went to the Crown 'and there supped at the club with my Lord Brouncker, Sir G. Ent, and others of Gresham College', and, after he had agreed to contribute £40 towards the building of a house for the Royal Society, 'thence with Lord Brouncker to the King's Head Tavern in Chancery Lane and there did drink and eat and talk, and above the rest did hear of Mr Hooke and my Lord an account of the reason of concords and discords in musique which they say is from the equality of vibrations; but I am not satisfied in it...'.

In the Diary of Robert Hooke from 1673 to 1680 the description of the meetings of the Royal Society on Thursdays is very frequently followed by a note of the names of some members who adjourned to Garaway's (a coffee-house in Cornhill where tea was first sold, and which was a centre for lotteries) up to 1675; or to Jonathan's, a house mentioned in the Spectator, up to the end of 1680 when Hooke's Diary ends. In January 1675 Hooke says 'we now began our new Philosophical Club, and resolved on engaging ourselves not to speak of anything that was revealed sub sigillo to anyone, nor to declare that we had such a meeting at all'. He often refers to these club dinners, and says that in August 1672 he 'dined at the Society's Club with Brouncker'; on three occasions in 1674 he records that he dined at 'the Club'.

Edward Ward, the author of a book entitled The Secret History of Clubs which was published in 1709, mentions that the Virtuoso Club of the Royal Society used to meet in 1701 but at the time when he wrote it consisted of a small number of men meeting nightly without organization.

Smythe in his history of the club (1860), says that Halley 'although para-lysed in his right hand came (1737) as usual once a week to meet his friends on Thursdays'. Thus it seems to be well established that for about twenty years the philosophers, and after that the Fellows of the Society, had carried on their old practice of informally dining together at a tavern to discuss scientific matters of common interest; when such meetings were held fairly regularly and were well attended it was spoken of as 'the Club', at other times it dwindled to occasional meetings of a few intimate friends.

As the number of Fellows had risen to three hundred by 1743 and was still increasing, there was now every prospect that such a club might become a permanent institution if suitably organized; and this has proved to be the case. It has formed a very useful addition to the Society itself since it provides opportunities for the Fellows to meet together for discussion of scientific subjects of interest more fully and more informally than is possible at the weekly meetings of the Society.

Sir Archibald Geikie has described the club, its members, the dinners and the numerous guests who accepted the hospitality of the club in his book *The Annals of the Royal Society Club*, so it need not be further discussed here. The club has for the past two centuries played an important part in the history of the Society, for a large number of its Fellows have belonged to it, and many statesmen, public men and foreign men of science have met its members at its dinners where they have learned to know the aims and activities of the Society. Candidates for the Fellowship were often invited to the club dinners by their proposers in order that they might meet other Fellows of the Society. The club still meets in the evening of each day on which an Ordinary meeting of the Society has been held. Its members now number seventy-five in addition to its honorary members, all of whom are Fellows of the Society. Until the end of 1780 the club met at the Mitre Tavern in Fleet Street and then transferred its custom to the Crown and Anchor in the Strand, where it continued to meet until 1847. As this tavern was then converted into a club-house the Royal Society Club had to move to the Freemasons' Tavern in Great Queen Street. In later years it has met at various hotels and restaurants as the Treasurers have found it convenient to arrange.

At a meeting of the Council on 10 November 1736 Mr M. Folkes, who had been for fifteen years a councillor and had acted often as a Vice-President, 'proposed a thought to render Sir Godfrey Copley's Donation (see p. 138) for an annual experiment more beneficial than it is at present; which was to convert the value of it into a Medal or other honorary Prize to be bestowed on the person whose experiment should be best approved; by which means he apprehended a laudable emulation might be excited among men of genius to try their inventions, who in all probability may never be moved for the sake of lucre'. The Council of that year thereupon resolved that a gold medal bearing the arms of the Society should be struck and awarded to the author of the most important scientific discovery or contribution to science, by experiment or otherwise. The awards were to be made on the nomination of Sir Hans Sloane and Mr A. Hill, the trustees under Sir Godfrey's will, and after Mr Hill's death by Sir Hans Sloane; on the death of the latter in 1753, the nomination devolved on the President and Council for the time being. The medal, which was the first gift to be made by the Society for any discovery of exceptional scientific importance, has been awarded annually except for a few years up to the present time. Until 1753 the grant of the medal was restricted to those of British nationality but from that year onwards it has been awarded without such limitation. By the selection of men who had carried out important scientific researches either published or communicated to the Society during more than two hundred years its award has come to be the most important scientific recognition which it is in the power of the Society to bestow, since the recipient thereby joins the company of all the most eminent men of science of the past two hundred years.

In 1743 the Society undertook a scientific investigation of fundamental importance when the Council appointed a committee to compare the Society's standard yard with that which was in the office of the Exchequer and other examples. The members of this committee were: the President of the Royal Society, Lord Macclesfield, the astronomer, Lord Charles Cavendish, a patron of science and at this time a member of the Council, John Hadley, a constructor of astronomical apparatus, William Jones, a mathematician at Shirburn Park Observatory, Peter Daval, a mathematician and later a Secretary of the Society, also Dr Cromwell Mortimer, one of the Secretaries. They met in 1743. An enquiry which led up to these metrological studies is described in the 42nd volume of the *Philosophical Transactions* where it is said that

Some curious gentlemen, both of the Royal Society of London and of the Royal Academy of Sciences at Paris, thinking it might be good Use for the better comparing together the success of experiments made in England and France, proposed some time since, that accurate Standards of the Measures and Weights of both nations, carefully examined and made to agree with each other, might be laid up and preserved in the archives both of the Royal Society here, and of the Royal Academy of Sciences at Paris: Which proposal having been received with the general approbation of both these Bodies they were thereupon pleased to give the necessary Direction for the bringing the same into effect. In consequence of which Mr George Graham (the horologist and instrument maker), Fellow of the Royal Society, did, at their desire, procure from Mr Jonathan Sisson, Instrument-maker, Two substantial brass Rods, well planed and squared and of the length of about 42 inches each, together with Two excellent brass Scales of Six Inches each, on both of which one Inch is curiously[1] divided by diagonal Lines and five Points into 500 equal Parts. And upon each of the Rods

---

[1] I.e., carefully.

Mr Graham did, with the greatest care, lay off the Length of Three English Feet from the Standard of a Yard kept in the Tower of London. He also at the same time directed Mr Samuel Read, Scale- and Weight-maker near Aldersgate, to prepare in the best manner he could, Two single Troy pound weights, with two piles of the same Weights decreasing from Eight Ounces to One Quarter of an Ounce respectively, Two Parcels of the lesser corresponding weights, that is to say, from Five Penny-weight to half a Penny-weight, and Grain Weights from Six Grains to one-fourth of a Grain; together with Two single Avoirdupois Pound Weights: all which when made were carefully examined and found to agree sufficiently with each other. Things being thus provided, the Two brass Rods, one of the Six-inch Scales, and one Set of all the Weights were sent over to Paris, one of the Rods to be returned and all the other Particulars to be presented for their Use to the Royal Academy of Sciences there: Who upon receipt thereof desired the late Monsieur Du Fay, and Abbé Nollet, both members of the Academy, and also Fellows (Foreign Members) of the Royal Society, to see the Measure of the Paris Half-toise containing Three Paris Feet accurately set off upon both the brass Rods, in the like manner as the length of the English Yard containing three English Feet, had already been set off on the same: After which those Gentlemen returned over one of the Rods to the Royal Society, together with a standard weight of Two Marcs or Sixteen Paris Ounces, accompanied with a Process Verbal or Authentic Certificate from the proper office, of the due examination thereof.

The Rod being returned, Mr Graham caused Mr Sisson to divide both the measure of the English Yard and the Paris Half-toise, each into Three equal Parts for the more ready taking off both the English and Paris Foot from the same: after which both the Rods and the Two-Marc Weight, sent over from France, were, together with the other Particulars before mentioned, carefully laid up in the Archives of the Royal Society where they now remain, as their Duplicates do in those of the Academy of Sciences at Paris: But as, before they were so laid up, an accurate examination and comparison of them was made by Direction of the Council of the Royal Society, the result of the same is here subjoined as follows:

Here follow the Paris Half-toise, the Paris Two-Marc Weight, the English Avoirdupois Pound and the Paris Foot expressed in terms of the English Foot or Troy Pound.

This was the first occasion when a careful investigation of our principal weight and measure had been undertaken, a subject which has since on several occasions called for the skilled assistance of many of the most eminent of the Society's Fellows.

In the following year an extension of the investigations was undertaken for in the words of the committee:

It was not at all the intention of the Society to determine what was the absolute legal length of the yard or the real legal weight of the several pounds, but to lodge and preserve two measures and two sets of weights sufficiently near to what were in common use, and well agreeing together, for the purpose of comparing together by some certain standard to which recourse might be had in either kingdom, the success of such experiments made either in England or in France in which measure or weight might particularly be concerned. But since some gentlemen have been desirous to know how far these standards really agreed with the original ones, as they are looked upon to be, in the Chamberlain's Office of the Exchequer, as well as those kept for public use at Guildhall, at Founders hall, with the Watchmakers Company, and in the Tower of London; Mr George Graham, F.R.S., was therefore requested, with such assistance as he should find necessary, to take upon him the comparison of the several standards which he has accordingly done and carefully viewed and examined the same at the Exchequer on Friday, the 22 April last, in the presence of the President, Lord Macclesfield, and other members of the Committee.

The Exchequer standards were two square rods of brass of the breadth and thickness of about half an inch, one being called the Yard, and the other the Ell. They had been deposited in the Exchequer in the reign of Queen Elizabeth, and their construction was of the rudest description; the other standards were also wanting in accuracy and no two of them agreed. The Society's yard was found to be 0·0075 inch shorter than that of the Exchequer. Full details of the results are given in the *Philosophical Transactions*, vol. 42, pp. 541–56.

Nothing more was done, though the matter was one of great national importance, until 1758, when a Committee of the House of Commons was appointed to enquire into the original standards of weights and measures, but even then no legislative action was taken until after the destruction of the Houses of Parliament by fire in 1834, when new standards had to be constructed. The Committee of the House of Commons had two brass rods made by John Bird, the instrument-maker, and these were compared carefully with the Royal Society's brass standard rods of 1742. One of the former pair was selected by the Committee and marked 'Standard-yard, 1758'. There the question was again allowed to remain without any legislative action being taken until 1814, when another House of Commons Committee was appointed (cf. *House of Commons Journal*, vol. 69, p. 414).

In 1729 Thomas Fairchild of Shoreditch, a gardener and clothworker of London, bequeathed the sum of £25 to the trustees of the Charity Children of Hoxton and their successors, and to the Churchwardens of the Parish of St Leonard, Shoreditch, and their successors in order that the income should be paid to a lecturer to preach on 'The Wonderful Works of God' or on the Certainty of the Resurrection of the Dead, at the Church of St Leonard on the Tuesday in Whitsun week. He was not a Fellow of the Society, but he and his work were highly esteemed by the botanists, Richard Bradley and Richard Pulteney, both of whom were Fellows: Sir Hans Sloane also was interested in his work and thought well of it. Fairchild, who corresponded with Linnaeus, was the first to succeed in scientifically producing an artificial hybrid: *Dianthus Caryophyllus barbatus*. He introduced the plants *Pavia rubra* and *Cornus florida*, and also carried out experiments which helped to prove the existence of sex in plants. In 1725 he joined a society of gardeners which was formed in the neighbourhood of London, and which later produced *A Catalogue of Trees and Shrubs both Exotic and Domestic which are propagated for Sale in the Gardens near London*. This has been attributed to Philip Miller who was at one time Secretary of the Society. It was published in 1730 (after Fairchild's death), but the topographical notes contained therein bear a strong similarity to those contributed by Fairchild to *The City Gardener*; so much so that the book is indexed under his name in the British Museum. Fairchild received the freedom of the Clothworkers' Company in 1704.

When his will was proved in 1729 the Council of the Royal Society considered that the income which the bequest would provide, twenty shillings, was too small to carry out worthily the wishes of the testator. Several of the Fellows therefore resolved to open a subscription in order to increase the bequest to one hundred pounds; among the subscribers were Sir Hans Sloane, then President of the Society, Lord Charles Cavendish, Dr Alexander Stuart and Dr James Douglas. At first the required total was not reached, but Archdeacon Denne, who was the first lecturer, added twenty-nine pounds from the money which he had received for preaching the sermon in several successive years. By 1746 the one hundred pounds had been collected and was invested in South Sea Stock, and in June of that year the trust and appointment of the lecturer were transferred to the Royal Society; from that time the dividend of three pounds was paid annually to the lecturer who was appointed by the President and Council. The President, accompanied by several Fellows of the Society, usually attended to hear the sermon. William Stukeley (1687–1765), a Fellow of the Society and Secretary to the Society of Antiquaries, has recorded in his journal that on

Whitsunday, June 4, 1750, I went with Mr Folkes, the President, and other Fellows to Shoreditch, to hear Dr Denne preach Fairchild's sermon on the Beautys of the Vegetable World. We were entertained by Mr Whetman, the Vinegar-merchant, at his elegant house by Moorfields; a pleasant place encompass'd with gardens well stored with all sorts of curious flowers and shrubs, where we spent the day very agreeably, enjoying all the pleasures of the country in town, with the addition of philosophical company.

No detailed accounts of expenditure in the eighteenth century have been preserved, but we may assume that the dividend of three pounds was paid each year to the lecturer. For 1830 however the payment has been recorded and also for each of the following years up to 1856. Professor C. Babbage, in his *Decline of Science in England* (1830), criticized the Council for appointing the same lecturer for so many years in succession that it almost became his perquisite. Mr Ayscough had delivered the lecture from 1800 to 1804, and after him the Rev. T. Ellis did the same for the next twenty-six years. In January 1873 the Council resolved to apply to the Charity Commissioners for authority to hand over the administration of the trust to the Gardeners' Company, who were willing to accept it, since they appeared to be a more suitable body for administering it. The necessary authority was granted and since that year the lecturer has been appointed by the Court of that Company. The lecture is still delivered in the parish church of St Leonard, Shoreditch, in Whitsun week annually.

From 1750 the hour at which the Anniversary meeting was held on 30 November was changed from 9 a.m. to 10 a.m., the later hour being more convenient for the attendance of the Fellows; the place at which the Anniversary dinner was held in the afternoon of the same day as the election was now changed from Pontac's Tavern in Abchurch Lane, where the Fellows had met for many years; Evelyn has noted in his Diary on 30 November 1696 'after the Anniversary meeting and its elections we dined at Pontac's as usual'. So long as the Society continued to hold its meetings at Gresham College, Pontac's Tavern was quite convenient, but when these took place at Crane Court, Abchurch Lane was found to be too far away, so the Devil Tavern near Temple Bar became the place for these dinners, where they were held for some years, probably until 1773 when the Crown and Anchor Tavern in Fleet Street became the meeting-place. It seems that at this time it was customary to make a collection after the dinner to meet the expenses, and if the amount collected was insufficient the deficiency was met from the funds of the Society, but this practice was discontinued some years later.

Professor J. Machin, who had been one of the Secretaries since 1718 and had been indefatigable in carrying out his duties on behalf of the Society, had been in indifferent health for the past two years and had not been able to attend the meetings of Council. He was not therefore recommended to the Anniversary meeting in 1747 for re-election and Mr Peter Daval was elected to replace him. The choice was an advantageous one for the Society for not only was the new Secretary regular in his attendance at the meetings but he was an astronomer and mathematician of much ability. As an officer he attended eighty-six out of the ninety-eight Council meetings which were held, and as a man of science he rendered valuable assistance to Lord Macclesfield in the computations and other work which the project for the reform of the calendar entailed. He had been elected a Fellow of the Society in 1740, and on being elected to the secretaryship was excused paying the annual subscription so long as he was holding that post as was then the custom; he also received the annual honorarium of £60 as a Secretary which had been authorized by the Council in 1732.

At the end of 1751 Dr Cromwell Mortimer, the other Secretary, was replaced by Dr Thomas Birch, one of the original trustees of the British Museum. He was a friend of Sir Hans Sloane and like him was an active and discriminating collector of manuscripts, of which he bequeathed a valuable collection to the museum. As a Secretary he was most regular in his attendance at Council meetings, having been present at 174 out of the 178 which were held during the period November 1751 to November 1765, but his knowledge of scientific matters can have been but slight. He published a history of the Society in four volumes but they only cover the period between 1660 and 1689, and are taken up for the most part with the proceedings of the Councils, and with the reports, communications and discussions which came before the Society at its Ordinary meetings.

The whole of the Society's own funds were still needed for administrative work so the provision of expensive instruments and the undertaking of scientific projects of importance were far beyond its means; it could, however, represent to the government the need for such investigations, and urge that the necessary financial assistance should be given. An institution which was much in need of help at this time was the Royal Observatory at Greenwich, where the brilliant abilities of James Bradley, who had been in charge since 1742, were greatly handicapped by the lack of suitable instruments. It was not surprising therefore that Lord Macclesfield, with whom he had been closely associated in astronomical investigations for several years, and who was on the Council in 1746 and 1748, should take the matter up. In 1748 Lord Macclesfield strongly advocated that new and better astronomical instruments should be provided for the Royal Observatory at Greenwich, where they were urgently required; and on such a subject he spoke with special authority having built and equipped a well-found observatory of his own at Shirburn Castle. This was agreed to by the Council who forthwith applied to the Government for a sum of £1000 to be spent on such instruments as were required. This application was approved and a mural quadrant of eight feet radius, as well as an eight-foot transit instrument, were constructed by John Bird and erected at Greenwich in 1750. The latter continued in use until it was superseded by one made by Troughton in 1816.

Lord Macclesfield, as an active and competent astronomer who had for years co-operated with James Bradley, must have been well aware of the growing dissatisfaction at the decline of scientific activity in the Society. By his scientific knowledge and by his position he was eminently suited to take action in order to improve the existing state of things, and he selected the *Philosophical Transactions* as the first object of his concern; these, it will be remembered, were first published in 1664, when the Council appointed Henry Oldenburg, one of the Secretaries, to edit the Society's scientific publications, which he continued to do until his death in 1677. Dr N. Grew completed the twelfth volume which was then in hand; and from 1679 to 1682 Dr R. Hooke edited seven parts of the *Philosophical Collections* which were of the same character. In 1683 the Council resumed the original arrangement and volume 13 was edited by Dr Plot, one of the Secretaries. He and other Secretaries did the same for volumes 14 to 46 which followed, in each case the editor being responsible for the selection of the communications which were printed and for the publication of them. Thus from 1664 until 1750 neither the Society nor the Council considered themselves as being primarily responsible for the *Philosophical Transactions* since they were produced and issued on the responsibility of the editor for the time being. But whatever might be the attitude taken up by the Society and the Council, foreign societies and men of science had from the beginning looked upon the *Transactions* as being prepared and issued on the Society's responsibility. It was high time that this duty should be acknowledged and that the Council should be charged with the selection and publication of such communications as the Society wished to give to the world of science. In February 1752 Lord Macclesfield, who was a member of the Council in that year, brought forward at a meeting of the Society a proposal to improve the character of the *Philosophical Transactions*. This was referred to the Council by whom it was considered at two meetings; the result was that the following resolutions were submitted to the Fellows at an Ordinary meeting:

That it is the opinion of the Council, that it would tend to the credit and honour of the Society if, for the future, they should so far take under their care and inspection the publication of such papers as shall have been read before, or communicated to them at their weekly meetings, as to appoint a Committee who should from time to time, as occasion should require, assemble together and select from the said papers (which should be referred to the said Committee for that purpose) such of them as they should think proper to be printed, and to order that no other papers should be published in the *Philosophical Transactions* than such as shall have been so selected by the Committee.

That it is the opinion of the Council that the President, Vice-Presidents and Secretaries should be constantly members of the said Committee; that several meetings should be appointed by the President, or in case of his sickness or absence, by one of the Vice-Presidents; and that due and sufficient notice of such meeting should be sent previously thereto to every member of the said Committee.

Not less than five members of the Committee shall constitute a quorum.

These resolutions together with others laying down the procedure of the committee were approved by the Fellows and adopted, thereby placing full responsibility for the Society's publications on the Council as representing the Society as a whole, and this 'Committee of Papers', as it was called, has dealt with them up to the present time. The next volume of the *Transactions*, No. 47, included papers which had been communicated from 1750, when the forty-sixth volume had been published by Dr Cromwell Mortimer, to 1753; after which the new scheme came into operation.

Until 31 December 1751 the Julian or Old Style calendar had been the legal calendar in this country, and as it was then eleven days behind the Gregorian or New Style calendar, which was in use in all European countries except Russia and Turkey, letters and documents in Great Britain had to be dated according to both systems of reckoning. In June 1751 Lord Chesterfield introduced a bill into the House of Lords for removing this anomaly and superseding the Julian calendar by the Gregorian in this country. Though the Royal Society in its corporate capacity had no responsibility for the bill, the draft of it was prepared and the necessary computations were carried out by Lord Macclesfield, who was then a member of Council, and by Mr P. Daval, one of the Secretaries, who were assisted by Rev. C. Walmesley and probably also by Mr W. Jones, both of whom were Fellows of the Society and eminent mathematicians. The President of the Society and Dr J. Bradley, the Astronomer Royal, both approved of the draft, the latter contributing three tables which were added at the end of the bill. The bill, which was entitled 'An Act (George II, Cap. 23, 1751) for regulating the commencement of the year and for correcting the calendar now in use', was introduced in the House of Lords by Lord Chesterfield in a clear and instructive speech; he was followed by Lord Macclesfield, who explained the involved nature of its preparation and the extreme care which had been exercised by those mathematicians who had assisted him in preparing the text of the bill and the tables which were included in it. The bill was adopted in Parliament without difficulty, but there was considerable opposition to it in the country among those who did not understand the need for it and were suspicious of its effects. Under the bill the day following 31 December 1751 became 1 January 1752, and in order to correct the error which had accumulated the eleven days between 2 and 14 September were omitted from the calendar for that year, the day after 2 September 1752 being called 14 September.

Though the President's interest in science may have been small, he was regular in attending to his administrative duties, having presided at forty-two out of the fifty-five meetings of the Council which were held, but in 1752 he had a stroke of paralysis and was too ill to attend any of the Council meetings; ten of them were held, both the Secretaries being present. Among other business it was ordered at the March meeting that the Charters and Statutes should be printed; and in November the annual remuneration of the two Secretaries was raised to £60 with an additional gratuity of £10. In 1752 Mr Martin Folkes informed the Council that in consequence of his failing health he would not stand for re-election to the presidency at the Anniversary meeting. His death took place in the following year.

Martin Folkes' presidency had not been so successful as to satisfy the scientific Fellows that literary and archaeological ability in a president would suffice to promote the Society's scientific aims as actively as they would wish, but it was not always easy to obtain the support of a majority for a man of science, when the non-scientific members on the Council and in the Society always outnumbered their scientific colleagues. Most fortunately at this time there was a candidate available who was a dis-tinguished scientific man and also well known both socially and adminis-tratively, this was George, the second Earl of Macclesfield; he had been elected a Fellow of the Society in 1722 and served on the Council in 1746, 1748, 1750 and 1752. He was distinguished both as a mathematician and as an astronomer, having studied these subjects at Oxford and later, under William Jones, F.R.S., the mathematician, at his residence, Shirburn Castle in Oxfordshire. While he was at Oxford he made the acquaintance of James Bradley, and these two co-operated frequently in their astronomical studies both at Oxford and later at Shirburn Castle. The Council therefore recommended to the Fellows at the Anniversary meeting that Lord Mac-clesfield should be the new President; that James West should be re-elected Treasurer; that Peter Daval should be re-elected as Secretary, and that Thomas Birch, one of the original trustees of the British Museum and a friend of Sir Hans Sloane, should be the other Secretary in the place of Dr Cromwell Mortimer who had filled that office for twenty-two years and had been most regular in his attendance at the meetings of Council.

The President was fifty-three years of age at the time of his election and gave twelve years of most valuable service to the Society, during which he devoted himself to raising the scientific standard of the Society's meetings and publications which, in his opinion, had fallen off greatly since Sir Hans Sloane's retirement. William Stukeley, F.R.S., records in his Diary that Lord Macclesfield, as soon as he had been elected to the presidency, asked the Fellows for their assistance in maintaining the dignity and the scientific prestige of the Society. Under him the Council meetings rose from an average number of five yearly to twelve—in 1760 there were no less than nineteen. He attended them with considerable regularity, being present at two-thirds of those which were held. The Secretaries were Peter Daval, an astronomer, who was a very capable mathematician and ably supported the President in his efforts to improve the scientific tone of the Society; his colleague Dr T. Birch, an historian but without scientific interests, was a regular attendant at all meetings but was not qualified to take an active part in promoting the Society's scientific aims. Financial questions early occupied the attention of Council and in order to meet the cost of issuing the *Philosophical Transactions* to those Fellows who wished to receive them the admission fee was raised from two to five guineas.

For several years past increased interest had been widely shown in the application of science in industry and in technological practice. In the seventeenth century a considerable number of the communications which the Society received had dealt with the practical application of science, but as time went on the character of the papers read and published became more academic, and those who were interested in industrial matters thought that the Society paid too much attention to theory, and had lost such touch with practical matters that it originally had. As early as October 1738 a certain Dr Peck wrote to Sir Hans Sloane, the President, putting forward 'a proposal for raising a stock of £1000 sterling by way of subscription, for the encouragement of arts and sciences'; this together with two other letters on 9 and 14 October were also read to the Council. 'And the contents thereof being considered, it was agreed that he should be informed "that this Society, as a Society, cannot assist in the establishment of such a foundation, nor will they give any interruption to the design of any other Society, which the proposer now seems to be in hopes may be formed thereon".' It will be remembered that at the time this decision was taken the Society had to extricate itself from a recurrence of its serious financial difficulties so that the Council would naturally be most unwilling to add to its liabilities. No society such as was proposed was formed at the time, but many had in their minds the importance and the practical value of such an institution; among these were Lord Folkestone, Lord Romney, Rev. Stephen Hales, F.R.S., a Mr Shipley, and others. In 1753 the Society of Arts, which had for its object the promotion of Arts, Manufactures and Commerce, was instituted, Mr Shipley being the actual originator of it; the others who have been mentioned actively supported him, as well as H. Baker, G. Brande, a chemist, J. Short, an optician, who were Fellows of the Royal Society, as well as N. Crisp, a watchmaker.

In 1753 Sir Hans Sloane died and left to the nation his collections on the condition that the sum of £20,000 should be paid to his executors for them. This was stated to be far less than their value and the government recommended to Parliament that the legacy and its conditions should be accepted. These collections together with the Cotton Library and the Harleian Manuscripts, which were already national property, formed the foundation of the British Museum which was at first housed in Montagu House in Bloomsbury. Nearly thirty years later, when the Royal Society transferred its quarters from Crane Court to Somerset House, its collections were offered to the trustees of the British Museum since the accommodation which was provided for the Society at Somerset House did not admit of their being properly exhibited there. They were accepted by the trustees in 1781.

In the year 1753 the Council awarded the Copley Medal to Benjamin Franklin for his 'Curious Experiments and Observations on Electricity'; these had been published in Vol. 47 of the *Philosophical Transactions*. The President, Lord Macclesfield, in his address to the Society at the Anniversary meeting, said that 'The Council could not overlook the merit of Benjamin Franklin of Pennsylvania; for though he be not a Fellow of this Society, nor an inhabitant of this island, he is a subject of the Crown of

Great Britain, and must be acknowledged to have deserved well of the philosophical world; and of this learned body in particular to whom he has at various times caused to be communicated many of the experiments he has made, and of which you have lately received a large collection'. In 1756 the Society elected Franklin to the Fellowship; he was recommended as a candidate by the Council, the President and seven of the members having signed his certificate. This was done without his having expressed a desire for election and as the election was voluntarily conferred on him by the Society the usual admission fees and liability for annual subscription were remitted by the Council.

The Mace, which had been presented to the Society by its founder and first Patron, King Charles II, had in the course of the ninety years that it had been in the Society's possession, become much tarnished, so that cleaning and regilding were necessary; the President therefore undertook to have this done at his own expense in 1756.

In October 1760 George III ascended the throne in succession to his grandfather, and in the following month the Council prepared an Address from the President, Council and Fellows, in which His Majesty was prayed to become the Royal Patron of the Royal Society as his two predecessors as well as the founder, Charles II, had been graciously pleased to do. Mr William Pitt, a Fellow of the Society and at the time one of the Secretaries of State, notified the President that His Majesty would receive him and a deputation of the Society. The Earl of Macclesfield as President, and several of the Fellows therefore waited on the king who received them graciously, and signed his name in the Charter-book as Patron of the Society. He was from the first greatly interested in its work and on several occasions gave his support to schemes which the Society's Council urged on the Government as being of national importance, but which its own resources were insufficient to carry out.

In 1714 Edmond Halley had published a paper in the *Philosophical Transactions*, vol. 29, calling the attention of astronomers to the transit of the planet Venus across the sun's disc which would occur in 1761, and urging that arrangements should be made in good time for the phenomenon to be carefully observed in that year. The Council discussed fully the most suitable places for observing the phenomenon and decided upon the island of St Helena, and if possible Bencoolen as well, as being the most likely to give satisfactory results. As the funds at the Society's disposal were quite insufficient to meet the expense which would have to be incurred in carrying out their design, for this would have amounted to £800 for each station, a Memorial was sent in the name of the President, Council and Fellows of the Society in July 1760 to the Lords of the Treasury praying for a grant of £1600 for this purpose; the President at the same time wrote to the Duke of Newcastle asking for his support in the consideration of their memorial, with the result that a grant of £1600 was approved. The

Council then took up the matter as one of urgency and held nineteen meetings in order to select observers, to draft instructions and to provide the necessary instruments; Dr N. Maskelyne, F.R.S., was appointed to take charge of the observations at St Helena but the weather there turned out to be most unfavourable, and he was only able to get two fair views of the transit. He did, however, avail himself of the opportunities furnished by his voyage to investigate the determination of longitude by the measurement of lunar distances, and his work has been of much value to navigators and others.

Messrs Mason and Dixon who had been appointed to go to Bencoolen in order to observe the transit from that place met with a series of misfortunes; their ship was attacked by a French frigate and had to put in to Plymouth for repairs. So much time however had been lost that they had to land at the Cape of Good Hope and make such observations as were practicable there. In England observations were made at the President's observatory at Shirburn Park by Messrs Hornsby and Phelps. The whole of the observations from the various stations were discussed by Messrs J. Short, an optician, and T. Hornsby, Savilian Professor of Astronomy, both Fellows of the Society, and published in vol. 52 of the *Philosophical Transactions*; the value of the solar parallax as determined by them was $8''.52$.

In 1754 Parliament approved the purchase of Sir Hans Sloane's collections and during the next five years they were removed to Bloomsbury where they were arranged so that they could be seen by the public, though for several years this was subject to considerable restrictions. A body of trustees had also been appointed; one of the trustees *ex officio* was, and still is, the President of the Royal Society. At this time there was a close relation between the two institutions, for in 1755 nine of the members of the Royal Society's Council were also trustees of the British Museum; whether this was done with the idea of ensuring co-operation between the two institutions is uncertain, but for the next ten years from six to nine members of the Society's Council were also trustees of the Museum, though after this the number fell off, and the practice, which did not increase the number of scientific councillors, was dropped.

In 1762 James Bradley, the Astronomer Royal, died and was succeeded by Nathaniel Bliss who held the post for two years only. For twelve years before his death Bradley had served as a member of Council of the Royal Society, having been appointed in 1742 when he was Savilian Professor of Geometry at Oxford, and then annually from November 1751 until his death in 1764.

Lord Macclesfield was taken seriously ill in the summer of 1763 and died early in the following year. Shortly before his death the Council had before them the question of the observations taken at the Royal Observatory, for the Society had adopted a resolution proposed by Dr Nevil

Maskelyne, the new Astronomer Royal, urging the Council to take steps to have these observations published regularly in the *Philosophical Transactions*. It had hitherto been the custom for Astronomers Royal to consider them as their own property so that they were taken possession of by their executors, and were published neither by the observatory nor by the Royal Society.

During the twelve years of his presidency Lord Macclesfield had done much to restore the scientific standing of the Society, and it was fortunate that at his death there was one of the Fellows who was well qualified to carry on his work. Lord Morton was both a competent mathematician and an amateur astronomer of distinction, and being a close friend of Professor Maclaurin, the mathematician, had assisted him in his scheme for enlarging a medical society which had been formed in Edinburgh in 1731, in order that it might include both philosophy and literature. In 1739 it was remodelled and was then known as the Philosophical Society of Edinburgh, Lord Morton being its first President. In 1783 the name was changed by royal charter to that of the Royal Society of Edinburgh. Lord Morton had been elected a Fellow of the Royal Society in 1733 and became a member of Council in 1758, 1759 and 1763; in 1764 he succeeded Lord Macclesfield as President. His Secretaries were Charles Morton, M.D., who had replaced Peter Daval, the mathematician, at the end of 1759, and Matthew Maty, M.D., who replaced Thomas Birch at the end of 1765. The former had taken up the medical profession early in his career and attained some distinction in it; he then devoted himself to literary work. Both were appointed heads of departments in the British Museum, the former in 1756 and the latter in 1753, and later each of them became in turn Chief Librarian of the institution. They attended the meetings of Council with considerable regularity, being present at about 80 per cent of them, but scientific initiative in these years was left to the President who was the only man of science among the officers.

It will be remembered that Queen Anne in 1710 authorized the issue of a Warrant appointing the President and Council of the Royal Society to be constant Visitors of the Observatory and giving them certain powers and authority in relation to the work carried on there. It had not been realized by the officers of the Society that the validity of the Warrant ended with the death of the sovereign who had authorized its issue, so they had taken no steps to obtain a new appointment on the accession of George I in 1714, or on that of George II in 1727. On the matter being raised the Law Officers of the Crown advised that these powers did not belong to the present Council and that a new authority had to be given by each successive sovereign in order that the Royal Society should be empowered to act as Visitors of the Royal Observatory during his reign.

Lord Morton lost no time in acting upon the proposal of Dr Nevil Maskelyne, who had succeeded Nathaniel Bliss as Astronomer Royal in

1765, and determined to memorialize the king not only to give the Royal Society power to make recommendations on the work and equipment of the observatory but also to pray His Majesty to give orders for the restitution to the observatory of the former observations. A draft Act containing regulations relating to the duties of the Astronomer Royal was also submitted. The king acceded to the prayer of the Society and appointed them to be Visitors of the Observatory by a Warrant dated 2 February 1765. The instruments there were repaired and in some cases new ones of a more modern type were supplied, a new ten-foot telescope being supplied by John Dolland. In response to another petition in 1767 the king ordered that the observations made at Greenwich should be printed by the government provided that the cost did not exceed £60 per annum.

In consequence of the increasing expenses, cost of publishing and issuing to Fellows of the Society the *Philosophical Transactions*, it had become necessary to amend certain of the statutes of the Society which related to the subscriptions of Fellows. The Council therefore decided in 1766 to raise the composition fees from twenty-one to twenty-six guineas, without however altering the rate of the annual subscription which remained at £2. 12s. 0d. This increase in no way checked the number of candidates for the Fellowship which at this time reached twenty-five admissions yearly on the average. The majority of these preferred to become life members by payment of the compounder's fee which even at twenty-six guineas was very moderate. Such fees were still being treated as income and were not invested to form a reserve fund until sixty years later. At the same time the admission fee was raised from three to five guineas. There were at this time 160 foreign members as compared with 345 Ordinary Fellows, a proportion which the Council recognized as being too high; they therefore resolved that in future only two foreign members should be elected annually until the number of them should be reduced to eighty, and that a certificate setting forth the qualifications of each candidate should be laid before the Council for their information before he was recommended to the Society for election; this maximum of eighty was increased to one hundred by the statutes of 1776.

The observations of the transit of Venus which took place in 1761 had not proved to be as satisfactory as it had been hoped that they would be; but fortunately the phenomenon would again be visible in 1769 but after that not for a century. The Council began early to make the necessary arrangements, taking steps to engage competent observers and to recall their instruments in order that they might be put in thorough order, and adjusted.

In February 1768 they submitted a memorial to the king setting forth the great scientific importance of the occasion and praying that the sum of £4000 should be made available in order that an expedition might be sent to the southern hemisphere and that sea transport should be provided, representing 'that the Royal Society are in no condition to defray this expense, their annual income being scarcely sufficient to carry on the necessary business of the Society'. This was no exaggeration, for the balance in the hands of the Treasurer at the end of the years 1764 to 1768 varied from a credit of £186 to a debit of £317; the average annual income for the decade 1761–70 was but £1100. On 24 March 1768 the President informed the Society that His Majesty had authorized the grant asked for and that the Astronomer Royal, Dr N. Maskelyne, had been directed to prepare instructions for the observers; these were to be: Lieut. James Cook and Charles Green in the Pacific; Messrs Dymond and Wales to Hudson Bay, for whom the Hudson's Bay Company granted a passage in one of their ships; and Mr Call to Madras. Mr J. Banks, a wealthy Lincolnshire landowner who had been in 1766 elected to a Fellowship on the ground of his knowledge of and interest in natural history, wrote to the Council requesting them to apply to the Lords of the Admiralty for permission for him to accompany Lieut. Cook in the ship *Endeavour* at his own expense. The Council, when communicating the names of Lieut. Cook and Mr Green as selected observers, added a request that Mr J. Banks together with his staff of seven assistants might form a part of the expedition. This was approved and the *Endeavour* sailed from Plymouth on 26 August 1768 having on board Banks and his staff which included Dr Solander, a Swedish botanist, two draughtsmen and four servants together with their baggage, stores and instruments.

As a boy at Eton, Banks had shown a keen interest in natural history, and as a gentleman-commoner at Oxford in 1760 he had continued the study of the subject, having engaged a lecturer in botany to instruct him in it. In 1763 on his father's death he came into a large estate in Lincolnshire, and besides continuing his scientific studies he devoted much of his attention to agriculture, the development of his property and the raising of stock. Here he made the acquaintance of Lord Sandwich who was First Lord of the Admiralty, a friendship which was to prove to be of much value to him in later years when he was advocating voyages of exploration in both northern and southern polar regions. Banks was elected a Fellow of the Society in 1766.

He was now a wealthy man and could follow any line of interest which appealed to him; this was natural history in all its branches, and at his house in New Burlington Street he met and entertained a wide circle of his friends who had similar interests. In the spring of 1766 an exceptionally favourable opportunity to extend his knowledge in this subject by travel occurred. An Oxford friend of his, Lieut. the Hon. Constantine Phipps, was serving in the Navy on board the ship *Niger* under the command of Captain Thomas Adams. Permission was given for Banks to accompany his friend in order to study natural history on the shores of Newfoundland and Labrador where the ship was to be stationed for the protection of the fisheries. Not only was the voyage a valuable experience in travel and exploration for Banks but he and Phipps became close friends, and when in after years the latter succeeded his father as Lord Mulgrave in 1775 and was appointed a Lord of the Admiralty in 1777, he was able to help Banks in some of his projects. He was elected a Fellow in 1771, and served on the Council of the Society for fourteen years.

The *Niger* sailed on 22 April 1766 and reached Newfoundland on 11 May. After five months in northern waters the *Niger* returned to Lisbon by the end of the year and reached London in the spring of 1767. In the following year Banks and his party joined the *Endeavour* under the command of Lieut. James Cook; they left Plymouth on 25 August and sailed for the Pacific by way of Rio de Janeiro and Cape Horn, reaching Otaheite on 13 April 1769. The transit was successfully observed, and the ship returned by way of Batavia and the Cape of Good Hope, reaching Deal on 12 July 1771. The voyage had been most successful, for besides the observation of the transit of Venus important additions were made to our knowledge of Australia as well as of the fauna and flora of the coasts and islands. The character and conditions of life of the primitive inhabitants were also studied. This voyage to the Pacific and the knowledge that Banks acquired of the coast of the Australian continent which the ship visited greatly influenced the whole of his future life. He never lost his interest in the lands of the Southern Pacific and applied his energies for many years to the colonization of Australia, and to improving conditions of life for the settlers as well as of those convicts who were later sentenced to transportation thither. With none of these projects was the Society concerned and they are only mentioned here as indicating the energy and the beneficent intentions of the man who was a little later to be its President for more than forty years.

On 12 October 1768 the President, Lord Morton, died and Sir James Barrow was elected to act in his place until the date of the Anniversary meeting. During the four years of his presidency Lord Morton had worthily carried on the work of his predecessor Lord Macclesfield, and had notably advanced the scientific activity of the Society. The Society had been reappointed Visitors of the Royal Observatory under the terms of a new Warrant and a grant of £1000 had been made for its equipment with modern instruments; the observations taken there were to be regularly printed by the government; and the transit of Venus of 1769 had been successfully observed with the aid of a grant which had been made to the Society. In the execution of his duties as President he was indefatigable, having presided at eighty-eight of one hundred and nine Council meetings which had been held during the four years of his presidency.

On Lord Morton's death the Society again turned to its non-scientific Fellows for a successor; their choice in 1768 fell on James West, an antiquary, who had been Treasurer for the past thirty-two years; so long a tenure of this office was probably not in the best interests of the Society. On Sir Hans Sloane's retirement the Society's income was about £1000 per annum, and was increasing; between 1753 and 1760 £1900 had been invested. West's presidency only lasted from 1769 to 1772 during which the Council were very fully occupied with the preparation of detailed instructions for Captain Cook's voyage to the Antarctic, on which he sailed in 1772. The Treasurer's duties were taken over by Samuel Wegg, who also was an antiquary; the two Secretaries were Matthew Maty, M.D., and Charles Morton, M.D. Of these two the former had just taken up the medical profession after taking his degree in Leyden, and practised in London from 1741; in 1750 he was publishing the *Journal Britannique* and in 1753 became an Under-Librarian at the British Museum and Principal Librarian in 1772. He became Secretary of the Society in 1765. The latter who had been Secretary since 1760 had taken his degree in Leyden and practised in London at first, but in 1756 he became an Under-Librarian at the British Museum and Principal Librarian in 1776. The President himself was a politician and a collector of books and manuscripts; he represented St Albans in the House of Commons from 1741 to 1768, and was Joint Secretary to the Treasury from 1741 to 1762. Thus it came about that at this time none of the officers of the Society were men of science; science was only represented on the Council by about half the members who had had a scientific training and none of these had the authority of an officer; there was therefore no one to initiate and promote scientific activity as the two former presidents had so effectually done. About seventeen Council meetings were held in each year and were well attended by the officers, and on the average about half the members were present. At this time the Society's average income is said to have been about £1450.

In November of this year it came to the President's knowledge that a large collection of letters and volumes of observations, etc., in manuscript which had been the property of Flamsteed, the Astronomer Royal, were at Islington in great disorder. He told the Council that he had informed the Commissioners of Longitude and they had agreed to pay £100 for the 'whole' collection. He had arranged that it should be brought to the Society's rooms and there sorted, those relating to the Society being retained there and the remainder sent to the Royal Observatory.

In 1771 the Council were informed that it was the intention of the Admiralty to send out an expedition to explore the high southern latitudes, and two ships, the *Resolution* and the *Adventure*, were commissioned and were to be under the command of Captain James Cook. As soon as Banks heard that this new expedition was contemplated, he again applied to the Admiralty for permission to accompany the expedition at his own expense, and made elaborate preparations which included engaging an artist and three draughtsmen besides two secretaries and nine servants, making fifteen in all besides himself. The Society agreed to lend two chronometers by Arnold to the expedition and these are still in the Society's possession. Ultimately the Admiralty refused their consent to Banks taking part in the expedition on the grounds that the alterations necessary to provide accommodation for Banks' large party and their stores and equipment were such as would materially affect the sailing qualities of the ship. He therefore changed his plans and chartered a vessel for himself and his party in which he made an expedition to Iceland, which was most successful in every way.

James West died on 2 July 1772 and James Barrow was again elected President until November when at the Anniversary meeting Sir John Pringle was elected in his place; Sir James Porter, a diplomatist who had no scientific qualifications, was proposed as an alternative candidate for the presidency but he was unsuccessful at the ballot.

The President, who was now sixty-five years of age, had been educated at the University of St Andrews, and in 1727 went to Edinburgh to study medicine; then, as many of the most promising students did at that time, he went to Leyden where Professor Boerhaave was lecturing. He took his degree of Doctor of Physic there in 1730 and returned to Edinburgh where he settled as a physician. In 1742 he was appointed physician to the Earl of Stair who was a commander of the army in Scotland and this changed the course of his life since he was afterwards appointed to the charge of the military hospitals in Flanders. In 1745 he was with the Duke of Cumberland in Scotland and in 1749 was appointed his physician-in-ordinary. He was elected a Fellow of the Society in 1745 and was a member of the Council in 1753. He greatly advanced the study and practice of army hygiene, and the treatment of wounds and sickness on military service.

In January 1773 the Council wrote to Lord Sandwich, who was the First Lord of the Admiralty, recommending that an expedition should be sent into the Arctic regions to explore them. The proposal was agreed to and the ships *Racehorse* and *Carcass* sailed in June 1773 under the command of Captain the Hon. Constantine John Phipps, afterwards Lord Mulgrave. Directions and scientific instructions for those taking part in the expedition were drawn up by the Council.

The government had now agreed to provide rooms for the Society for which no rent would be charged, and had instructed their architect, Sir W. Chambers, to prepare plans for the information and consideration of the Council; in July 1776 therefore discussions began between Sir William Chambers and the Council on the subject of providing accommodation for the Royal Society, and some other Societies, at Somerset House.

In 1772 Dr N. Maskelyne, the Astronomer Royal, communicated a paper to the Society in which he strongly urged that the deflection from the vertical of a plumb line by the attraction of a mountain mass of ascertainable size and composition should be determined experimentally and named certain hills in Yorkshire and Cumberland which might be found suitable. The Council agreed that such an experiment should be carried out by the Society and appointed a 'Committee of Attraction' to go into the details, the members of which were Dr S. Horsley, a Secretary, and Dr N. Maskelyne, the Astronomer Royal, together with the Hon. Daines Barrington, Hon. Henry Cavendish, Dr Benjamin Franklin, Mr Raper, Dr Watson and, later, Mr Joseph Banks. They employed a Mr Charles Mason to carry out the preliminary surveys, who reported that the hills in England which he examined were not suitable, but that one in Perthshire, called Schiehallion, seemed by its situation, size and figure to be the most suitable for the proposed experiment. The Council adopted Mr Mason's report and added a strong recommendation that Dr Maskelyne should be invited to direct the observations on the spot. This he agreed to do and the Society authorized that the cost of the work as well as Dr Maskelyne's personal and travelling expenses and those of Dr Hutton, who carried out the mathematical investigations, should be at its charge. The Astronomer Royal and his assistant Mr Burrow arrived at the locality in June 1774 and completed the work in the following November. The cost was £600 which was paid by the Society from its funds (cf. *Philosophical Transactions*, vol. 65). The attraction of the mountain mass on the north and south sides was found to be 11".6. Dr C. Hutton, F.R.S., a well-known mathematician, undertook the calculations for determining the density of the earth from the observations made at Schiehallion and found it to be 4·481; the details of his investigation were published in the 68th volume of the *Philosophical Transactions*. The Copley Medal was awarded to Dr N. Maskelyne in 1775 for this work. Dr C. Hutton was the Copley Medallist in 1778 for mathematical investigations in ballistics.

Some of the statutes which had been originally approved in 1663 had been modified from time to time as conditions had changed; the supply of the *Philosophical Transactions* to the Fellows free of charge had added to the Society's expenses, so that the admission and composition fees had to be increased; candidates had formerly to present themselves for admission within four weeks after their election, but later, when many did not live in or near London but in more distant parts of the country, the time had been extended to ten weeks; the Treasurer's balance which was at first kept in his chest had for several years past been invested; since 1710 the Society had had a house of its own and a resident officer had become necessary; foreign members were limited to one hundred. It was therefore agreed by the Council that it was high time to revise the statutes and republish them for the information of the Fellows; this important task was taken in hand in the year 1774 when the Council met seventeen times, and was completed in 1776 after nineteen meetings had been held in that year.

Again the all-important question, how the proportion of the scientific members of the Society to those who were not interested in the promotion of Natural Knowledge could be increased, was not dealt with; nor was the control of the finances made more efficient.

In 1775 the Society received from the Executors of Mr Henry Baker, F.R.S., an antiquary and naturalist, the sum of £100 which he had bequeathed to the Society 'for an oration or discourse to be spoken or read yearly by some one of the Fellows of that Society on such part of Natural History or Experimental Philosophy at such time and at such manner as the President and the Council for the time being shall please to order and appoint'. Should however a year pass without such a discourse having been read or spoken the capital of £100 was to be returned to the executors of his estate. The first lecture was given by Peter Woulfe; his name and those of all the subsequent Bakerian lecturers are given in the *Record of the Society* (4th ed. 1940, pp. 364-74).

In 1769 the Dean and Chapter of St Paul's Cathedral had asked the Council to advise them on the best and most effectual way of fixing electrical conductors to the cathedral; the Church of St Bride having recently been struck by lightning and seriously damaged, the cathedral authorities wished to assure themselves that all possible precautions had been taken to protect their charge from harm. A committee was therefore appointed consisting of John Canton, an electrical expert, Edward Delaval, a chemist, Benjamin Franklin, Sir William Watson, a physician and writer on electricity, and Benjamin Wilson, a painter to the Board of Ordnance and a student of electrical physics, who examined the cathedral carefully and reported to the Society on 7 June that the principal object was 'to make a complete metallic communication between the cross placed over the lanthorn and the leaden covering of the great dome; as from its height, if any lightning struck it, it would probably affect the cross'. This was done and the water pipes were utilized to serve as conductors from the roof to the ground. In 1772 the government received intelligence of the destruction by lightning of several powder magazines at Brescia; and as their own magazines at Purfleet were totally unprotected, they determined on taking measures to protect them. The Society was requested to give its opinion on the best form of lightning-conductor for the protection of the magazines. The Council appointed a committee of the same composition as that which had recently reported on St Paul's Cathedral except that Hon. Henry Cavendish replaced J. Canton, and John Robertson, a mathematician, took the place of E. Delaval. The report of the committee, which was drawn up by Benjamin Franklin and published in vol. 63 of the *Philosophical Transactions*, strongly recommended pointed conductors, B. Wilson alone dissenting. He published a long paper in the *Philosophical Transactions* setting forth his reasons for the opinion which he had expressed though these were strongly contested by Franklin, Nairne, Henley, Swift and other electricians in the Society. The government however accepted the Society's advice and used pointed conductors.

Unfortunately in May 1777 the magazines at Purfleet were struck by lightning and slightly damaged but no explosion of their contents took place; a house at Tenterden fitted with pointed conductors was also struck. The Board of Ordnance therefore wrote to the Society requesting their advice and a second committee consisting of the President, the Secretaries and Messrs Henley, Lane and Nairne was appointed. They examined the magazines carefully and recorded their unanimous opinion in favour of pointed conductors. Wilson then sent a long report to the Board of Ordnance quoting a series of experiments which he had carried out and maintaining his views; this was referred to the Royal Society which appointed Hon. Henry Cavendish, Lord Mahon, Messrs Henley, Lane, Nairne and Dr Priestley for consideration and report; they reported against Wilson's conclusions. Wilson then informed the Board of Ordnance that the committee's report which was laid before the Society was not to be considered as the opinion of the Society as a whole; the Secretary of the Board asked therefore what the opinion of the Society as a whole on the matter was. In reply the Society explained that it had never been its practice to give an opinion as 'a body at large' but always by committees specially selected and appointed for the purpose in each case; and that in this instance the Society had no reason to be dissatisfied with the report of its committee.

At this time England was at war with the American colonies, and Benjamin Franklin, who was now their representative in Europe, had introduced and advocated the pointed conductor. The matter now passed out of the hands of scientific men and became a political question, since those who advocated pointed conductors were identified with the supporters of the insurgent colonists. The king is said to have taken the side of Wilson in the dispute on political grounds; he had blunt conductors installed at the palace, and even endeavoured to make the Royal Society rescind its resolution in favour of pointed conductors. In an interview with Sir John Pringle at the end of 1777, His Majesty is said to have urged him to use his influence in supporting Mr Wilson. The President however replied: 'Sire, I cannot reverse the laws and operations of nature.'

At the Anniversary meeting on 30 November 1778 Sir John Pringle did not stand for re-election. His biographer, Dr Kippis, denies the report that his resignation was the result of the royal displeasure, but this dispute, which had lasted already for three years, and had passed from the scientific field into the more complicated political arena, may well have influenced his decision; the President was now seventy-one years of age and had spent a very busy life; it was quite reasonable therefore that he should seek release from a position which showed no signs of becoming easier, and the retention of the presidency by him might even prejudice the future of the Society now that it had definitely ranged itself in opposition to its Patron's wishes.

The choice of a successor to Sir John Pringle presented many difficulties for the Council, who naturally wished to put an end to the Society's estrangement from the court, and at the same time to press forward with such plans for scientific investigations as they may have had in mind, but which they could only finance to a very moderate extent from their own resources. Finally they decided to recommend Mr Joseph Banks, who had been elected a Fellow in 1766 and had served on the Council in 1774, 1775 and again in 1778; he was elected by the Society to the presidency at the Anniversary meeting of 1778 and was re-elected annually until his death in 1820 forty-two years later. This was a remarkable departure from previous practice, for he was only thirty-five years old at the time, whereas former presidents had not usually been less than fifty years of age, and in some cases as much as sixty-five.

During the period between the election of Martin Folkes to the presidency and the retirement of Sir John Pringle the Fellowship had increased from 300 to about 450 and was still rising; the scientific Fellows accounted for about 29 per cent of the total at the end of the period as compared with the 71 per cent of the non-scientific majority. In consequence of the action taken by the Council in 1758 and 1776 the number of foreign members was falling off but there were still 125 of them in 1780. The special elections (peers) remained about the same at 10·5 per cent of the total Fellowship or about 45, but as usual few of them took any active interest in the scientific aims of the Society.

The financial crisis of 1740 had been overcome with little difficulty by employing the powers which Sir Hans Sloane had obtained for the Council in 1728-30 and which only needed to be regularly and firmly applied whenever the need arose. On Martin Folkes' retirement in 1753 Weld records that the Society possessed £3000 invested in various stocks, the result of the steps which had been taken to see that the subscriptions of Fellows did not fall into arrear. The property which had been bought at Acton and that which had been bequeathed to the Society at Mablethorpe and Coleman Street, as well as a few gifts, had all contributed to make the financial position easier and more stable; all that was needed was a careful control of the administration. This Lord Macclesfield had provided and under him £1900 was invested in Consol Annuities or Bank Annuities. In the decade 1751-60 the average annual receipts were about £1300 and the corresponding expenses were about £1000. When James West resigned the Treasurership in 1768 the income is stated by Weld to have been £1450 having risen to this figure from £760 in 1736.

The Anniversary dinners are not often referred to during the eighteenth century though they were held regularly on each St Andrew's Day. There is in the Society's library a small book in which the names of the Fellows who dined at these dinners were recorded for most of those which took place from 1768 to 1809. The date, the place at which the dinner was held, and the names of those who dined are all that it contains, so there is not much information to be extracted from it. The average attendance of Fellows was about seventy and the President of the day usually presided; private guests of Fellows do not seem to have been present but once or twice guests of distinction are included in the list as though they were guests of the Society; the practice of inviting official guests was not introduced until about the middle of the nineteenth century. These dinners which were held at Pontac's Tavern in the seventeenth century, and at the Crown and Anchor in the Strand from 1772 until 1848, still take place regularly at some hotel or restaurant selected by the officers.

During the presidency of Martin Folkes the Councils do not show evidence of much activity; there were frequently but three meetings in the year and the average number for the first decade is but 4·8. This was changed at once when Lord Macclesfield became President and the improvement which he initiated was continued up to the end of his presidency. The average number of meetings which were held annually reached nineteen in the third decade and fifteen in the last.

The number of scientific members of Council showed no sign of increase, the average for the whole period being nine, too low a number to indicate that the Council's outlook had become more scientific. The effect of the presidency of the Society having been held for sixteen years by antiquaries who had little knowledge of natural science (Martin Folkes and James West) was definitely unfavourable to any proposal to strengthen the representation of science on the Council as had at one time, under Sir Isaac Newton, seemed to be possible.

Under Lord Macclesfield and Lord Morton there was a temporary increase in the activity of the Council which is shown by the number of meetings. Some of the statutes were revised in 1753; instructions for observing the transits of Venus were prepared in 1760 and 1768, and a thorough revision of the statutes was undertaken in 1776; in all of these years meetings were more frequent than of late, there having been forty-three in 1768.

The average attendance at the meetings also increased slightly, being usually between nine and twelve and occasionally as high as fifteen (cf. Appendix II C).

In the middle of the eighteenth century the Councils were active in promoting expeditions for observing astronomical phenomena, and for exploration, but the scientific councillors still remained too few in number.

Lord Macclesfield [1722][1] and Lord Morton [1754], as presidents and

[1] The dates in square brackets indicate the year of election to the Fellowship.

astronomers, endeavoured to promote research, as also did Sir William Herschel [1781] and Nevil Maskelyne [1758] later in the century.

In mathematics Lord Macclesfield drafted the bill for the reform of the calendar, and was assisted by Peter Daval [1740], a Secretary of the Society.

In chemistry and physics Henry Cavendish [1760] and Joseph Priestley [1766] were rapidly advancing our knowledge in these branches of science. James Black's discoveries should of course find a place in this period, but though he was a Fellow of the Society of Edinburgh he never became a Fellow of the London Society.

In medicine John Pringle [1745] and James Lind [1777] were improving the conditions of hygiene in the army and navy; the former introduced many improvements into military hospitals, and the care of the wounded, while the latter in his treatise on scurvy showed how this scourge of seamen could be avoided, or at least much mitigated by supplying fresh fruits and vegetables of suitable kinds.

At the end of the eighteenth century and in the early years of the nineteenth the scientific efficiency of the Councils had not increased, indeed it had fallen off to some extent. There was in fact a feeling growing amongst the Fellows of the Society that steps must shortly be taken to introduce greater scientific efficiency into its procedure but there was no general consent how this was to be done; there was indeed among a considerable number of the Fellows the conviction that matters might as well remain as they were.

REFERENCES

KIPPIS, Dr A. *Biographica Britannica*, 1778–1795. (Sir John Pringle and others.)
Royal Society, Record of. 4th ed. 1940.
SMITH, E. *Life of Sir Joseph Banks, Bt.* London, 1911.
WELD, J. C. *History of the Royal Society.* Vol. II. London, 1848.
WILLIAMS, B. *The Whig Supremacy 1714–1760,* vol. XI of Oxford History of England. 1939.
WOLFE, A. *A History of Science, Technology and Philosophy in the Eighteenth Century.* London, 1938.

CHAPTER VI

SIR JOSEPH BANKS: 1778–1820

As SIR JOHN PRINGLE did not wish to be nominated for the presidency in 1778–9, the Council had to select some other suitable Fellow, and to submit his name to the Society at the Anniversary meeting. In 1778 only eight councillors out of the twenty-one were men of science so it was unlikely that one of them would be selected by the Council.

The names of two candidates for the presidency had been proposed in the Council, Mr Alexander Aubert and Mr Joseph Banks. The former was a wealthy London merchant and an active member of the Society of which he had been elected a Fellow in 1772; he was a keen amateur astronomer who had built three observatories, one at his house at Austin Friars, another at his house at Highbury, and a third near Lewisham; he was much respected in scientific circles and had served on the Council for three years, so that he could probably count on the support of many of the scientific Fellows. Banks had been elected to the Fellowship in 1766 and had served on the Council in 1774, 1775 and again in 1778. He was well known to the Fellows of the Society as being a wealthy landowner and an ardent naturalist who was ready to finance scientific expeditions as well as to take part in them, and to undergo the many risks and discomforts of foreign exploration which were inseparable from them. He had already been a member of an expedition to Newfoundland where he had made considerable collections, and later had accompanied Lieut. James Cook on his first voyage (1768–71) when Banks had borne all the expenses of the staff and equipment necessary for carrying out botanical and other biological investigations at the places which the expedition visited. On the return of the expedition the President, Sir John Pringle, had presented him to the king who accorded him a long interview and for several years showed much interest in his work. He had since organized and carried out a scientific and exploratory expedition of his own to Iceland. He was in favour with the king as a wealthy landowner who devoted much time to the development of his land and the improvement of his stock; he took no part in the politics of the day, and identified himself with neither of the principal parties, nor would he stand for election as a Member of Parliament, all of which qualities were in his favour.

After a lengthy discussion the Council decided to recommend Mr Joseph Banks for the presidency; he was still a young man, being only thirty-five years of age at the time of his selection whereas former presidents had usually been between forty-five and sixty-five. His election to the presidency was not a matter of chance; in 1778 the Council met thirteen times and the average attendance of the members was fifteen out of twenty-one, an exceptionally large number; at the meetings in October and November, when the new officers and members of Council to be recommended to the Society for election were probably selected, fifteen or sixteen members were present, and at the last of these, which was held shortly before the Anniversary meeting when the final decision would be taken, nineteen out of twenty-one councillors attended. It may therefore be accepted that the Council as a whole considered him to be the more suitable candidate; at this time eight of the twenty-one members of Council were scientific men. In August 1778, Dr Solander wrote to Banks telling him that Aubert would be a candidate for the presidency, but that most of the Fellows would, he believed, support Banks. At the Anniversary meeting Joseph Banks was elected President, 'unanimously to appearance, with 220 votes', as Rev. Sir John Cullum wrote to Rev. Michael Tyson a few days later.

There was no lack of subjects to which the new President's zeal could be usefully directed; the membership was increasing steadily, the financial position gave no cause for anxiety, and the rapid growth of industrial activity together with the growing introduction of machinery in production showed clearly the larger part which scientific knowledge was already beginning to play in the life of the country. Banks, who was a man of much self-confidence and of great energy, at once devoted himself to improving the Society's administration. He was most regular in presiding at the meetings of Council and of the Society, rarely missing any of them; in fact during his long presidency of forty one and a half years he presided at 417 meetings of Council out of the 450 which were held—or 93 per cent. A man of so forceful a personality and such exceptional ability would soon make himself familiar with the business of the Society and would then control to a great extent the activities of the Council; it was not long before it was rumoured that he alone selected those who were to form the Council, and that he took a leading part in accepting or rejecting candidates proposed for election. There was no doubt some ground for such views for he had announced to the officers and the members of Council soon after his election that unsuitable candidates would not be recommended for election unchallenged. It was time that some such action was taken for it was freely stated that the Secretaries had been in the habit of selecting candidates for election without consulting either the President or the Council. A writer in Tilloch's *Philosophical Magazine* says that Dr D'Alembert of Paris was in the habit of asking his friends who were about to visit England whether they wished to be made foreign members of the Royal Society, humorously saying that he could easily arrange it. This of course was wholly untrue, for since 1766, when the statutes were revised, only two foreign members could be elected in any year until the number of them had been considerably reduced, but that it should have been said even in jest shows that the Society's discrimination between worthy and unsuitable candidates had not been as strict as it should have been. During the four decades of Banks' presidency the average number of candidates elected as Fellows annually ranged from twenty-one to thirty, which did not differ much from the numbers in the preceding decades, so that during his administration admission to the Society was not restricted in any way. It was also stated that he packed the Council with his own friends who would in all cases support his view, but a careful examination of the Council minutes does not bear this out. The allegation of favouritism in the selection of councils or committees is easy to make and is always difficult to refute; an analysis has therefore been made of the membership of his councils for each year of the two periods 1778–1800 and 1801–20 in order to see whether there was any basis for the criticism. Without taking into account the Secretaries, who were usually re-elected for several years, out of 171 Fellows who served on the Council in the first period 88 per cent served for one or two years only; of the rest, nine served for three years and five for four or five years. In the second period, out of 169 Fellows 89 per cent served for one or two years, four for three years, and four for four or five years.[1] In the first period there were four who served for more than ten years: Lord Mulgrave with whom, as Lieut. Phipps, Banks had sailed to Newfoundland in 1766 and who was now a Lord of the Admiralty, for fourteen years; Hon. Henry Cavendish, the eminent physicist, for seventeen years; Dr Nevil Maskelyne, the Astronomer Royal, for twenty years; and Sir William Musgrave, a Commissioner of Customs, for twenty-two years. In the second period Hon. Henry Cavendish served for ten years more; the Astronomer Royal for eleven; Dr Blagden, an eminent physician, for nineteen; and Lord Morton for eighteen years. Of those who served for exceptionally long periods, there were rarely more than three of them on the Council in any one year. The Astronomer Royal seems to have been included as though he were an *ex officio* member. Half of these were eminent scientific men while the others may well have been able to render valuable service as administrators of long experience; and this may have been a matter of some importance since seldom more than half the members of Council attended the meetings. They could have exercised more control, had they wished to do so, by more regular attendance, but they were content to leave things as they were and for the Society's business to be carried on by the officers and by the few councillors who were the most regular in attendance. The criticisms were for the most part ill-founded or based on exaggerated accounts of what did happen. It was no doubt true that the organization and the administration of the Society were far from being as satisfactory as they might have been; not much im-

[1] A similar analysis for the years 1701–40 may be compared with this one; see p. 125.

provement had resulted from the revision of the statutes in 1774–6 for the committee engaged on it had avoided dealing with the three most important matters: the restriction of the Fellowship; the introduction of some form of selection; and a thorough overhaul of the financial system. As it was, the Council were satisfied with the report furnished to them and a few desirable changes were made, but the really important ones, which would probably have aroused considerable opposition at that time, were left untouched for another fifty years.

It cannot be denied that Banks made the Society, both at home and abroad, much better known than it had been hitherto; he entertained generously at his house in Soho Square which he had bought in 1777. Here he gave weekly receptions at which all were welcome and free to inspect his exceptionally complete library of works on natural science and his collections; they heard the latest scientific news, and the accounts of travellers who had lately returned from foreign lands. He also invited many of his friends, foreign men of science and travellers or explorers who had returned from their journeys, to breakfast with him, and afterwards in his study he discussed with them their experiences and heard their news to more advantage than was possible at the evening receptions. A man of such wide interests and endowed with such ample wealth was of course the object of many appeals for support or assistance. These he examined carefully and was a generous patron of those who could satisfy him that they were capable of carrying out their plans, and that these were worth attempting. Banks always said that he valued the presidency of the Society above any other distinction which he received, and he certainly spared no pains to maintain the standard of dignity and efficiency to which he considered that it was entitled. He made it his regular practice to preside at its meetings in court dress and wearing his decorations so as to honour suitably the institution whose patron was the sovereign.

THE OFFICERS

The Treasurer, S. Wegg, was an antiquary who doubtless carried out his duties to the best of his ability, but since the financial procedure had not been dealt with by the Statutes Committee in 1776 many practices were allowed to continue which should have been drastically amended. He was replaced by W. Marsden, an orientalist, in 1802, whose successor in 1810 was S. Lysons, an antiquary; reforms were left for a later Treasurer, J. W. Lubbock, to introduce, which he did in a series of illuminating annual reports presented to Council in 1831 and the following years.

The Secretaries during Banks' presidency were at first Dr M. Maty and J. Planta, both librarians of the British Museum; the former resigned in 1784 and was replaced by Dr C. Blagden, M.D., who retired thirteen years later in favour of Dr E. W. Gray, M.D. In 1804 Dr W. H. Wollaston, the chemist, replaced J. Planta and in 1806 Humphry Davy, who was also a distinguished chemist, replaced E. W. Gray; from then until 1812 both Secretaries were scientific men, but in that year Sir Humphry Davy married and resigned his post, being replaced by Taylor Combe, a numismatist. It was not until 1825 that both the Secretaries were again men of science.

It may be thought that the President, who must have recognized that a higher standard of administrative efficiency was attainable, would have taken steps to ensure its introduction, but he does not seem to have taken this view. In the administration of his own affairs, his correspondence, his library, his collections, he exercised the most detailed control; he kept copies of all his letters, which were regularly indexed, as were all the reports, letters and information which he was continually receiving from all parts of the world. As a large landowner and one greatly interested in the yield of crops and the improvement of his stock he exercised much intelligent foresight and demanded from his employees careful and faithful service. It remains a mystery why he should have tolerated the financial irregularities and administrative shortcomings which existed in the Society and could not have been unknown to him. They were already being much criticized and the dissatisfaction due to them materially hastened the reorganization which took place during the twenty or thirty years following his death. It should be remembered that the presidency of the Royal Society was only one of many responsibilities and duties which he undertook. He never lost his interest in Australia and its colonization; he promoted in many ways the exploration of Northern Africa and of Arabia; he gave much time and help to the development of the Royal Gardens at Kew and sent collectors far and wide to bring back plants, etc. which might be useful to this country or its colonies. As President of the Royal Society he was a member of the Board of Longitude which had been established in 1718 but latterly had not shown much activity; so long as he was on the Board a marked improvement took place which was due to his zeal and energy. In 1797 he was made a Privy Councillor and service on the committees of this Council took up much of his time.

In 1783 Banks met with a serious carriage accident and was laid up for several weeks at Revesby, his Lincolnshire home; until then he had been active and fond of an open-air life, but after this he began to suffer frequently from attacks of gout. In 1808 his health was already beginning to fail and he had to be carried from his coach to the committees and councils which he attended. In his later years it would seem as though he realized that considerable changes in the organization of the Society could not be long deferred and that a greater part must be played by science in the future, for shortly before his death he expressed the opinion that Dr W. H. Wollaston, the chemist, would be the most suitable of the Fellows to succeed him.

COUNCILS

Exceptional activity in the administrative or scientific work of the Society was usually indicated by an increase in the number of meetings of the Council which were held; but it is not so evident why the number of those members who attended the meetings of Council should have been so small and have varied so little. The period from 1741 to 1780 which has been discussed in the preceding chapter was too short to provide any explanation of this uniformity so the matter was left until the data for an additional period, 1781–1820, could be utilized. As Sir Joseph Banks was President throughout the whole of this latter period and no marked changes of policy took place it should be very suitable for the purpose.

In Appendix II C statistics relating to the Councils and their meetings have been given for each of the twenty-year periods between 1663 and 1820. The results are both instructive and interesting. There is no indication of any gradual increase in the number of the scientific councillors, or in the average annual number of meetings, or in the number of councillors who attended them. The variation throughout the whole period is very small. During the presidency of Sir Joseph Banks the average number of men of science who were members of Council was lower than during the presidencies of Lord Macclesfield, Lord Morton or Sir John Pringle, but did not differ much from the corresponding figures for earlier periods. The Council meetings were markedly fewer than in the previous period when the Councils had been more active than usual in preparing instructions for expeditions, in revising the Society's statutes, in preparing various reports and other matters; but the average number of councillors who attended the meetings of the Council varied very little from year to year, and hardly at all from decade to decade. It was not a satisfactory state of things that less than half of the twenty-one councillors should on the average have attended Council meetings even though the four officers, the President, the Treasurer and both the Secretaries, were usually present; but that only seven out of the seventeen other councillors could be depended upon to take part in discussing the Society's business suggests that they were not taking their duties seriously in the period preceding Sir Joseph Banks' presidency; while he was President the attendance was even lower.

When the whole period of 157 years from the Anniversary meeting of 1663 to that of 1820 is considered, the activity of the various Councils who administered the affairs of the Society for more than a century and a half do not impress us with any high opinion of their zeal or efficiency. During the last thirty-five years of the seventeenth century it was probably difficult to find more than eight scientific men to serve on the Council year by year, but it is incredible that any such difficulty existed during Sir Joseph Banks' presidency when such Fellows were about four times as numerous in the Society as they had been in 1663. In the two intervening periods the average number of scientific councillors was 9·5 and 9·0, a small improvement, but there is nothing to suggest that at any time during the eighteenth century any attempt was made to increase the number of the scientific councillors and thereby to raise the scientific standard of the whole administration of the Society. The average number during the presidency of Sir Isaac Newton was only 9 though he certainly appreciated the importance of eliminating 'useless' members, a course which he had once advocated.

The suggestion has been made in Chapter IV that from about 1712 to 1715 it might have been possible for him to prevail upon the Society to adopt a policy whereby the number of scientific councillors would be gradually increased up to fourteen, or two-thirds of the total number. But if he had reason to believe that such a change would be strongly opposed by the literary and archaeological group the prudent course may have been to leave it alone. Martin Folkes, who succeeded Sir Hans Sloane as President, would certainly not have favoured any such change as his interests were historical and archaeological, and the same may be said of James West who succeeded Lord Morton in 1768. Soon after the house in Crane Court had been acquired and the number of Fellows was steadily increasing was the moment when something might have been done by so distinguished a President as Sir Isaac, but the opportunity passed and so favourable a one did not recur until a century later.

Some of the anomalies which have been noticed can be understood. If the Councils took but a slight interest in the scientific progress of the Society they would not be surprised at three or even four of the officers' posts being occupied by non-scientific Fellows, as often happened. In the four years of James West's presidency and the first six of those of Banks not one of the officers was a man of science. No signs of any permanent' improvement are noticeable, and those which had taken place between 1741 and 1780 under Lord Macclesfield, Lord Morton and Sir John Pringle soon disappeared, and the average figure during Sir Joseph Banks' presidency differs little from those of the century which preceded it. Whatever individual men of science may have done in the advancement of science by their own researches in the eighteenth century it was a period of comparative stagnation so far as the Councils' activities were concerned, for the councillors were content to accept things as they were rather than to stimulate their colleagues to promote scientific research with all the zeal and energy at their command especially by recommending that the number of men of science among the councillors should be materially increased by the Society. It may be argued that the four officers who were certainly well acquainted with the Society's affairs together with half a dozen of the councillors who were interested would probably deal with the agenda as effectively as a full Council of which the majority were not scientific either by their training or their interests; and this may be so, but when we come to 1821 and the following years when the majority of the Council were men of science, it will be seen that the average attendance at a Council meeting rose quickly to twelve and later to sixteen and even more; and that a considerably larger amount of business was transacted at their meetings than had been usual in the eighteenth century. For the present it is enough to say that the most probable cause of the indifference of councillors to the zealous performance of their duties as the administrators of the Society resulted from the majority of the Fellows on the Councils and also in the Society having little knowledge of or interest in science. Dr A. B. Granville[1] in 1830 criticized the meagre attendance of the members of Council at their meetings since it rarely exceeded half their number; but he did not realize that this had been the general practice ever since the Society had been founded, and was not peculiar to the years when Sir Joseph Banks was President.

THE FELLOWSHIP

During the presidency of Sir Joseph Banks the Fellowship increased from 450 to 640, rather more rapidly than during the preceding forty years; but the proportion of non-scientific to scientific Fellows rose by a small amount only. There was a marked falling off in the number of foreign members, which was the result of the action taken by the Council in 1766 and 1776. The number in this category had reached a maximum of 170 in 1766 and by 1778 had fallen to 120; at the time of Banks' death there were only forty so that their numbers which had been excessive were by then being effectively controlled. Nothing had been done to remedy the predominance of non-scientific Fellows in the Society, which remained a constant source of difficulty by maintaining the large number of Fellows who had no real interest in promoting the aims of the Society, and were inclined, whenever the question of requiring a higher standard of qualifications from candidates was raised, to oppose it or at least to favour inaction. The scientific Fellows were increasing in numbers but only slowly, and it was not until about the third decade of the nineteenth century that their number grew materially.

Before 1847 there was no statute or by-law limiting the number of candidates which might be admitted to the Fellowship in any one year; it is therefore remarkable that the proportion of non-scientific to scientific candidates so elected should have varied as little as it did. The number of ordinary Fellows in each of the seven years 1663, 1671, 1698, 1740, 1770, 1800 and 1830 have been analysed (see Appendix II A) and the non-scientific Fellows in the Society in these years ranged from 67·0 to 71·0 per cent except in 1671 when it was 75·0; it would appear therefore that the

[1] *History of the Royal Society in the Nineteenth Century*, 1836.

councils aimed at keeping the ratio of the two groups of scientific and non-scientific Fellows to approximately what it was at the end of 1663, namely 67·8 to 32·2 per cent or 2·3 to 1·0. The possession of fortune and leisure was considered as being of more importance than a good education or wide knowledge. John Howard, well known as a wealthy man and later as a strong advocate of reforming the prisons of his day, was elected a Fellow in May 1756 before he had begun his philanthropic work. At this time, according to his biographer Aiken, he was unable to speak or write his native language correctly.[1] His name does not occur among the benefactors of the Society as having made either a gift or a bequest to it.

After 1848, when the number of candidates to be recommended for election was limited to fifteen, this ratio fell to 1 to 0·9 in 1860, and before the end of the century practically disappeared. The number of peers who became Fellows under the statute giving them the privilege of special election remained at about 10 per cent of the total number, but hardly any of them took any active interest in the Society or its doings. Charles, the third Earl Stanhope, who was elected a Fellow in 1772, was an exception; he invented two calculating machines, supported the proposal for a unification of weights and measures in 1816 and bequeathed the sum of £500 to the Society in his will.

The scientific Fellows were now beginning to be drawn from a wider field than hitherto; the medical profession still provided the largest number, and in most years there were three and even four of them among the members of Council. It was not until after the statutes governing the election of candidates to the Fellowship had been amended in 1847 that the number of scientific men increased steadily, and after 1860 it always exceeded that of their non-scientific colleagues. Already those of the biological sciences, of whom previously there had been but few, were increasing though they did not yet equal those of the mathematical and physical sciences. The Society was becoming more fully representative of all branches of natural science (see Appendix II B).

FINANCE

No detailed account of the Society's financial position during this period of forty years can be given since few contemporary records relating to it have been preserved. Weld gives a list of the total receipts and of the total expenditure of each of the years 1781 to 1800, and A. B. Granville does the same for those from 1800 to 1829, but they are only quoting the very meagre statements which were presented to the Fellows at each Anniversary meeting in the same form as that which had been employed during the previous century and a half, no details of the income or of the ex-

[1] Communicated by Rt Hon. Lord Justice Sir Frank MacKinnon, P.C.

penditure being given. Granville, when criticizing the wholly inadequate character of this information, lays the blame on the directions contained in the 'ill-digested statutes under which the treasurers had to work'; but there was no reason whatever why the Council should not at any time have amended the statute prescribing the form in which the financial report should be made annually to the Auditing Committee of the Council, or have introduced a new one designed to give the Fellows a full and intelligible report on the financial position of the Society at each Anniversary meeting, but this had never been done. As early as the seventeenth century accounting in business circles was well advanced, and a book dealing with the subject had been published by one Richard Dafforne in 1635 of which three editions had been issued before the end of the century. Early in the eighteenth century an accountant, one Charles Snell of Foster Lane, London, was employed to investigate the transactions of a director of the South Sea Company who had been dealing in its stock, and in carrying out this work he examined the accounts of at least two firms of merchants. In the Society books of account had always been kept though none of them dating from the earlier times are now in existence; ledger accounts of the fees and subscriptions paid by Fellows must have been posted regularly in order that from them the amount of arrears due from those who were in default could readily be ascertained and these were often called for and reported to the Council. Weld in 1846–8 records the income received at certain periods of the eighteenth century, the costs of publication, receipts from rents, etc., which could only have been taken from the registers which, though available to him, are no longer in existence. The information was therefore to be had at the time and it must have been the settled practice of the Council to keep the details of the Society's financial position to themselves as being the authority entrusted with the administration of the Society, reporting to the Fellows only the total amounts received and expended, and the balance remaining in the Treasurer's hands at the end of the year. It was in such a form as this that the Earl of Bedford[1] received the annual statement of his income and expenditure after the year's accounts as prepared by the heads of his various departments had been checked by his auditor; but the earl had also month by month carefully checked and initialled each item of his receipts and expenses so that he had a full knowledge of how his stores and funds were being employed. To what extent the Treasurers of the Society controlled its daily expenditure is not known, nor can it now be ascertained how much freedom of action was allowed by them to the staff. The final stage of the accounts in much the same form as they were usually prepared for an owner of great estates in the seventeenth century was laid before the Anniversary meeting, but the much fuller statement which must have been prepared for and communicated to the Council was never transmitted

[1] G. S. Thomson, *Life in a Noble Household*, 1641–1700. London, 1937.

by it to the main body of the Fellows for their information until 1830, as will be described later. Under such conditions the control which the Auditing Committee could exercise can have had little effect. Such a procedure when it had been in operation for a century and more would not be easy to alter, and the Councils would probably have regarded any attempt by the Fellows to demand a fully detailed statement as being an invasion of their privileges.

James West and Samuel Wegg were the Treasurers during a period of sixty-six years, from 1736 to 1802, and they could certainly have produced annually from the books which were in their charge a statement of the Society's financial position as informative as that prepared for 1830 and the following years by J. W. Lubbock, if the Council had expressed any desire that one should be provided. They should certainly have urged on the Council the desirability of such a statement being prepared and communicated to the Society by whom the funds had been subscribed and to whom they belonged, for they must have been well aware of what the practice in business firms and houses of repute was at the time. It is difficult to avoid the conclusion that the Councils of the eighteenth century were unwilling to acquaint the Fellows with the details of the Society's financial position, preferring to keep the knowledge of it to themselves, and administering the funds as their members saw fit.

At the end of Martin Folkes' presidency in 1753, there was a reserve fund of £3000 which had been invested; this had increased to £11,361 by 1813, and to £16,500 by 1830, though there were then outstanding liabilities to the amount of £2500 to be met; the reserve had been increased therefore by about £140 per annum in the former period and about £240 in the latter.

Though it is not known when the registers, etc. relating to the accounts were lost or destroyed, some may have disappeared when the Society moved from Crane Court to Somerset House in 1778, and again when it moved to Burlington House in 1853. Panizzi, in his report on the library at Somerset House, reported in 1833 that many of the books there had been seriously damaged by damp, so it may be that the archives also had suffered though no mention is made of them. When the land owned by the Society at Acton was being sold in 1882 the would-be purchaser raised the question whether the Society had a legal right to dispose of the land; but the Treasurer of that day was still able to produce evidence which satisfied the Court that the land had not been received by gift or bequest but that it had been bought in 1731 with funds derived from the fees and subscriptions paid to the Society by the Fellows and therefore the Council of that time were within their rights in purchasing the land and in selling it later. It would seem that in 1882 more information may have been available than we have to-day in the minutes of Council of 1731.

From time to time the Council have authorized the destruction of papers, etc. of no value, since, in the office of an institution such as this with operations continuing over many years, and often on a considerable scale, there must always be a constantly accumulating mass of documents which are of only transient interest. These it is both reasonable and desirable to dispose of from time to time, but for all except those well-defined types of documents of merely temporary importance, such as petty cash receipts and others of a similar nature, lists of those which have been passed for destruction by a properly constituted Board of Survey should be prepared and filed for permanent reference. This does not seem to have been done in the eighteenth century, or even at the beginning of the nineteenth century when a committee of the Council criticized the administration of the library and the way in which the archives were being cared for.

All the descriptions which we have of Sir Joseph Banks and his methods of administration, as well as the numberless letters of his which have been preserved, bear witness to his businesslike and orderly habits; all the books and documents which were in his library at his country seat of Revesby in Lincolnshire were entered in a catalogue, and a duplicate of this always accompanied him when he was away from home. It is the more remarkable therefore that with his characteristic zeal for efficient administration he should not have insisted on a more businesslike treatment of the Society's funds and records. The Society had to wait until 1830 for a president who realized the importance of careful accounting and appreciated the Council's duty to keep the Fellows informed of its financial position.

Though the annual totals of receipts and expenditure which Weld and Granville quote from the minutes of the Anniversary meetings are but a poor substitute for the more detailed reports which we should like to have, we can draw a few interesting facts from them. Starting at about £1000 per annum in 1781, the yearly income forty years later had risen to more than £4000 though the Fellowship during the same period had only increased by 180 which would but increase the amount received from this number of annual subscriptions by about £470. Gifts and bequests do not seem to have been either large or numerous, and the £500 bequeathed by Lord Stanhope in 1786 has not been included in the table of annual receipts given below. In the first decade the average sum received from the composition fees amounted to £802 according to Weld, but this increased considerably later when the practice was introduced of exercising pressure on candidates to pay the composition fee of £40 in addition to the admission fee of £10 instead of the annual subscription of £2. 12s. 0d. Such composition fees should have been invested but no rule to this effect was made until 1833, when it was also laid down that the payment of a composition fee in place of the annual subscription fee was purely optional.

While the average annual income rose from £1288 to £2850 (121 per cent) the corresponding figures for the annual expenditure rose from £1262 to £1930 (54 per cent). At the end of 1820 £2000 were invested but no other indications are given of how the rest of the funds received were used. In the ten years 1781–90 the balance carried forward by the Treasurer and reported to the Anniversary meetings was as often a debit as a credit; over the ten years there was but £53 in favour of the Treasurer. In the period 1801–10 his average credit balance was £342; and for the following decade was £600 after investing £2000 to 1820. He gave no explanation of the change by which this improved state of things had been brought about.

Annual Receipts: 1781–1820

Year	£	Year	£	Year	£	Year	£
1781	891	1791	1,516	1801	2,337	1811	2,370
1782	1,101	1792	1,393	1802	1,754	1812	2,578
1783	1,104	1793	1,544	1803	2,089	1813	2,342
1784	1,106	1794	1,809	1804	1,799	1814	2,541
1785	1,168	1795	1,557	1805	2,052	1815	2,301
1786	1,933	1796	1,491	1806	2,342	1816	2,896
1787	1,486	1797	1,917	1807	1,707	1817	2,939
1788	1,631	1798	1,956	1808	1,789	1818	2,006
1789	1,160	1799	1,342	1809	2,094	1819	4,117
1790	1,298	1800	1,652	1810	2,200	1820	4,408
Total	12,878		16,177		20,163		28,498
Average	1,288		1,618		2,016		2,850

Annual Expenditure: 1781–1820

Year	£	Year	£	Year	£	Year	£
1781	727	1791	1,047	1801	2,129	1811	1,622
1782	1,130	1792	1,121	1802	1,601	1812	1,795
1783	958	1793	1,119	1803	1,680	1813	1,726
1784	1,309	1794	1,299	1804	1,585	1814	1,997
1785	1,337	1795	1,486	1805	1,392	1815	1,765
1786	1,914	1796	1,557	1806	2,356	1816	1,949
1787	1,169	1797	1,730	1807	1,483	1817	1,905
1788	1,455	1798	1,697	1808	1,515	1818	2,061
1789	1,300	1799	1,563	1809	1,705	1819	2,359
1790	1,318	1800	1,535	1810	1,416	1820	2,117
Total	12,617		14,154		16,862		19,296
Average	1,262		1,415		1,686		1,930

After the foregoing sketch of the administration during Sir Joseph Banks' exceptionally long tenure of the presidency, the historical occurrences of these forty-two years demand our attention.

The Society had for sixty-five years been in possession of the two houses in Crane Court which Sir Isaac Newton had bought for it in 1710 when Gresham College could no longer provide accommodation adequate for its meetings, its library and its museum. The objections which were raised at first by some of the Fellows to leaving Gresham College had long since been forgotten and Crane Court had served the Society well; but the number of Fellows was now two and a half times what it had been in 1710, and was steadily increasing, so that more ample accommodation would soon become necessary. By this time the Royal Society was generally recognized as being an institution of national importance, and was being frequently asked by the government to draft instructions for the scientific work which should be carried out by such expeditions as those led by Lieut. James Cook and others; it was therefore with considerable hopes of receiving a favourable reply that the Society had been urging on the government its claims for more suitable and more commodious accommodation. On 10 April 1775 Parliament was recommended in a message from the Crown to settle upon Queen Charlotte the house in which she then resided, formerly called Buckingham House but then known as the Queen's House; if this plan was adopted Somerset House which was already settled upon her should be given up and appropriated 'to such uses as shall be found most useful to the public'. The demolition of the old buildings was commenced as soon as an act could be passed to give effect to the royal message and Sir William Chambers, the architect, was appointed to carry out the new building, the street front of which was completed in 1779. In 1776, before Sir John Pringle retired from the presidency he received an intimation from the government that it was their intention to provide the Society with suitable rooms in the new Somerset House which was about to be built.

The government's offer to provide accommodation for the Society in Somerset House was carefully considered by the Council to whom it seemed that the space which was to be allocated was not sufficient to provide room for the Society, its offices, library and museum. They therefore represented this to Sir William Chambers who replied explaining why more ample accommodation could not be provided, but how by certain alterations that portion which it was proposed to place at the disposal of the Society could be made more convenient for its use. Those suggestions were very carefully considered and finally the Council came to the conclusion that the advantages of the government's offer which provided a fine building, convenient proximity to other scientific societies, a favourable position and remission of any charge for rent outweighed such objections as there were. The offer was therefore accepted, and to facilitate the allocation of the space available, the Council in 1779 offered the Society's collections to the trustees of the British Museum. The Society's repository had been for about a century the most important and largest institution of its kind in London, but since Sir Hans Sloane's collections had been added to those already in the possession of the nation the collections now assembled at Bloomsbury far surpassed those of the Society. It was therefore a wise and graceful act on the part of the Society to offer its collections to the trustees of the British Museum to be incorporated with those in their charge. In June 1781 the trustees replied accepting the collection and returning their thanks to the President, Council and Fellows for their 'considerable donation'. A committee had been appointed to confer with Sir William Chambers on the work which had to be done to make the Society's portion of Somerset House convenient for its requirements, and so expeditiously was their work carried out that the Anniversary meeting of 1780 was held in the new rooms at Somerset House, in many respects a more suitable home for it than the house at Crane Court which had however served the Society well for seventy years. That house and the smaller house adjoining it were sold in 1782 for £1000.

A certain amount of friction was to be expected when a young and energetic president assumed the direction of the Society's affairs. He fully realized the importance of the institution over which he now presided and such experience as he had gained as a councillor during three years showed him that the regulations governing the admission to the Fellowship had been relaxed to an undesirable extent. Soon after his election Banks had announced to the Secretaries and members of Council his determination to watch over the admission and the election by ballot. Previous to an election he gave his opinion freely on the merits of the candidates put forward, and, according to Lord Brougham, had laid down two principles for his own guidance: first, that any person who had successfully cultivated science, especially by original investigations, should be admitted whatever might be his rank or fortune; secondly, that men of wealth or station disposed to promote, adorn and patronize science, should with due caution and deliberation be allowed to enter. It was only to be expected that such views should be construed as an attempt to control the Fellows' rights to elect whom they would as granted by the Charter, but if we examine the number of candidates elected the result gives no support to this allegation; the average number of Fellows elected in the years 1771 to 1780 was 23; and the corresponding figure for the following decade was 19·7. The number in individual years varied a good deal, being from 15 to 29 in the first decade and from 13 to 29 in the second. The number of foreign members elected was 27 and 26 in the two decades respectively.

An attempt to introduce a higher efficiency into any administration always produces some discontent and usually a crop of unfounded allegations, but these in most cases die down as those concerned become accustomed to a better organized system; and this would probably have been the Society's experience also had it not been for an unfortunate occurrence which took place in 1783. It is very fully described by Weld in his *History of the Society* (vol. II, ch. VI) from which the following account is mainly taken.

It will be remembered that the post of Foreign Secretary was not established until 1823; before that time a Fellow, usually not a member of Council, was appointed by the Council to deal with the Society's foreign correspondence and was called Assistant to the Secretaries. P. H. Maty had carried out these duties from February 1772 until the end of June 1774; J. Planta replaced him until 14 January 1779 and Dr Charles Hutton then took them over. Maty was not a member of Council in 1772–4; Planta was a councillor from the time that he became a Secretary in November 1778; and Hutton was one from 30 November 1778 until 30 November 1780 when he was not re-elected to the new Council. There was nothing invidious in his omission from the Council since by the Charter ten out of the twenty-one members of the Council have to be replaced each year; at this time Dr Maskelyne, the Astronomer Royal, was usually but not continuously one of its members, but from 1785 to 1811 he was however always a member. During the two years when Dr Hutton was a member of Council he had been present at twenty-four out of the twenty-eight meetings which were held, so he had been very regular in his attendance. At the meeting of Council on 24 January 1782 it was reported that the foreign correspondence had not been carried out with sufficient punctuality and that the remuneration of the officer charged with the execution of it was not adequate to his duties; the President therefore proposed that the duty of translating papers and extracting from foreign books should be dispensed with. A committee consisting of the President, Mr Pownall, an antiquary, Dr Blagden, M.D., and the two Secretaries, P. H. Maty and J. Planta, was appointed to define his duties. They recommended that these duties should be: to notify foreign members of their election; to return thanks to foreigners for presents; to reply to all letters from foreigners and to write such letters to foreigners as the President or the Council may direct. Dr Hutton attended a meeting of the Council on 25 April 1782, agreed to the duties proposed and proceeded to carry them out. No other entry occurs in the minutes until 20 November 1783 when it was resolved (Dr Maskelyne and P. H. Maty dissenting) that the business of the Foreign Secretary should be done by a person constantly resident in London; this was approved by Council at their meeting on 17 December, when Dr Hutton's resignation was reported. This caused great dissatisfaction, and at a meeting of the Society on 11 December 1783 Mr Poore moved that the thanks of the Society be given to Dr Hutton for his services as Secretary of Foreign Correspondence. This was opposed by the President who maintained that only the officers and the members of Council could know if the duties had been performed efficiently or not; and that he, who did know, was of opinion that the duties had been neglected. A motion that the question of returning thanks of the Society to Dr Hutton having been put to the vote was passed by a majority of five, and the President returned thanks to Dr Hutton.

At the next Ordinary meeting on 18 December Dr Hutton put in a written defence of his action, showing that he had dealt with all the letters which the Clerk had passed to him; that he had attended about 75 per cent of the Ordinary meetings since he had ceased to be a member of the Council; and that he had incurred much expense by doing so, since he had to stay at least one night in London on each of these occasions. The Society accepted his explanation as being satisfactory by a majority of thirty, forty-five out of the sixty Fellows who were present voting in favour of it.

The matter might well have ended here but during the discussion Dr S. Horsley, who had been a Secretary from 1773 to 1778, spoke with much violence against the President and his administration of the Society; Dr N. Maskelyne and Mr Maseres supported this view. In response to a special memorandum which the President sent out notifying that matters of importance would be discussed at the next Ordinary meeting, one hundred and seventy Fellows attended it early in January 1784; Mr Anguish then moved that this Society do approve of Mr J. Banks as their President, and mean to support him in that office; Hon. Henry Cavendish seconded this. The opposition included Mr Poore, Mr Maseres, Dr N. Maskelyne, Mr P. H. Maty, Dr S. Horsley, Dr Glenie and Mr Watson. After a stormy discussion the motion was carried by a majority of 77 on a vote of 161. At the next Ordinary meeting a motion that the resolution of Council requiring the Foreign Secretary to reside in London be rescinded and for Dr Hutton to be requested to resume his duties was negatived by a majority of 38.

On 26 February two notices were brought forward by Dr Horsley and eight of his friends which implied that the President interfered in the elections of councillors and of candidates for the Fellowship; these were debated with much heat. The first was negatived by a majority of 88 out of 142, and the second, after being amended, by 79 out of 125. This ended the meeting. A month later P. H. Maty offered an anonymous pamphlet entitled 'An authentic Narrative of the Dissensions and Debates in the Royal Society'. The President then said that as this pamphlet contained many misrepresentations and statements discrediting the Society he would not propose the usual thanks to the donor unless a majority of the meeting desired it. Dr Horsley then proposed that a vote of thanks to the donor be accorded but only the seconder, Dr Hutton, and P. H. Maty supported it, while several Fellows expressed the wish that the motion should be withdrawn. Mr Maty then resigned his office of Secretary. On 5 May the Society met to elect a new Secretary; Dr C. Blagden, M.D., having declared himself a candidate and being supported by the President was elected by a majority of 100 out of 139 votes cast.

The view seems to have been widely held that Dr Horsley was aiming at ousting Sir J. Banks in order that he might replace him as President, and as this became more generally known and they heard his intemperate speeches, the general opinion of the Fellows hardened against him. It was the calm and judicial attitude taken by the Society generally, as well as its refusal to be rushed into a rash decision, which allayed the dissatisfaction that had been aroused, and enabled the normal working of the Society to be resumed. Lord Brougham[1] severely criticizes his conduct and dilates on his unfitness for such a post. Dr Kippis too[2] criticizes Dr Horsley's manner and violence in debate. Banks is said to have shown neither jealousy nor prejudice after the settlement of the dispute, and at his address at the following Anniversary meeting passed over it lightly. The accounts of this unfortunate affair, which have been preserved in the Minute-books and Journals of the Society, lead us to the conclusion that the Council had not dealt with it wisely or tactfully. In the first place they treated Dr Hutton with scanty consideration and even with injustice; and secondly, they gave to Dr Horsley the opportunity for which he had been waiting of attacking the President and his administration of the Society's affairs. Dr Hutton was an eminent mathematician who had discussed the observations which the Astronomer Royal had made at Schiehallion in 1774 to determine the deflection of that mass on a plumb line in its neighbourhood; his results had been published in volume 68 of the *Philosophical Transactions*. He had also been awarded the Copley Medal in 1778 for his researches in ballistics. He was well known to the officers and to several of the Council since he had been a member of it in 1779 and 1780, and had then attended most regularly. At the Council meeting of 24 January 1782 when the duties of the Fellow charged with the Society's foreign correspondence first came under discussion the President proposed that the duties should be lightened, and a committee appointed to consider the details agreed to this. The Council's decision was communicated to Dr Hutton on 25 April, who accepted it. No further reference to this appears in the minutes until more than eighteen months later when, on 20 November 1783, the Council decided that whoever was charged with the foreign correspondence should live in London, a condition which was impossible for Dr Hutton to fulfil since he was Professor of Mathematics at the Artillery College at Woolwich. The Council do not appear to have given him any opportunity of replying to the complaints of his work which had been made, and when he was able to do this before an Ordinary meeting the Society completely exonerated him from all blame. On this occasion the officers did not attempt to disprove his statements nor did they bring forward any specific instances of neglect of duty. The President's sympathetic attitude in January 1782 is in marked contrast to the opinion which he expressed

[1] *Men of Letters and Science*, II, p. 370.
[2] *Observations on the late Contests in the Royal Society.*

to the Society on 11 December 1783 to the effect that Dr Hutton had neglected his duties, when he may possibly have felt it incumbent on him to support the decision reached by his colleagues on the Council.

Dr C. Blagden, M.D., was elected Secretary in May 1784 in the place of P. H. Maty who had resigned; and on 17 June 1784 Rev. C. P. Layard was appointed to take charge of the foreign correspondence, but without a seat on the Council; this duty he carried out until the end of 1803. The inconvenience caused by the foreign correspondence not being dealt with by a member of Council was not removed until about forty years later when at Sir Humphry Davy's suggestion the Council in 1823 appointed a Secretary for Foreign Correspondence who should rank with the other two Secretaries and like them should be elected annually at the Anniversary meeting.

About 1783 Cassini, who had been directing the trigonometrical survey of France, moved the Académie des Sciences in Paris to approach the Royal Society to obtain its support to a project for connecting Greenwich Observatory by triangulation with the Paris Observatory in order to determine the difference between the longitudes of the two institutions. The Council readily agreed and early in 1784 submitted a petition to the king for the necessary funds which were estimated at £3000; it was favourably received and on 24 June the President was able to report that the undertaking had been approved, the grant authorized, and Major General Roy, who had been a Fellow of the Society since 1767, had been appointed to carry out the work under the direction of the President and Council of the Royal Society.

J. Ramsden, the well-known instrument maker, was directed to construct a theodolite of the highest class with a divided circle of 36 inches in diameter, but though the instrument had been ordered early in 1784 it was not delivered until the summer of 1787; Ramsden was well-known for the excellence of his work, but he was sadly unpunctual in fulfilling his engagements and General Roy complained bitterly of the loss of time caused by these dilatory habits. In the meantime a base had been measured on Hounslow Heath, an account of the work being published in the *Philosophical Transactions* for 1785. The triangulation was begun in August 1787 and completed by the end of the year. A base of verification was measured on Romney Marsh as a control of the accuracy of the work. The Copley Medal was awarded to General Roy in 1785 in recognition of his services to science.

In 1785 the President at the desire of the Council and on an application by Sir William Herschel submitted to the king a scheme for the construction of a large reflecting telescope, and prayed for His Majesty's assistance in providing it. This he not only agreed to do but provided the whole of the necessary sum, which amounted to £4000, from funds at his own disposal. It was completed, mounted and brought into use by August 1789.

In 1791 a triangulation of a similar character to that carried out by the Society in 1787 was put in hand for the whole country under the direction of the Master General of Ordnance in order to provide a control for the complete mapping of the whole country; this however was a State under-taking for which the Royal Society had no responsibility.

About 1783 James Edward Smith came to London as a young student of medicine, and when breakfasting with Sir J. Banks he learned that the whole collection of books, manuscripts and natural-history specimens of Linnaeus had been offered to him for the sum of a thousand guineas but that he (Banks) intended to decline it. Smith thereupon decided to purchase it himself and with his father's assistance acquired the whole collection which arrived in London in 1784. On his death the collection was purchased by the Linnean Society, in whose possession it still is. This Society was founded in 1788 by Sir James Smith, Sir Joseph Banks, the Bishop of Carlisle (Dr Goodenough) and other Fellows of the Royal Society, and was the first specialized scientific society to be established. In this case no objection was raised by Sir Joseph Banks to its foundation on the ground that it was likely to encroach on a field of knowledge hitherto within the activity of the Royal Society; it may perhaps have been fortunate that its aims were similar to those in which the President of the Royal Society was himself keenly interested. It was not long before the new Society was able to demonstrate that there was a real need for its activities; for during the first twenty years of its existence nineteen Fellows were elected annually on the average, in the next decade twenty-three, and between 1820–30 thirty-two, by which time it was well established as an important scientific society; by 1800 its Fellows numbered 209 and by 1820 there were 463.[1]

Botany and zoology had for many years been two of the branches of science which the Society had promoted, and in which its Fellows had made notable advances. At this time the President apparently did not anticipate that the new society would compete with the Royal Society or encroach upon its field of activity, though Granville says in 1830 that for some years past the Royal Society had received no botanical communications of importance, all these having been sent to the Linnean Society.

A few years later however, when other groups of scientific men desired to take similar action in fields of science which interested them, his policy changed and he opposed their foundation as independent institutions. According to Francis Baily, the astronomer, who was one of the first secretaries of the Astronomical Society, Sir J. Banks at first opposed the formation of the Royal Institution. In 1820 when the Astronomical Society was

---

[1] For this information I am indebted to the Secretary of the Linnean Society.

being formed, he wrote: 'A similar attack was made by Sir Jos. Banks on the Astronomer Royal, who, if the report be true, made a very spirited reply. As a similar, and indeed a more violent attack was made at the establishment of the Geological Society, and also of the Royal Institution, and which only tended to unite more firmly the original members, we hope that a similar result will be produced here.' No record of such difference of opinion is to be found at the Royal Society or at the Royal Institution, so by some means or other the dispute seems to have been settled to the satisfaction of both parties, and the respective fields of action of the two bodies were so defined as not to interfere with one another, for on 9 March 1799 a meeting was held at the house of Sir Joseph Banks in Soho Square for the purpose of founding the Royal Institution, at which were present: Sir Joseph Banks, the Earl of Morton, Earl Spencer, Count Rumford, Richard Clark and Thomas Bernard. Its title and purposes as then laid down were:

Institution: For diffusing the knowledge and facilitating the general intro-duction of useful mechanical inventions and improvements; and for teaching, by courses of philosophical lectures and experiments, the application of science to the common purposes of life.

The Institution was incorporated on 13 January 1800, and has ever since maintained close relations with the Royal Society, many of whose Fellows have always been among its members.

In 1800 the Council appointed a committee to consider whether the amounts paid to the Secretaries and to the Clerk were adequate in view of the increased cost of all necessaries of life at this time, and on their re-commendation the Council raised the honorarium paid to each of the Secretaries to £105, and the salary of the Clerk to £280.

The formation of the Linnean Society had shown that there was already a real need for additional scientific societies at which those who were interested could meet and discuss matters of common interest more fully than could be done at the weekly meetings of the Royal Society; also a large number of communications dealing with these younger and growing branches of science were now being contributed. In 1807 Dr William Babington, M.D., a Fellow of the Royal Society, invited to his house a number of those who were interested in mineralogy and might therefore be disposed to subscribe to a fund for publishing Count Bournou's monograph on carbonate of lime. Other meetings followed and among those who were most active in working for the formation of a Geological Society was Sir Abraham Hume. In 1809 Hon. Charles Greville, a coun-cillor and Fellow of the Royal Society, proposed that the Geological Society should become an assistant association of the Royal Society in which the members, who were also Fellows of the Royal Society, should conduct the affairs of the new society. This was supported by Sir Joseph

Banks, Sir Humphry Davy and others, but it was definitely rejected by a special General meeting of the Geological Society held on 10 March 1809, since the members preferred to develop their society in such a manner as might best suit their own purposes.

In this same year some Fellows of the Royal Society who were interested in promoting the study of Chemistry and Physiology, established a Society for the promotion of Animal Chemistry, and requested the President of the Royal Society that their body should be constituted as Assistant Society affiliated to the Royal Society. It seems doubtful if this could be done under the Charter but the Council approved the scheme and the President spoke warmly in favour of the arrangement at the Anniversary meeting. Sir Everard Home was the most active member in bringing about its foundation, but when he ceased to support it a few years later the society came to an end.

The difference of Sir Joseph Banks' attitude to the Society of Animal Chemistry and to the Geological Society is instructive. So long as the new bodies were prepared to be definitely subordinate to the Royal Society and to be in effect controlled or administered by its Council he had no objection to them; but if they demanded an independent existence and complete freedom in the management of their affairs he opposed them. He took a similar line in opposing the foundation of the Royal Astronomical Society in 1820, but he was then an old man and in very poor health, so he cannot be considered as then expressing a considered opinion on the matter but rather as following a policy which he had supported for some time past and from which he was not prepared to depart.

In 1781 the king had conferred a baronetcy on Sir Joseph Banks, and in 1795 created him a Knight of the Order of the Bath in recognition of his numerous public activities. Such recognition was quite exceptional at the time and he was the first President of the Society to be so honoured since Sir Hans Sloane.

During the eighteenth century the gifts and bequests which the Society had received had not been numerous, but in 1794 Sir Clifton Wintringham, a physician and physiologist of considerable distinction resident in York, bequeathed to the Society the sum of £1200 subject to a life interest to his wife. His wife died intestate in January 1805, and letters of administration were taken out by Mr Clifton Wintringham Loscombe, by whom the bequest should have been then paid over to the Society. This he did not do, and it was not until Mr Loscombe's affairs were being settled by his solicitors that the Society was asked if the bequest had ever reached them. The matter was then put into the hands of the Society's solicitors, but as a claim was advanced the Governors of the Foundling Hospital set up a counter claim on the grounds that the dividends not having been applied according to the terms of the bequest this necessarily passed to the hospital. A lawsuit to decide this point was eventually settled by a compromise by which the £1200 Consols should be transferred to the Society as well as six years' dividends, and that out of these and the accruing dividends £100 and the costs of the trial should be met. This compromise was accepted by the Council and was sanctioned by the Master of the Rolls in 1842. It was then found that the conditions of the will were so stringent, and involved so much expense, that it was practically impossible to carry them out even when the rate of interest paid on Consols was 3 per cent. As there was a provision in the will that in the case of failure on the part of the Society to fulfil the intentions of the testator the income of the fund should be paid over to the Governors of the Foundling Hospital, that institution has received in each year the interest accruing from the fund; this is now about £34 which is paid over to the hospital yearly.

In July 1796, Benjamin Thompson, who had been elected a Fellow in 1779 and later made Count von Rumford in 1784 when he was in the service of the Elector of Bavaria, wrote to the President of the Royal Society to inform him that he had purchased and transferred £1000 stock in the funds of this country to the Society. He wished that 'the interest of it should be given once every second year as a premium to the author of the most important discovery or useful improvement which shall be made or published by printing, or in any way made known to the public in any part of Europe during the preceding two years on heat or on light; the preference always being given to such discoveries as shall, in the opinion of the President and Council, tend most to promote the good of mankind'. The premium was to take the form of two medals, the one of gold and the other of silver, to be together of the value of two years' interest on the £1000 or £60 sterling. In the case of there being no new discovery in heat or light during any term of years which, in the opinion of the President and Council, 'is of sufficient importance to deserve the premium, direction is given to invest its value in the purchase of additional stock in the English Funds, and the interest of this additional capital is to be given in money with the two medals at each succeeding adjudication'. In a subsequent letter Count Rumford suggested 'that the premium should be limited to new discoveries tending to improve the theories of fire, of heat, of light and of colour, and to new inventions and contrivances by which the generation and presentation and management of heat and light may be facilitated'. Chemical discoveries and improvements in optics, so far as they answer any of these conditions, are to be within the limits of the award, but the count wished to encourage especially such practical improvements in the management of heat and light as tend directly and powerfully to increase the enjoyments and comforts of life, especially in the lower and more numerous classes of society. The first recipient of the medals was Count von Rumford himself. Before 1846 it not infrequently happened that no medal was awarded for some years; and between 1818 and 1832 the only recipient was M. Fresnel. Consequently the value of the invested funds have increased and they now amount to about £1800. The first dies were engraved by John Milton, assistant engraver at the Royal Mint.

In 1801, at the request of the government, the Council appointed a committee to enquire into the causes of an explosion at the gas works which had just been established in Westminster; the members were: Sir J. Banks, Sir C. Blagden, Colonel Congreve, Mr Lawson, Mr Rennie, C.E., Dr W. H. Wollaston and Dr T. Young. They produced a full report which is to be found in the ninth volume of the Council minutes, and recorded their opinion that if gas lighting was to become general the works in which it was produced should be placed at a distance from other buildings, and that the gasometers for storing it should be small and numerous. Later, as technical knowledge of the industry grew and experience was gained, it was seen that the apprehensions of the committee were exaggerated.

Dr T. Young, who was one of the members of this committee, was a man of great ability not only in physical science but in medicine also; he also attained an exceptional knowledge of the language of ancient Egypt and in deciphering the hierographic and demotic scripts, in which he was one of the earliest pioneers. He was elected a Fellow of the Society in 1794 at the age of twenty-one after he had communicated a paper in which he attributed the accommodating power of the eyes to a muscular structure acting on the crystalline lens. In 1804 he was entrusted with the Society's foreign correspondence, becoming a member of Council two years later. In 1823, when the statutes were revised, he became the first Secretary for Foreign Correspondence and continued to hold the post until his death in 1829. The results of his researches in physical optics were communicated to the Society as the Bakerian Lecture in the years 1800, 1801 and 1803, which dealt with the undulatory theory of light. In 1818 he was appointed Secretary to the Board of Longitude, an appointment which was criticized by Professor Babbage and others who considered that the post would be more fittingly held by an astronomer.

In 1811 Dr Nevil Maskelyne, the Astronomer Royal, died and was succeeded by William Pond who served as a member of Council for the next thirty years. There had been no ruling to the effect that the Astronomer Royal should be an *ex officio* member of Council but it had been the practice to re-elect Dr Maskelyne annually for many years and the custom was continued throughout Pond's tenure of the post. Maskelyne's death gave to the Council as the Board of Visitors an opportunity of approaching the Treasury on the subject of the upkeep of Greenwich Observatory. The principal instrument designed by the late Astronomer Royal was not yet complete and for its effective use one additional assistant at least would be indispensable; the cost of the observatory's administration had increased, and the salary of the Astronomer Royal was wholly inadequate to attract an astronomer competent to take charge of the observatory unless he had considerable private means of his own, as had been the case with Dr Maskelyne. The salary of the post had indeed been increased recently, but with more instruments and longer hours of observing the cost of heating, lighting and other incidental charges had also grown, and these were charged to the Astronomer Royal. The Council therefore addressed a memorial to the Lords of the Treasury setting forth the financial position of the post of the Astronomer Royal, and of the observatory in general. They urged that an additional assistant should be given to the Astronomer Royal and that the cost of heating, lighting and other incidental charges should be met by the State.

As has been related in Chapter v, the Council of the Royal Society and the Académie des Sciences of Paris had in 1742 and 1743 undertaken a comparison between the standard yard and pound troy of this country and the half-toise and marc weight which were in the possession of the Académie. This was carried out with scrupulous care, and after the verification of them examples were exchanged between the two institutions in order that measurements made for scientific purposes in one country might be accurately expressed in terms of the weights and measures of the other. One important result of this investigation was that the old so-called standards of this country were found to be so roughly made and their divisions so crudely marked that no great confidence could be placed in them. The matter was of very great national importance but no action was taken in Parliament until 1814. In the meantime Sir George Shuckburgh-Evelyn, Bt., took up the question of accurate standards of measurements on account of the importance of the subject to men of science in their researches, and has discussed the matter fully in a paper which was published in the *Philosophical Transactions* of 1798. Having obtained from Edward Troughton, the instrument maker, a five-foot scale of brass he compared it with various other measures of accepted or reputed authority. This scale, which is known to metrologists as Sir George Shuckburgh's scale, afterwards came into the possession of Hon. Charles Cecil Jenkinson (afterwards the Earl of Liverpool) who gave it to the Royal Society in 1828. A few years ago the Society presented it to the nation and it is now preserved at the Science Museum in South Kensington in the metrological collection there. The results which he obtained in his comparisons in terms of 36 inches on his five-foot scale were:

	Inches	Difference
Standard yard, Henry VII, 1490	35·96700	− ·03300
Standard yard, Elizabeth, 1588	35·99300	− ·00700
Royal Society's Scale, marked E	36·00130	+ ·00130
Royal Society's Scale, marked EXCH	35·99330	− ·00670
Bird's Standard, 1758	36·00023	+ ·00023
General Roy's Standard, First Yard	35·99990	− ·00010

In 1814 the question of standard weights and measures was taken up by the House of Commons and on 10 May a committee of the House was appointed consisting of twenty-three members including the law officers and Mr Davies Giddy, F.R.S. On 1 July they reported firstly, that the brass rod made by John Bird in 1758, currently known as 'Bird's standard', should be called a yard though no new comparison of it with other standards had been made, and secondly, that the length of a pendulum vibrating seconds of time in the latitude of London had been ascertained to be 39·13047 inches of which Bird's standard contained 36. The scientific witnesses examined by the committee were Professor L. Playfair and Dr W. H. Wollaston.

In the following year it was resolved in a committee of the whole House that a bill should be brought in for establishing and preserving a uniform system of weights and measures and that Mr D. Giddy and two other members should prepare a bill for this purpose. On 15 March 1816 the House of Commons resolved that

A humble Address be presented to H.R.H. The Prince Regent that he will be graciously pleased to give direction for ascertaining the length of the pendulum vibrating seconds of time in the latitude of London, as compared with the Standard Measure in possession of this House; and for determining the length of the said pendulum at the principal stations of the Trigonometrical Survey extended through Great Britain, and also for comparing the said Standard Measure with the ten-millionth part of the Quadrant of the Meridian, now used as the basis of linear measure on the Continent of Europe.

In 1816 the Council of the Society received a letter from the Secretary of State requesting them to afford all the assistance in their power to ascertain by experiments the length of a pendulum vibrating seconds of time in the latitude of London and at different stations of the Trigono-metrical Survey. This brought the questions of standards of length and of geodetic measurements officially before the Society in continuation of what had been done by some of its Fellows in 1742. The Council resolved to take every measure in their power to comply with the request of the Secretary of State for the Home Department, and appointed a committee consisting of Sir Joseph Banks, Dr W. H. Wollaston, T. W. Combe (Secretary), Mr D. Giddy, Mr J. Pond, the Astronomer Royal, Mr E. Troughton, Captain H. Kater, R.E., Sir Henry Englefield, Mr Henry Browne and Colonel W. Mudge, R.A.; Sir John Barrow, F.R.S., and Sir Humphry Davy, Bt., F.R.S., were added to the committee later.

In order to comply with the request of the Home Secretary, Captain Kater employed for his observations a bar of plate brass $1\frac{1}{2}$ inches wide and $\frac{1}{8}$ inch thick. To this were attached knife-edges 39·4 inches apart which were made of Indian wootz steel; these knife-edges were hardened and rested on agate planes. The experiments were made in the house of

H. Browne, Esq., a member of the committee, in Portland Place, London, W. I. They were carried out in June and July 1817, every precaution to guard against changes of temperature being taken and micrometer microscopes being used to read off the distance between the knife-edges. Captain Kater's results gave, when the necessary corrections had been applied, the length of a pendulum vibrating seconds in a vacuum at this station—situated in 51°, 31′, 8″·4 north latitude—in terms of

Sir G. Shuckburgh's Standard	39·13860 inches
General Roy's Scale	39·13717 ,,
Bird's Parliamentary Standard	39·13842 ,,

(*Phil. Trans.* for 1818)

The Society awarded the Copley Medal for 1817 to Captain H. Kater for his work.

The Commission, with Captain Kater's observations before them, recommended in their report of 24 June 1819 that General Roy's Scale should be made the legal Standard Yard, but on 13 July 1820 they altered their opinion and decided to recommend that J. Bird's Parliamentary Standard should be adopted in preference. Nothing further was done however until 1819, when the prince regent at the request of the House of Commons appointed a Commission of Inquiry to go into the whole subject. So far, action had lain with Parliament but the Commission which the prince regent appointed brought the matter to the notice of the Royal Society since five of the six commissioners, Sir Joseph Banks, Davies Gilbert,[1] Dr W. H. Wollaston (chemist), Dr T. Young (physicist) and Captain H. Kater, R.E., were Fellows of the Society; the seventh, Sir George Clark, had just been appointed a Lord of the Admiralty.

The destruction of the Houses of Parliament by fire in 1834 involved the loss of the standard yard. A Commission was appointed to consider the steps to be taken for the restoration of the standards of weights and measures, the members of which were all Fellows of the Royal Society. They consisted of Mr Airy, Mr Baily, Mr Bethune, Mr Davies Gilbert, Sir J. F. W. Herschel, Mr Lefevre, Sir John Lubbock, the Dean of Ely and Rev. R. Sheepshanks.[2]

In 1817 Sir Joseph Banks wrote to Viscount Melville, then First Lord of the Admiralty, to urge on the Admiralty that the time appeared to be exceptionally favourable for an Arctic expedition, and an attempt to make the North-West Passage since whalers in these seas reported that they were exceptionally free from ice. In December the Admiralty replied that it had been decided to fit out four ships which were to sail at the beginning of March 1818, and that their Lordships would be happy to avail themselves of the assistance of the Society in preparing detailed instructions for

---

[1] He had changed his name from Giddy to Gilbert in 1817 on his marriage.

[2] Weld, *History of the Royal Society*, II, p. 267.

the guidance of the officers on board the vessels. The Council referred this request to the Pendulum Committee who prepared a list of the necessary instruments as well as the instructions for their use. They also recommended that Lieut.-Colonel E. Sabine, F.R.S., should accompany the North-West Passage expedition, and that Mr Fisher should accompany the Polar expedition. The ships returned at the end of October 1818 without having accomplished what they had hoped though the scientific observations were of value. The Admiralty however sent out two other ships in the following May, and Sabine, on the recommendation of the Council, was allowed to accompany them as astronomer and magnetic observer.

During the last twelve years of his life Sir Joseph Banks' health frequently left much to be desired; he had much difficulty in walking and often had to be carried from his carriage to the committee room where his work required his presence. Attacks of gout which became more frequent as the years went by troubled him and made it very painful for him to attend the Council and Ordinary meetings of the Society; nevertheless he continued to be present at them with his customary regularity. In September 1817 he wrote to his friend Sir Everard Home to say that he was starting on a visit to Revesby; he would not travel more than forty miles a day and would stay at the houses of friends for the three nights which he would spend on the road. He returned to his town house at Spring Grove in time for the first autumn meeting of the Society and of the Royal Society Club. In the following August he was in his carriage when it was overturned and though he received no actual injury the shock affected him considerably. Nevertheless he continued his work and attended ten out of twelve Council meetings in 1817–18 and eight out of nine in the following year. Early in 1820 he was present at two out of the first four Council meetings which were held, but on 18 May Sir Everard Home informed the members of Council that Banks felt himself to be unable to carry on the duties of President and therefore desired to resign that honourable office. The Council unanimously desired him not to withdraw from the Chair, and he therefore withdrew his resignation.

A few days later, when a deputation of the Council waited upon King George IV with the Charter-book in order to obtain His Majesty's signature as Patron of the Society, the king congratulated the deputation on Sir Joseph Banks' determination to retain the office of President. On 19 June however he died.

On 29 June an Extraordinary meeting of the Council was held, Davies Gilbert the Treasurer being in the Chair. He moved, and Captain Kater seconded the proposal, that Dr W. H. Wollaston should be appointed President and this was carried. Dr Wollaston therefore became President until the next Anniversary meeting, when the officers and councillors for the coming twelvemonth would be recommended by the Council to the Anniversary meeting for election.

At the next meeting of Council on 7 July it was resolved that a monument should be erected to the memory of the late President, the cost being met by subscriptions from the Fellows of the Society.

For many years Sir Joseph Banks had occupied a very exceptional position in the country; he had identified himself with no political party, he had declined invitations to stand for a seat in Parliament, and would take no part in political disputes. When he was elected a Fellow of the Society he was very wealthy and was giving his attention to the development and improvement of his estate at Revesby in Lincolnshire; he had been so fortunate as to have made three voyages to explore distant parts of the globe and had thereby greatly developed his knowledge and his ideas for British colonial expansion. His election to the presidency of the Royal Society came to him at a time when he was in a position to work for it to the best advantage, and he himself said that of all the distinctions which came to him this was the one which gave him the greatest pleasure and satisfaction. His interests were very wide and he was free to develop them as he might desire. These lay primarily in the field of natural history, and especially in botany, horticulture, agriculture and the raising of live stock in this country and its oversea possessions in order to increase their productivity and the welfare of their inhabitants. He had never devoted much attention to mathematics, astronomy, physics or the related branches of science, and this formed one of the complaints made against him by Dr S. Horsley in 1784. Nevertheless no evidence was ever produced of his having unfairly discriminated against those who were interested in these subjects. His three years of service as a member of Council must have made him familiar with the aspects of the Society's administration which were sorely in need of reorganization, and he gave notice in Council soon after his election that he would oppose the selection of candidates whom he considered to be inadequately qualified for the Fellowship. It may be thought that he might have done more when we see in the next chapter what a determined and talented body of reformers were able to accomplish within a generation after Sir Joseph's death; but it must be remembered that only two years before his election the Councils of 1774 and 1775, of which he was a member, had discussed the statutes and in 1776 had presented the results at which they had arrived to the Society which had accepted them; in them the limitation of the number of Fellows, the introduction of some kind of selection of them, and the introduction of a more efficient control were not touched upon, as though the non-reforming majority in the Society and on its Councils were too strong to make their adoption practicable. Banks may well have made up his mind that a conflict on so wide an issue, on which the Society had so recently expressed its opinion, would be neither wise nor conclusive and therefore decided to take no action.

As its President he spared no efforts to do all that he could for the Society; his house in Soho Square and his evening receptions were open to every man of learning, serious student and traveller of note; and he entertained many of them at the dining club of the Royal Philosophers. But it has to be remembered that from the first he had many calls upon his time besides those of his estate in Lincolnshire and the collections which he had amassed during his travels between 1766 and the end of 1772. As President of the Royal Society he was chairman of the Board of Longitude and also of the Board of Visitors of the Royal Observatory, Greenwich; he was also *ex officio* a trustee of the British Museum; all these he attended regularly and took a keen interest in the business. His wider activities included a keen and active interest in the colonization of Australia, which dated from his visit to the Southern Pacific with Lieut. James Cook between 1768 and 1771. His botanical and horticultural interests led to a long and intimate friendship with the king and to the development of the Royal Gardens at Kew. The acquisition of rare and curious flowers and the introduction and acclimatization of economically useful plants for these gardens, and for this country and its colonies, became one of Banks' most useful activities. He saw to it that collectors were trained and sent with suitable equipment to various foreign lands, and his generosity as a patron of these important explorations resulted in the introduction of several thousand varieties of plants being brought to this country. He was also a leading supporter of the African Society and contributed generously to the cost of giving Burckhardt and others training in the oriental customs and languages which were indispensable for the successful execution of their proposed explorations. At home too the government made frequent calls upon him for advice and assistance, and after he had been made a Privy Councillor he was often appointed a member of the Privy Council's committees. With all these occupations on his hands, and much of the responsibility for their successful execution left to him, he may well have found it impossible to devote more time than he did to internal affairs of the Society. There were a Treasurer and two Secretaries to supervise them, all of whom had had some years' experience in their duties, and the sums involved were not large, needing only reasonable care and methodical control, which he was entitled to expect his officers to exercise; this may have influenced his decision to make no bequest to the Society.

In his later years he seems to have realized that the time was near when the scientific aims of the Society would have to be accorded a more prominent place in its organization for he expressed his opinion that Dr W. H. Wollaston, the distinguished chemist, would be the most suitable of the Fellows to succeed him.

Sir Joseph Banks, by his personal ability, his energy, and the influence which he exercised over all with whom he came into contact, made the Society while he was its President far more widely known than it had hitherto been and showed how vast a field of scientific activity lay before it in all branches of knowledge.

It will be remembered that Bishop Sprat expected that those who were elected to the Fellowship without having a knowledge of natural philosophy would be ready to give financial support to the Society. This did not prove to be the case in the later years of the seventeenth century, but then the Society's future seemed to be far from being assured. By 1700 the position began to improve, and from 1740, when there were 202 Fellows in this category, to 1820 when there were 641, the increase was continuous. During these one hundred and twenty years many hundreds of such Fellows had been elected but not one of them is recorded in the list of the Society's benefactors as having contributed to its funds[1] except Sir Godfrey Copley whose gift of £100 was devoted to providing the Copley Medal.

Among the men of distinction at this time were Dr Nevil Maskelyne [1758],[2] Sir William Herschel [1781], astronomers; John Dalton [1822], W. H. Wollaston [1793], Sir Humphry Davy [1805], chemists; Henry Cavendish [1760]; Thomas Young [1794]; Sir William Hooker, botanist [1812]; John Hunter, anatomist [1767]; James Watt, engineer [1785].

[1] *The Record of the Royal Society*, 4th ed., 1940, pp. 141, 142.
[2] The dates in square brackets indicate the year of election to the Royal Society.

REFERENCES

BABBAGE, Professor C. *The Decline of Science in England*. London, 1830.
GRANVILLE, A. B., M.D., F.R.S. *History of the Royal Society in the Nineteenth Century*. London, 1836.
Royal Society, Record of. 4th ed. 1940.
SMITH, E. *Life of Sir Joseph Banks, Bt*. London, 1911.
WELD, J. C. *History of the Royal Society*. Vol. II. London, 1848.

CHAPTER VII

THE SCIENTIFIC REVOLT: 1820–1860

AFTER the death of Sir Joseph Banks it was inevitable that many changes would be introduced in the policy and administration of the Society. He had been its President for forty one and a half years and maintained a close control over the Council's discussions and their decisions except perhaps during the last few years of his life; he had presided at 417 of the 450 Council meetings which had been held during his presidency and undoubtedly had guided them in the path which he considered to be the best and wisest for the Society. He took a pride in making it the great intellectual centre of the country, and this he promoted by the weekly receptions which he gave at his house in Soho Square, where not only science but literature, history, archaeology, art, as well as travel and exploration were represented by the most notable of their votaries, foreign as well as British.

During his presidency great advances were being made, notably in physical and biological science but also in many branches of industry and technology, though these did not appeal to him so much. He did not as a rule favour the foundation of the specialized scientific bodies which many scientific men at that time were desirous of forming in order to facilitate more detailed discussions of branches of science in which they were interested than the meetings of the Royal Society could provide; the number of such men was growing rapidly and the hundred and fifty out of the total Fellowship of the Royal Society who were scientific men were far from adequately representing those who were doing valuable work in science. While Banks would not purchase the collections of Linnaeus when they were offered to him in 1783, he encouraged J. E. Smith, F.R.S., to do so, and with other Fellows of the Royal Society he supported the foundation of the Linnean Society in 1788, for its activities were to a large extent such as interested him. On the other hand, he opposed the formation of the Geological Society in 1807, and also that of the Astronomical Society in 1820 when the men who were making notable advances in these fields felt it to be essential to their work that they should have the fullest freedom to develop their subjects in the manner which they regarded as the most suitable. Once these younger societies had started on their own careers the rapid growth of their membership showed that they were meeting a need which already existed and that a brilliant future awaited them; the Linnean Society, founded in 1788, had a membership of 209 by 1800, and of 463 by 1820; the Geological Society was founded in 1807 with thirteen Fellows, and by 1833 its Fellowship exceeded five hundred; by 1863 it had grown to more than a thousand. The Astronomical Society was founded in 1820 and ten years later there were 243 Fellows; these had increased to 500 by 1870, and have now reached 900. It has long since been recognized that Banks' fears lest these new societies should compete with the Royal Society to its detriment were without foundation; on the contrary, the contributions to research made by the Fellows of such specialized societies have greatly promoted the advance of science and have raised its standing in this country. Banks may have regarded the Royal Society as a trust committed to him to preserve with as few material changes as possible rather than as a living institution ready to seize upon any opportunity of improving Natural Knowledge and of promoting the advance of technical industry by every aid that science could give. Both by education and temperament he was inclined to regard that which came up to his ideals as only to be modified or recast under exceptional circumstances.

The result of the policy which had been followed throughout the latter half of the eighteenth century and persevered in throughout Banks' presidency was that science was not being actively promoted by the Society as an institution, though much was done by the researches carried out by several Fellows who were eminent as men of science; little more than one-third of the councillors and rather less than 30 per cent of the Fellows were men of science. This state of things was recognized by many as being unsatisfactory but a remedy was difficult to find.

After the unsuccessful attempt by Dr Samuel Horsley to oust the President in 1784, successive Councils renominated Banks for the presidency year after year, and the Fellows at each succeeding Anniversary meeting re-elected him. He had a longer experience of the Council's work than any of his fellow-councillors, and no proposal which did not commend itself to him had much chance of being carried. If he did not select the names which were to be put forward at the Anniversary meetings for election to the new Council he certainly saw and approved the list shortly before the meeting and before it was available to the Fellows, sometimes only at the meeting itself. The result was that at the time of Banks' death there was a strong feeling in the Society that the time had come for science and especially mathematics, physics and chemistry, to be brought much more to the front instead of being kept in a subordinate position by the inadequately informed opinion of a majority. Banks himself seems to have realized this during the later years of his life for he expressed the opinion that Dr W. H. Wollaston would be the most suitable man to succeed him. This was perhaps in the minds of the Council when on 29 June 1820 they elected Dr William Hyde Wollaston, the distinguished chemist, to succeed him as President of the Society and to hold office until the next Anniversary meeting.

Wollaston had been a Secretary from 1804 to 1816 and had served for seventeen years on the Council so that he was well acquainted with the procedure and policy of the Society. It was soon clear, however, that Sir Humphry Davy considered that he had the better claim and that he intended to stand as a candidate for the presidency at the coming annual election in November 1820 whatever recommendation the Council might make. Wollaston had only accepted the position until then and had no wish to be involved in a contested election. Sir Humphry Davy was therefore elected by a large majority, and once again the Society had a man of science in charge of its affairs; Davy had been Secretary from 1807 to 1812 and had served on its Council for five years more, including the years 1818 and 1819, so that he had an intimate knowledge of the matters which were now awaiting the attention of the officers.

Besides the work which awaited the new Council in devising an administrative system under which the number of Fellows elected annually could be controlled, and their scientific qualifications ensured, there were many matters which called for its early consideration. The statutes which had been adopted in 1776 now needed revision for many changes had taken place; the Society's quarters in Crane Court had been closed in 1780 when the government provided free accommodation for the Society and other institutions in Somerset House, and in 1782 the two houses in it which the Society had owned since 1710 were sold for £1000; also as there was not sufficient room at Somerset House to accommodate its collections these had been presented to the British Museum in 1781. The financial position was not altogether satisfactory, and several other matters needed the attention of Council.

The first sign of a change of policy was that on 30 November 1820 twelve out of the twenty-one members elected to form the new Council were men of science instead of only seven, eight or nine, as had been the custom for many years past, and that Sir Humphry Davy had been elected President. This was but a beginning; twenty-seven years were to pass and considerable opposition had to be overcome before the way was clear for the continuous and effective progress of the Society from being an institution in which the scientific members were a minority to one in which they were to form the great body of the Society and to determine its policy.

Sir Humphry Davy was forty-two years of age when he was elected President of the Society. He had become a Fellow in 1803 and a Secretary in 1807 but resigned this post in 1812 on his marriage. In 1801 he had been appointed director of the chemical laboratory at the Royal Institution where he soon built up a reputation by his researches and discoveries in chemistry, and by the lectures and demonstrations which he gave there. He was knighted by King George IV, in 1812, and was created a baronet in 1818 for his work on coal-gas and for designing his Miners' Safety Lamp. His wife, whom he married in 1812, possessed a considerable fortune which enabled him to give up the routine work of lecturing and to devote his time and energy to original research in chemistry which it had always been his intention to do as soon as it became practicable. Life under these new conditions is said to have changed him considerably and led him to regard social position and recognition as being of greater importance than the distinction which he had won for himself by his scientific researches. He aimed at carrying on the duties of his position in the Society much as his predecessor had done, but he did not always exercise the tact and courtesy which had been so characteristic of Sir Joseph Banks. Davy was offended when some of Faraday's friends wished to propose him as a candidate for the Fellowship of the Society because his own consent as President had not been obtained first; he even told Faraday that he would oppose his election. He did not however maintain his opposition, and Faraday was elected in 1824. He continued the practice of drawing a distinction between the qualifications expected from scientific candidates and those of others, which Banks had laid down for his own guidance when he took over the presidency, for when Sir Roderick Murchison was elected a Fellow in 1826 Sir Humphry Davy explained to him that he had not been elected for his scientific work but because he was a man of means and leisure and so might be expected to be a profitable Fellow for the Society. Murchison at that time had been for several years an active geologist and had been a Fellow of the Geological Society for at least eight of them. These and other inconsiderate acts diminished Davy's popularity in the Society.

THE OFFICERS

During Sir Humphry Davy's presidency the Treasurer was Davies Gilbert who had replaced S. Lysons in 1819. The Secretaries were W. T. Brande, a chemist, who replaced Dr W. H. Wollaston in 1816, and was succeeded by J. F. W. Herschel, the astronomer, in 1824; the other was Taylor Combe, a numismatist and a member of the staff of the British Museum, who had replaced Davy at the end of 1812 and was succeeded by J. G. Children, a chemist, in 1826. All of these officers were very regular in their attendance at the meetings of Council.

This frequent change of officers was not to the advantage of the administration but the Fellows who were brought in to fill the vacancies were now men of science, and Taylor Combe was the last of the Secretaries who was not one.

COUNCILS

Dr W. H. Wollaston having been elected President in succession to Sir Joseph Banks at the end of June certainly influenced the choice of the new councillors who were recommended to the Society at the Anniversary meeting to act for the coming year. As a distinguished chemist Wollaston might be expected to increase the number of scientific councillors who had of late been fewer than formerly; actually his Council recommended the election of twelve scientific councillors out of twenty-one. For the next one hundred and twenty years the scientific Fellows in each Council numbered from twelve up to twenty-one. The result was that the Councils' work had now a definitely scientific character and the councillors felt it incumbent on them to attend regularly at the meetings. By the middle of the century the average number of scientific men serving on the Council was eighteen and the average number of councillors who attended a meeting was from twelve to fourteen. Sir Humphry Davy, when he had been elected President at the Anniversary meeting on 30 November 1820, followed this practice and Councils consisting almost wholly of men of science became customary.

The Society was now on the eve of important changes, and the most important of them was this change in the character of the Council from a body in which only a minority of from six to ten were men of science to one which consisted almost wholly of such men; this meant that the administration of the Society, and the planning of its policy, passed into the hands of those who were competent to advance the original aims of the Society, the improvement of Natural Knowledge. The change seems to have come about quietly, starting with the election of a Council the members of which were recommended to the Anniversary meeting on 30 November 1820. Dr W. H. Wollaston had been President since Sir Joseph Banks' death; it was known that Sir Humphry Davy was the candidate who was most likely to succeed him, and this was in accord with the strong feeling that existed among many of the Fellows of the Society that its scientific activities should be encouraged. The increase of the number of scientific councillors to twelve, and even more in subsequent years, was the first sign of this movement which the recommendations of the committees of 1823 and 1827 strongly supported.

THE FELLOWSHIP

The number of Fellows in the Society in 1820 was 641, besides 40 foreign members; these numbers continued to mount up steadily until 1846, when it consisted of 771 Fellows, but in February 1847 the number of candidates who might be recommended by the Council for election annually in future years was fixed at fifteen, so that from that time onwards the number began to decrease; by 1880 it had fallen to less than 500.

When the statutes were revised in 1776 the limit for foreign members was fixed at one hundred; the selection of candidates for the foreign members' list became much stricter until by 1810 there were only fifty. The result of limiting the foreign members to this number was soon visible in the higher standard of those who were selected to this special group. Throughout the nineteenth century as the scientific standard of the Fellowship rose, so also did that of the foreign members, and now for many years past only the most eminent scientific men of different countries have been admitted to it. In 1776 the proportion of foreign members to Ordinary Fellows was one to three; by 1810 it had fallen to one to eleven; and by 1823 it was one to seventeen. In these days when the number of Ordinary Fellows has fallen to about 450 the proportion is about one to nine.

The maximum number of foreign members was fixed at fifty by the statutes of 1823 but this made no difference as they had not exceeded this number since 1810. The percentage of the Fellows which was contributed by the nobility was ten or eleven up to 1830 but by 1860 it had fallen to 4·6 and was still decreasing. By 1860 men of science were for the first time in the majority, and before many years had passed they had replaced their non-scientific colleagues altogether except for those who were specially elected and these were strictly limited in number. Between 1830 and 1860 the scientific Fellowship increased by 117 while the non-scientific Fellows, who had hitherto always formed the majority, fell off by 146 and were thirty fewer than their colleagues. The tide had turned and the Society was moving steadily towards an almost exclusively scientific Fellowship after two centuries of frustrated effort. The way in which this important change was brought about and the change which it caused in the composition of the Society will be described later.

Fellows of different professions in 1830

Professions	No. of Fellows	No. of communications in *Phil. Trans.*	Remarks
Bishops	10	9	All by one Fellow
Noblemen	63	None	
Naval officers	27	7	
Army officers	39	28	Two officers together contributed 25
Clergymen	74	8	
Legal men	63	28	16 contributed by one Fellow
Physicians	79	66	
Surgeons	21	137	109 contributed by one Fellow
	376	283	
Fellows not belonging to these professions	286	205	

In 1836 Dr A. B. Granville, F.R.S., published an analysis of the Fellow-ship of the Society according to their professions, and added the number of communications which had been published in the *Philosophical Trans-actions* in each case.[1] As he compiled this in 1830 when he had ready access to the records of the Society his results are of considerable interest, and are summarized in the table on p. 233. If the case of Sir E. Home, the surgeon, who contributed one hundred and nine papers, is omitted as being exceptional, the average number of papers contributed by members of the professions is about three-fourths of one apiece. The category of those Fellows who are not included in the professions referred to includes both scientific men as well as those who had little or no interest in the aims of the Society, and so little is known of many of them that a detailed analysis is impracticable; it can however be said that seven contributed from five to ten papers, four contributed between eleven and twenty papers, and one had twenty-four papers accepted and published (cf. Appendix II A, B).

FINANCE

During the decade 1811–20 the receipts as reported to the Fellows at the Anniversary meetings exceeded the expenditure by nearly £1000 per annum on the average. This was probably due to the practice which had been introduced of requiring candidates to compound for their annual subscriptions by the payment of twenty-six guineas or later of forty pounds which was included in the receipts of the year and was not as a rule invested. It is not clear therefore why the Council should have found it necessary in 1823 to raise the annual subscription from £2. 12s. to £4, and the com-position fee from twenty-six guineas to forty pounds, unless there were outstanding liabilities which the statements laid before the Anniversary meetings did not include.

Early in 1827 on the proposition of James South, the astronomer, the Council appointed a committee to consider the administration of the Society and the election of Fellows. Davies Gilbert, the Treasurer, was a member of this committee and agreed to its report which was ready for Council by June; by the time it was received by the Council Sir Humphry Davy's resignation had been received, and Gilbert had been elected President for the remainder of the year. At the Anniversary meeting he was re-elected President for the year 1828 and Captain Kater became Treasurer. Nothing more was done about the committee's report which was suppressed by the new Council at its first meeting.

In 1829 J. W. Lubbock was elected a Fellow of the Society, and though this may have seemed at the time to be a small matter it led to results of the greatest importance. He was an eminent mathematician and had carried out important investigations on the oceanic tides; but besides

[1] *History of the Royal Society in the Nineteenth Century.* London, 1836.

being a man of science he was a director of the family bank of Lubbock and Forster,. and his election to the treasurership at the Anniversary meeting of 1830 enabled him to make himself fully acquainted with the financial position of the Society and with the way in which its funds had been administered in past years. This was of inestimable benefit to the Society, since his annual reports gave to those Fellows, who had been urging the need for a complete reorganization, the evidence which they needed and had hitherto been unable to obtain. His opportunity was not long in coming, for at the first Council meeting which he attended as Treasurer on 9 December 1830 the newly elected President, the Duke of Sussex, was in the Chair. At this first meeting of the new Council, the President introduced a resolution, which was adopted, to the effect that the report of the last audit and the Treasurer's accounts for the twelvemonth ending 30 November 1830 should be printed and distributed to the Fellows forthwith, and that this practice should in future be continued annually. It seems that this was not on the agenda of the meeting but that the President made the proposal on his own initiative, and if that was the case he must have learned from some well-informed source that the decision which the Council then took would be of the utmost importance to the Society. Proposed, as it apparently was, from the Chair by the newly elected Royal President, no one ventured to oppose the motion. This statement of the accounts for the year 1 December 1829 to 30 November 1830, which was prepared and circulated to all the Fellows by the order of the President, was the first which had ever been issued to the Fellows, and which informed them of the financial position of the Society, what sums had been received and how their funds had been expended. The accounts for 1830 had been prepared by Captain Kater, the Treasurer, who had replaced Davies Gilbert in 1827; but as Gilbert had stated that he wished to resume the treasurership as soon as possible the statement of accounts was doubtless in the same form as that in which it had been rendered by him to the Councils during his tenure of the treasurership and by his predecessors for many years past. This Receipts and Expenditure Account is therefore of exceptional interest and is reproduced on p. 236.

There does not seem to be any sufficient reason why such statements should not have been furnished to the Fellows in each of the past years. A year later in accordance with the instructions of the Council the Treasurer submitted a financial statement which was presented and circulated to the Fellows before the Anniversary meeting of 1831. He was then able to show a balance in hand after one year's administration of the Society's funds of £775, or £650 more than at the corresponding date of the preceding year, but this was due mainly to a considerable payment made by the British Museum for manuscripts which the Society had transferred to it; this payment was spent in the purchase of scientific books for the Society's library.

Accounts: 1 December 1829—30 November 1830

Receipts	£	s.	d.
Balance with Treasurer	296	15	3
Contributions at £2. 12s.	218	4	0
Contributions at £4	145	0	0
Admission fees at £10	350	0	0
Compositions for annual contribution	1306	0	0
Rents	177	4	0
Dividends	654	12	0
Miscellaneous receipts	444	11	3
	£3592	6	6

Expenditure	£	s.	d.
Bakerian lecture fee of previous year	4	0	0
Chronometer Stove	6	10	0
Engraving	130	13	9
Invested in stock	1062	18	0
Salaries	600	5	0
Fairchild and Bakerian lecture fees	7	0	0
Gratuity to Mr Coppard	10	0	0
Solicitor's bill	2	3	0
Levelling Commission on Thames	92	0	0
Annual bills	1388	11	11½
Rates and Petty Cash	166	4	7
Balance with Treasurer	122	0	2½
	£3592	6	6

	£	s.	d.
Arrears of subscriptions	78	12	0
Rent due (Mablethorpe land)	107	0	0
	£185	12	0

Accounts: 1 December 1830—30 November 1831

Receipts	£	s.	d.
Balance in hand 30. 11. 30	122	0	2½
Weekly and Quarterly Subscriptions	286	0	0
29 Admission fees at £10	290	0	0
27 Composition fees at £40	1080	0	0
Rents	381	9	0
Dividends on:			
£16,500 Reduced Bank 3% Annuities	495	0	0
Other investments	181	11	1
Cash receipts	32	15	9
Sales of *Philosophical Transactions*	379	0	4
From British Museum on account of Arundel MSS.	956	0	3
	£4203	16	7½

Expenditure	£	s.	d.
Instruments and Scientific books	292	5	2
Purchase of stock for Donation and Rumford Funds	262	2	9
Salaries	638	5	0
Special payments for work	23	0	0
Bills for House and Library	1464	9	3
Books bought for Library from British Museum payment	528	12	3
Rates and Petty charges	219	14	5
Balance with Treasurer	775	7	9½
	£4203	16	7½

During 1831 when the Treasurer had had an opportunity of examining the accounts of the previous years carefully he found that considerable liabilities were outstanding or would mature shortly, so in 1833 he had to ask the Council to authorize the sale of £2500 of the Society's invested funds to meet the debt, thus reducing the Society's reserve to £14,000 of 3 per cent Reduced Bank Annuities. It is not stated for what the debt had been incurred, and in the former Treasurer's reports no reference was made to its existence.

In November 1833 the Treasurer presented to the Council a very useful report on the Society's financial position. There were then 11 Royal Fellows, 46 foreign members and 690 Ordinary Fellows; of the last category 595 had compounded for their annual contributions, 44 were paying the old subscription of £2. 12s. and 51 paid the new rate of subscription of £4.

He also gave the following estimate of the ordinary income and expenditure of the Society for the following year:

Estimated Income and Expenditure for 1834

Ordinary Income	£	£	Ordinary Expenditure	£	£
Rents	284		Salaries	645	
Dividends on Stock	501		Lighting, Heating and Petty Expenses	364	
Weekly and Quarterly contributions (about)	270		Miscellaneous and House Bills	200	1209
Sale of *Philosophical Transactions*	350	1405	*Philosophical Transactions* on basis of past five years:		
			Printing	350	
			Paper	259	
			Engraving	285	894
			Total		£2103
			Non-recurring expenditure incurred but not yet paid (part recoverable)		£1600

There were also the admission fees of £10 from each candidate elected to be included; the average number of these during the last five years had been twenty-six. The number who paid a composition fee instead of their annual subscriptions varied considerably from year to year. On this subject he says: 'But it being now optional for members to compound or not for their annual payments the compositions will most probably go on decreasing, or may cease altogether; and until the amount of the present annual subscription of four pounds has come into full operation a temporary inconvenience will be experienced from this circumstance as well as from the falling-off in the compositions.' Professor C. Babbage, writing in 1827, says:[1] 'It should be observed that all members (of the Society) contribute equally and that the sum now required is fifty pounds. It used until lately to be ten pounds on entrance and four pounds annually.' The practice of putting pressure on candidates to compound for their annual payments seems to have been in operation for longer than Babbage thought, for of 576 Ordinary Fellows who were elected during the twenty years 1814 to 1833, inclusive, 463 had paid composition fees and only 113 were paying their subscriptions annually. After this time the number of those who compounded for their subscriptions steadily decreased. During the twenty years 1814 to 1833 a total of £15,256 had been received on account of the composition fees paid by 463 Fellows, but only about one-third of this sum had been invested, £5138. 6s. 8d. 3 per cent Reduced Bank Annuities having been purchased. The average amount of the Society's annual expenses for the preceding twenty years had been £2330, so the Society's financial position at this time was by no means satisfactory.

In 1834 the Council resolved that the composition fee should be raised from £40 to £60 except for those Fellows who had contributed a paper which had been printed in the *Philosophical Transactions*, but as the number of candidates who were prepared to pay the composition fee was already falling off the benefit which the Society would derive from this alteration was likely to be inconsiderable.

In 1833 the Council had under consideration the honoraria to be paid to the officers and resolved that the two principal Secretaries of the Society should each receive £100, and that the Foreign Secretary should have £20, which was provided by the Keck bequest. When the Treasurer's report was received in November he asked that £2500 of the Society's invested funds should be sold in order to meet liabilities which had been incurred before 1831 but had not been paid. At the beginning of the following year it was resolved, on his recommendation, that all composition fees should be invested instead of being treated as a part of the year's revenue; this injunction has remained in force ever since. The payment of such fees instead of an annual subscription was now declared to be optional, and the number of those who compounded began to fall off more rapidly.

The health of the President, the Duke of Sussex, was now failing and his eyesight was seriously affected so that for the next three or four years he was often unable to attend the Council and Ordinary meetings. Whether this had anything to do with a tendency to relax or vary some of the resolutions which Council had lately adopted it is impossible to say with certainty, but in 1834 candidates' certificates were to be suspended for five months only instead of ten as had been the rule; and the composition

[1] *The Decline of Science in England.* London, 1830.

fee was raised from £40 to £60 except for those who had had a communication accepted for publication in the *Philosophical Transactions*.

In 1831 a reference appears to £200 which was held by the Society as the Pulteney Fund. Richard Pulteney was a botanist of Bath who died in 1801, but it is not clear for what purpose the trust was constituted or how it was dealt with. It disappears from the accounts after 1835. In this same year the two houses in Coleman Street which had been bequeathed to the Society by Dr Thomas Paget in 1717 were expropriated and sold to the city since the space was required for improving the access to London Bridge. The purchase money, £3402. 1s. 1d., was invested in Consols and held by the court on the Society's behalf.

At the end of 1835 J. W. Lubbock resigned the treasurership though he continued to be a member of the Council. During the President's illness he had acted for him at nearly all the meetings, which entailed a considerable amount of additional work for one who was a busy director of a bank. He also objected to the action of the Council in continuing to recommend for re-election a president who, they knew, was unable to perform the duties of his office owing to ill-health. It is also probable that he was opposed to the action taken by a majority of the Council to relax or rescind several of the resolutions which had only recently been adopted in order to improve the administration. At the time of his resignation he was able to report that the balance in the Treasurer's hands was £218. The number of Ordinary Fellows was then 737 of whom 591 had compounded for their subscriptions. He was succeeded by George Rennie, the civil engineer, who however only held the post for three years, for on the resignation of the Duke of Sussex at the end of 1838 and the election of Lord Northampton to the presidency J. W. Lubbock was again elected to the treasurership, which he then held until the end of 1845. Sir John Lubbock, who was then living at some distance from London, found it extremely inconvenient to be present at the meetings of the Council and of the Society. He therefore told the Council that he did not wish to be re-elected to the treasurership, and was again replaced by George Rennie.

During his tenure of the treasurership Sir John Lubbock had rendered most valuable service to the Society; it was due to him that the accounts had been printed and circulated to all the Fellows before the Anniversary meeting since the beginning of 1831; he had reorganized the accounts and had introduced a much more economical administration of the funds, and had shown that for many years past far too little care had been exercised in dealing with the Society's finances. Since he had been the Treasurer he had been able to report a balance in hand at the end of each year's working, and had moreover been able at the end of 1843, and again at the end of 1844, to invest £1000 in government funds. What was even more important was that by supplying irrefutable evidence that for a good many years the administration of the Society had been carried on by successive Councils with insufficient care, he had given a great impetus to the movement which had for its object the reorganization of the Society on the lines proposed by Dr W. H. Wollaston's committee in 1827. When the annual report of the Treasurer was circulated to every Fellow from 1831 onwards before the Anniversary meeting, the secret which had been most closely guarded by successive Councils of many years past was published in November of each year and the details of the receipts and expenditure were presented for discussion. Granville complained in 1830 of the impossibility of getting any information about the financial position of the Society from the Treasurer's office or from his registers, since access to these was restricted to members of the Council, and even they were not encouraged to press their investigations far.

At the beginning of May 1846 a 'Charters Committee' was appointed by the Council to consider what modifications were desirable in them or in the statutes. The Committee reported at the end of June recommending that only fifteen candidates should be elected to the Fellowship in each year, after having examined the relative advantages of fifteen, seventeen and twenty annual elections. As this raised several financial questions of importance, the Committee had examined the probable effect of their proposals on the Society's finances and submitted the following estimate based on the average of the preceding five years:

Estimated Receipts	£	*Estimated Expenditure*	£
Rents	235	Salaries	635
Dividends on Investments	602	Printing, etc.	749
Sale of Publications	244	Library	233
Annual Contributions	989	House Bills	154
Admission fees (15)	150	Petty expenses	143
Composition fees	440	Rates, Insurance, etc.	72
		Balance	674
	£2660		£2660

In December 1846 the Council appointed a Finance Committee 'to inquire into the Prospective Financial Condition of the Society'. This Committee rendered their report on 14 January 1847 to the effect that if not more than seventeen candidates were elected annually the Society's income would be sufficient to meet expenditure estimated on the basis of the preceding twelve years, and attached the estimate shown on p. 241. The Committee also reported that at the date of their report the Society possessed £14,000 3 per cent Reduced Annuities and £8852 3 per cent Consols which represented the capital of the trust funds, yielding in all £685. 11s. 2d. gross income. The Mablethorpe estate, the Acton estate and the Sussex fee-farm rents produced a total rental of £214. 4s. The income derived from the trust funds amounted to: Donation Fund £145, Rumford Fund £73, and the Fairchild Fund £3. This report was considered by the Council to be sufficiently satisfactory to justify them in adopting the report of the Charters Committee.

Estimated Income	£	s.	d.	Estimated Expenditure	£	s.	d.
Rents	214	4	0	Salaries, Bills and General	1033	13	0
Dividends	685	11	2	Expenses			
Annual Contributions	1041	8	0	Printing *Philosophical Trans-*	800	0	0
Admission fees (17)	170	0	0	*actions*			
Composition fees	320	0	0	Purchase and Binding of	400	0	0
Annual Contributions of new Fellows	41	4	0	Books			
Sale of *Philosophical Trans-actions*	273	0	0				
	£2745	7	2				
Deduct for Income Tax	26	5	0	Balance	485	9	2
	£2719	2	2		£2719	2	2

In October 1857 the Treasurer of that year presented a report on the finances of the Society for the nine years which had elapsed since the new statutes regulating the admission of Fellows came into operation, during which period the number of Ordinary Fellows had fallen from 764 to 660 while those who had compounded had decreased from 480 to 375, or by about the same amount; the annual subscribers had increased by one. For the ensuing year he estimated:

Estimated Income and Expenditure for 1858

Estimated Income	£	Estimated Expenditure	£
Subscriptions	1126	Printing for *Philosophical Trans-actions* and *Proceedings*	1052
Entrance fees	150		
Composition fees	360	Salaries	738
Rents	195	Purchase of books	158
Dividends	830	House and other expenses	492
Sale of *Philosophical Transactions*	268	Balance	489
	£2929		£2929

This the Council considered to be sufficiently satisfactory to justify them in continuing the new procedure at least for a further period.

Having summarized the various changes in administration which the introduction of control by men of science had brought about, their influence on the affairs of the Society up to 1860 may be described.

From 1821 a marked change took place in the policy of the Society, and a more active spirit began to show itself in the Councils. This coincided with the election of Sir Humphry Davy to the presidency and with the appearance for the first time of a majority of scientific Fellows on the Councils. The minutes show a greater desire to improve the administration and to keep the Fellows informed of the Councils' intentions than there had been in the past. All this was the outcome of the steady growth of interest in scientific knowledge and in its application to technical industry which had been in progress for several years past and had led to the demand for a reorganization of the Society itself. The principal supporters of the movement were to be found among the scientific Fellows, the 'reformers' as they were called, but so long as they were in a minority both in the Society and on its Councils they were unable to accomplish much. Now their opportunity seemed to have arrived, but there were still many of the senior Fellows who looked on the proposed reforms with extreme disfavour and used every means in their power to hinder them or to prevent their adoption; one of the most persistent of these was Davies Gilbert, who had been Treasurer since the end of 1819 and was elected President eight years later.

In 1821 a letter was received from the Académie des Sciences and the Board of Longitude in Paris asking that the triangulation connecting the observatories of Paris and Greenwich, which had been carried out by General Roy in co-operation with MM. de Cassini, Mechain and Legendre in 1790, should be re-observed. The Council agreed and adopted a resolution in which the great importance of the work and the desirability of carrying it out as early as possible were emphasized. General Colby and Captain Kater were nominated as the British Commissioners by the Society to act in co-operation with MM. Arago and Mathieu, who had been appointed by the Académie des Sciences. The angles were measured with the Society's 36-inch theodolite which had been constructed by Ramsden in 1787 and the stations on the coasts of England and France were marked by special lamps designed by M. Fresnel, the French physicist. The stations which were occupied inland were marked by stones one foot square and four or five feet long, sunk in the ground until their ends were level with the surface of the ground.[1] The original observations are preserved with the archives of the Society.

The Council's next step was to appoint a committee in 1823 to review the statutes and to report what additions and alterations were required in those which the Council of 1776 had approved. Much had happened since then; the Society's house had been transferred from Crane Court to Somerset House in 1778, and the Society's collections which had been brought together by gift and purchase from 1666 had

[1] Cf. *Philosophical Transactions*, 1828.

been accepted by the trustees of the British Museum for incorporation with the national collections.

The committee recommended that the annual subscription of Fellows which had been one shilling a week since 1660 should be raised to four pounds a year, and that the obligation to sign a bond binding a candidate under a penalty to pay his fee to the Society regularly should be revoked: the admission fee was to be raised from five guineas to ten pounds, and the fee for which the payment of annual subscriptions could be compounded was raised from twenty-six guineas to forty pounds. They also recommended that the maximum number of foreign members should be reduced from one hundred to fifty, and that in future candidates for foreign membership should be nominated by the Council, the opinion of each member being taken by ballot before the candidate's name was brought before the Society for election.

The Fellow who had hitherto been called the Assistant for the foreign correspondence of the Society was now to be given the rank of a 'Secretary'; he was classed with the other two Secretaries of the Society, and was to be a member of Council, receiving the income of the Keck bequest as before. Dr Thomas Young, the physician and physicist who had performed the duties since November 1803, now benefited by the new statute and became the first Foreign Secretary, holding the post until his death in 1829. It was also proposed to abolish the post of Clerk, which had been in existence since the Society was formed, and to create that of Assistant Secretary in its place, since it was considered that his work was primarily that of assisting the Secretaries in carrying out the instructions of the Council, and in preparing the business for the meetings. The Clerk had by the Charter to be appointed by the Society as a body and they alone could dismiss him, but the Council, whose servant he had come specially to be, should, it was thought, have the power to engage and to dismiss such an employee. This distinction between the two posts was easily lost sight of and accounted for a resolution adopted by the Council in January 1835 which laid down that 'the denomination of Clerk be adopted instead of that of Assistant Secretary'; a month later the reason for the change made in 1823 was explained to the Council who thereupon rescinded their resolution of January 1835. The committee of 1823 did not however deal with the most urgently needed reform in the constitution of the Society whereby the number and the qualifications of its Fellows might be regulated. This question was not a new one for it had been first considered by the Council of the day a hundred and fifty years earlier; the minutes of a meeting of Council on 27 August 1674, at which Sir W. Petty, Sir J. Lowther, Sir J. Cutler, Sir C. Wren, Sir P. Neile and Mr Oldenburg were present, record that 'to make the Society prosper good experiments must be in the first place provided to make the weekly meetings considerable, and that the expenses for making these experiments must be secured by legal subscriptions for paying the contributors; which done the Council might then with confidence proceed to the ejection of useless Fellows'.

About this time the departments of the government, especially the Admiralty, on several occasions requested the opinion of the Council on matters which were related in one way or another to science. Such action on their part had often been taken, but had been limited usually to a request that instructions should be prepared for scientific or exploratory expeditions; now their enquiries were more often of a physical or chemical character. In 1823 the advice of Council was desired by the government on a project of Professor C. Babbage for making a calculating machine of a new type. The Council supported his proposal but the outcome of the scheme was not a complete success. After he had spent much of his own money on it, and a considerable sum which the Treasury had granted, further support was refused and in 1842 the project was abandoned. So much of the machine as was completed, together with the plans, etc., are now in the Science Museum at South Kensington. The Admiralty also asked for a report on Mr Snow Harris' scheme for affixing lightning conductors on ships and to this a favourable reply was sent.

In January 1823 the Admiralty asked for a report on the copper sheathing which was then being adopted for ships of the naval service, and a committee was appointed consisting of Sir H. Davy, Mr Brande, Mr Combe, Mr Hatchett, Sir J. Herschel and Dr Wollaston, four of whom were chemists, to examine the problem and to report. Sir H. Davy took upon himself all the work of this enquiry and devised a method of protection which he recommended should be adopted. Copper sheathing so protected was not corroded but it became so foul by the growth upon it of seaweed, shell fish, etc., that the ships' sea-going qualities were seriously affected, and the use of the protectors had to be abandoned. This was announced just after Sir H. Davy had read an elaborate paper on the subject before the Society, and the failure of his plan caused him the deepest mortification. Soon after this his health began to deteriorate; he only delivered his address to the Anniversary meeting of 1826 with the greatest difficulty, and was unable to be present at the dinner. Early in 1827 he went abroad in the hope of recovery, but he was disappointed in this; recognizing that he was unable to face the work of another session he wrote to the Treasurer, Mr D. Gilbert, on 1 July to say that he must resign the presidency.

It was at this time common knowledge, not only within the Society but also outside it, and even abroad, that nearly every candidate who was proposed for the Fellowship of the Society was admitted; and many of the Fellows realized the extent to which its prestige had been lowered by this practice. Some drastic reform in its statutes and procedure to deal with this state of things, which neither the committee of 1774 nor that of

1823 had reported on, was urgently needed. Early in 1827 the Council, on the proposal of James South the astronomer, appointed a committee consisting of

Dr W. H. Wollaston	Mr C. Babbage
Mr D. Gilbert (*Treasurer*)	Captain Beaufort, R.N.
Mr J. F. Herschel (*Secretary*)	Captain Kater
Dr T. Young (*Foreign Secretary*)	Mr J. South

to 'Consider the best means of limiting the members admitted to the Royal Society, as well as to make such Suggestions on that subject as may seem to them conducive to the welfare of the Society'. The committee reported to the Council at the beginning of June 1827[1] to the following effect:

i. The Committee are satisfied that the increase of the Fellowship has been in a much higher ratio than that of the population, or of the general growth of knowledge, or the extension of science:

ii. The utility of the Society is directly proportional to its respectability which can only be secured by including men of high philosophical eminence, and this demands that the public should be convinced that membership of the Society is an honour. From studying the return of the Fellows who have contributed papers to the *Philosophical Transactions* the Committee recommend that the number of Fellows should be fixed at 400 exclusive of foreign members and that until this number is reached only four new members annually should be admitted.

iii. Candidates should be elected once only in each year, and printed lists of them should be circulated beforehand to all Fellows.

iv. The Committee realized that such a reduction in the number of Fellows would lead to a loss of income; in 1826 composition fees brought in £1200 besides what was received for subscriptions, dividends, rents, etc., but the balance in hand at the end of the year was small although no money had been invested. The Committee drew attention to the expenditure of more than £1200 on paper, engraving and printing in that year and expressed the view that expenditure should be controlled by a standing Finance Committee.

v. Seeing that the policy and affairs of the Society are entrusted to the Council the Committee recommended that nominations to a future Council should be subject to the diligent and anxious deliberation of the whole Council, and not be limited to the acceptance of a list presented to them.

vi. A point which the Committee regarded as of great importance was the selection of papers for the *Philosophical Transactions*; they considered that each one should be referred to an appropriate Committee which would report its opinion, and the grounds on which it was based.

[1] The Report is printed in full by Professor Babbage in his *Decline of Science in England*. London, 1830, pp. 160–5.

The committee further recommended that their report, after full discussion by the Council, should be remitted to another committee who would draft the requisite alterations in the existing statutes in order to give effect to it.

The Council at their meeting of 25 June 1827 received and read the report of their committee, and ordered it to be entered on the minutes. Usually the Council met once in July to deal with business likely to require attention during the next two or three months when the Society did not meet, and their meetings in October were mostly occupied with matters relating to the Anniversary meeting and with the membership of the new Council; it was not unreasonable therefore that the consideration of their committee's report which raised several issues of great importance should have been postponed, and 'recommended to the serious and early consideration of the Council for the ensuing year' which would be elected at the Anniversary meeting on 30 November.

When Sir Humphry Davy resigned the presidency in July 1827, Davies Gilbert, who had been the Treasurer of the Society for eight years, was elected by the Council to act as President until the Anniversary meeting in November. Owing to the Council meetings being interrupted by the annual vacation Sir Humphry Davy's letter of 1 July announcing his desire to retire from the presidency was not brought before the Council until 6 November, but even before this steps had been taken by Gilbert to seek for a successor among the politicians rather than from the scientific men, of whom J. F. Herschel, who was a Secretary at this time and had received the Copley Medal in 1821, was one of the most prominent. On 25 November 1827 Adam Sedgwick wrote from Cambridge to Sir Roderick Murchison saying that he had heard of a suggestion to elect Robert Peel, then Home Secretary, to the presidency, and adding:

> The republic of science will indeed be degraded if the Council of the Royal Society is to become a mere political junta and we are to sit under a man who *condescends* to be our patron....Why don't some of you propose Herschel? He is by far the first man of science in London, and would do the work admirably.

There was not time to proceed with the proposal to choose Peel, and the certainty of a contested election if it were persisted in caused Peel's name to be dropped. Gilbert was elected at the Anniversary meeting, but he made it clear that he wished to hold the position for one year only, after which he desired to resume the treasurership; he did however continue as President for three years. Gilbert in 1785 became a gentleman-commoner at Pembroke College, Oxford, where he devoted himself to the study of the physical sciences; he was elected a Fellow of the Society in 1791 and soon after returned to his home in Cornwall where he occupied himself with various branches of science and maintained a correspondence with Davy, Hitchins, Trevithick, Hornblower and others. In 1804 he was elected Member of Parliament for Helston and two years later he was elected Member for Bodmin, for which borough he continued to sit until 1832. In the House of Commons he was considered to represent scientific interests, and thus was able to serve usefully on many committees at which scientific and technical matters were considered; in this way he was brought into contact with many statesmen and politicians perhaps to the detriment of his scientific interests. During his presidency Gilbert did not attempt to entertain the Fellows of the Society on the scale that either Sir J. Banks or Sir H. Davy had done, but he maintained contact with many of them by giving dinner parties at the Thatched House Tavern in St James' Street and breakfasts at the Royal Society in Somerset House during the session. His place as Treasurer was taken by Captain H. Kater, the geodesist.

An unexpected act of royal recognition of the Society's increasing activity and national utility came to it in December 1825 when Robert Peel, the Home Secretary, wrote to the President saying that he had the King's Command to acquaint him of His Majesty's proposal to found two Gold Medals of the value of fifty guineas each to be awarded as honorary premiums under the direction of the President and Council of the Royal Society in such manner as shall, by the excitement of competition among men of science seem best calculated to promote the objects for which the Royal Society was instituted. The President and Council were invited to submit a draft of the regulations under which the award of the medals would be made. These regulations were adopted by the Council in January 1826, and later were approved by His Majesty. The medals for that year were awarded to John Dalton for the development of the chemical theory of definite proportions, and to James Ivory for his work on Refractions.

The practice has always been that the names of the proposed recipients of the royal medals are submitted to the sovereign for approval before the award becomes effective and the medals are struck. On accession each sovereign has intimated through the Home Secretary to the Society that these medals will be provided annually, and they have been awarded to eminent men of science in each successive year up to the present time. The designing of the obverse of the medal was entrusted to Mr Chantrey, the sculptor, and the reverse was to be designed by Sir Thomas Lawrence, R.A., who unfortunately delayed its execution from year to year until at his death not even a sketch of his ideas for it was to be found. Though ten medals had been awarded during the five years 1826 to 1830, not even the dies from which they should have been struck were in existence. On this being brought to the notice of the Duke of Sussex in 1833 when he was President, he took the matter up and arranged for the completion of the dies and the provision of the medals which had been already awarded. The regulations under which these medals have been awarded have been revised from time to time; at first they were given for 'the most important

discoveries or investigations completed and made known to the Royal
Society in the year preceding the day of award'. This was soon found to
be very inconvenient and the time limit of one year was extended to five
years. The rules under which these medals are awarded are now published
each year in the Year-book of the Society.

Dr W. H. Wollaston, the chemist, died in 1828 just after he had made
a gift to the Society of £2000 in order to found a fund, the income of
which was to be devoted to scientific research. The President contributed
£1000 to the fund and other Fellows also contributed generously until
a sum of £3410 had been collected. This donation, which has always been
known as the Donation Fund, was the first which the Society had received,
the dividends of which were to be applied from time to time to promoting
experimental researches, or in rewarding those by whom such researches
may have been made, or in such other manner as shall appear to the
President and Council for the time being most conducive to the interests
of the Society in particular or science in general. The application of the
funds extends to individuals of all countries but not to members of
the Council of the Society. The dividends are not to be hoarded parsi-
moniously but expended liberally. The gross income of the fund is now
about £480.

The first meeting of the Council which had been elected at the Anniversary meeting of 1827 was held early in December, and among other
business was the report of the committee which had been appointed in the
early part of that year at the suggestion of James South; their report had
been sent to the Council in June, and at the Council's meeting of 25 June
it was remitted to the new Council of 1827-8. In this new Council
Davies Gilbert, as the President for that year, presided over the Council's
meetings. One of the new members of Council was Lord Colchester, a
distinguished lawyer of long experience, who as Charles Abbot had been
Speaker of the House of Commons, where Gilbert had made his acquaintance; he was made Lord Chief Justice in 1818, and was raised to the
peerage in 1827. He had been elected a Fellow of the Society in 1793 but
had taken no active interest in its affairs. According to Professor Babbage
the President had included Lord Colchester in the Council in order to have
his support in the discussion of the committee's recommendations and
counted on him as an advocate of great legal dexterity and experience
with whom such of the councillors as were in favour of the recommendations would be at a manifest disadvantage in any discussion. He
succeeded, with the support of the President, in prevailing on those
members of Council who attended the meeting to reject the committee's
report without further discussion. Thus the first attempt which the
scientific members of the Council made to initiate improvements in the
methods and practice of the Society was successfully stifled by a few of
their colleagues who were opposed to every kind of reform whatever its
merits might be. That a President should employ such a device to defeat

the recommendations of a committee of which as Treasurer he had been a member and in favour of whose report he had voted a short time before was strongly condemned by many.

In 1830 the consent of the Duke of Norfolk was obtained to transfer to the trustees of the British Museum the manuscripts (except the Hebrew and Oriental ones) included in the Arundel Library which at John Evelyn's suggestion Henry Howard, afterwards the sixth Duke, had presented to the Society in 1667. The manuscripts were valued at £3559, the whole of which sum was expended in the purchase of English and foreign scientific works for the Society's library where they were of the greatest value. The manuscripts, which related principally to archaeological, philosophical and ecclesiastical matters, were much more suited to the museum where they would be more readily accessible to students than at Somerset House, and where reference to them would be much more frequently made.

About this time an important duty devolved upon the President of the Society under the will of the Earl of Bridgewater, who bequeathed the sum of 'eight thousand pounds to be invested in the funds, which sum with the accruing dividends thereon was to be paid to persons nominated by the President'. The persons selected were to write, print and publish a thousand copies of a work 'On the Power, Wisdom and Goodness of God as manifested in the Creation'. The President requested the assistance of the Archbishop of Canterbury and of the Bishop of London in carrying into effect the testator's intentions, and the following were appointed to write separate treatises:

The Rev. William Whewell, F.R.S. Astronomy and General Physics considered with reference to Natural Theology.

The Rev. Thomas Chalmers, D.D. On the Adaptation of External Nature to the Moral and Intellectual Constitution of Man.

John Kidd, M.D., F.R.S. On the Adaptation of External Nature to the Physical Condition of Man.

Sir Charles Bell, F.R.S. The Hand: its Mechanism and vital movements as evincing design.

Peter Mark Roget, M.D., F.R.S. On Animal and Vegetable Physiology.

The Rev. William Kirby, F.R.S. On the History, Habits and Instincts of Animals.

William Prout, M.D., F.R.S. On Chemistry, Meteorology and the functions of Digestion.

The Rev. William Buckland, F.R.S. On Geology and Mineralogy.

At the time of his election Gilbert had stated that he had no wish to occupy the position of President for a long period, and he soon began to look for a suitable successor. His choice fell on the Duke of Sussex, the sixth son of George III, but why he should have thought that a member of the royal house would be especially suitable is not clear. George III had

on more than one occasion supported the Society's petitions for considerable grants in order to carry out scientific schemes which it considered
to be of national importance, but which its own resources could not
adequately provide for. George IV had recently founded two royal medals
to be awarded annually for scientific work of exceptional merit, but these
benefactions do not seem sufficient to justify such a departure from the
Society's well-established practice of keeping the administration of the
Society wholly in the hands of its Fellows; he may have been influenced
by the fact that the Royal Academy of Arts, which had been founded by
George III in 1768, was in a special sense the King's Academy. Alternatively he may have thought that with a royal duke as President, whose
knowledge of the Society and its aims could not be extensive, he as Treasurer
could count upon the duke's support and so strengthen Gilbert's opposition
to the reforms which the scientific Fellows had in view. All this is but
conjecture, but the President had been in touch with members of the duke's
household for some months before he allowed any mention of it to reach
the Council, by which time the matter was well advanced; and by the
middle of 1830 it became known that the President and certain of his
friends were prepared to propose that the duke should be invited to
become the Society's next President. That a member of the royal house
should allow his name to be put forward as a candidate at an election which
would be hotly contested was quite unprecedented, but it may be that
knowledge of the strong opposition which had been aroused did not reach
the duke or members of his suite until shortly before the Anniversary meeting. A representative group of the scientific Fellows being indignant at
what they regarded as an interference with their independence on the part
of the Court, persuaded J. F. Herschel, the astronomer, to allow himself
to be nominated in opposition to the duke, and Sir R. Murchison took an
active part in getting up the formal request to Herschel which was signed
by Adam Sedgwick and most of his friends in Cambridge. Sedgwick,
who was at the time a member of Council, wrote to Murchison on
21 November 1830: 'I intend to take a private step and a very strong one.
By this post I shall write to the Duke of Sussex and explain my views to
him very plainly. I shall then have liberated my conscience.' At that time
the duke was a frequent visitor at Cambridge where he greatly enjoyed the
hospitality of Trinity College, two of whose Fellows, A. Sedgwick and
G. A. Browne, had been appointed his chaplains. He was not displeased
with Sedgwick's boldness, and wrote in reply: 'I thank you for your
candour, and whether our opinions may differ on this or any other subject I
know how to respect the talents as well as the motives of any individual.'
   By this time it was too late to avoid a contested election. Canvassing
had been carried on by both parties with the rancour that was then usual
in a political contest, and all available influence was brought to bear on
doubtful voters.

During 1830 meetings of several of the most influential Fellows of the Society had been held and the procedure for selection of a President and a new Council was discussed at one of these. It was moved by J. F. Herschel and seconded by M. Faraday that

the President and Council be recommended to take into consideration the propriety of making out a list of fifty Fellows out of whom they would advise the Council and Fellows for the ensuing year to be chosen; and that such list contain: First the names of the existing Council stating whether any and which thereof are dead or absent from the country; Secondly, the names of twenty-nine other Fellows, out of whom they would advise the Society to select persons to fill up the vacancies on the day of election.

These resolutions were acted on by the Council who also recommended persons for the offices of Treasurer and Secretaries; but the selection of a President was left to the Society itself.

On the day of the election only two candidates had been proposed, the Duke of Sussex and Mr J. F. Herschel. When the ballot was closed it was found that 119 votes had been given for the Duke of Sussex and 111 for Mr Herschel. Mr J. W. Lubbock was elected Treasurer and Dr Roget and Mr Children were to be the Secretaries. Dr C. König, a mineralogist, became Foreign Secretary but he was not included in the twenty-one members of Council. Davies Gilbert served for three more years on the Council but not as an officer.

The non-scientific majority had for the second time scored a success.

In 1830 Dr A. B. Granville, a Fellow of the Society, published anonymously a strongly worded criticism of the Councils of the Society in which he drew attention to numerous omissions and shortcomings, as well as to cases of negligence and maladministration. Among the suggestions which he made was that the Duke of Sussex should be considered for the presidency of the Society. Seeing the stress that Granville had laid on the need for promoting its scientific aims more energetically his advocacy of a non-scientific president seems strange; but an explanation may be found in his early history: Augustus Bozzi Granville, the son of the postmaster-general at Milan, was born in 1783: he took his degree of M.D. at Pavia in 1802, and was well known as an Italian patriot. By his mother's wish he assumed the name of Granville, and settled in London as tutor to the sons of William Richard Hamilton, an antiquary and diplomatist, in 1813. In the following year he brought warning of Napoleon's expected escape from Elba; and in 1815 he headed a deputation from Milan to offer to the Duke of Sussex the crown of Italy. In 1836 he republished his earlier pamphlet of 1830 under his own name, having made numerous additions to it to include the improvements which had been introduced during the first five years of the duke's presidency, and also indicated what might still be done by the Society to improve its administration.

The three publications which appeared between 1827 and 1836, *The Decline of Science*, by C. Babbage, F.R.S., *Science without a Head*, 1830, and *The Royal Society in the Nineteenth Century*, by Dr A. B. Granville, F.R.S., 1836, and a pamphlet by James South, F.R.S., the astronomer, 1830, all criticized severely the way in which the administration of the Society had been carried on during the previous thirty or forty years. In several cases the criticism goes beyond what the evidence would seem to justify, and this is especially the case in J. South's publication, but many facts which they quote seem to be correct and well supported so that they are of historical value. Not a few of their suggestions were adopted by the Councils of 1833 and in subsequent years with advantage.

The Duke of Sussex accepted the presidency to which he had been elected by the Fellows at the Anniversary meeting with a majority of eight votes only in a total of 230 votes cast by a membership of 659. Looking at the figures more closely the very small support which he received from the scientific Fellows of the Society who numbered 213 is clearly seen: 111 of the scientific Fellows (52·1 per cent) voted for J. F. Herschel, the scientific candidate; and 119 (26·7 per cent) of the non-scientific Fellows voted for the Duke of Sussex; about 430 of the Fellows (65 per cent) were not present at the ballot. The Council which had been elected at the Anniversary meeting of 1830 consisted of twelve scientific Fellows and nine who had various other interests; among them were two, D. Gilbert and Captain H. Kater, who had been members of the committee of 1827 whose report the Council of 1827–8 had passed over without discussing it.

Immediately on taking up his duties the President at the first meeting of Council proposed on his own initiative that the Treasurer's report for 1829–30 should be printed and circulated to all the Fellows, as has already been mentioned. At the next Council in January a committee was appointed to report on the statutes, and this led to the formation of a special committee of forty-two members who went into the whole question of the statutes which were then in force recommending such modifications as they considered to be desirable; this in effect brought up again before the Council the recommendations of the former committee and certain other matters. The result was that the Council adopted the following resolutions:

1. The certificate of each candidate was in future to be signed by six, instead of only three Fellows, as supporters;
2. Elections were to be held only at the first Ordinary meeting of the Society in January, February, April and June;
3. The annual subscription was to be raised to £4, but no bond was to be demanded from candidates in future;
4. A list of the Council and Officers was to be prepared and circulated to all Fellows previous to the Anniversary meeting;

5. A single balloting list for the Council and Officers was to be prepared;

6. An abstract of the Accounts was to be prepared by the Treasurer, and printed for circulation to all the Fellows before the Anniversary meeting;

7. Provision was to be made for calling a Special General meeting when necessary.

At the same Council meeting it was ordered that all the papers and documents of the Society should be properly arranged and classified, an instruction which suggests that until then no orderly system for filing or indexing had existed. Apparently not very much was done, for Granville notes in 1836 that though the documents had been tied up in bundles nothing more had resulted; he does however agree that between 1830 and 1836 the library 'was put into decent order'.

The misgivings of the scientific group of Fellows had been amply justified, and their criticisms were now supported by the production of irrefutable evidence which had not previously been accessible to the general body of the Fellows. The services which Lubbock rendered to the Society during his tenure of the treasurership proved the urgent necessity for a thorough reorganization of the administration, which had yet to be undertaken. It will be remembered that in February 1752 Lord Macclesfield, who was at the time a member of Council, proposed at a meeting of the Society that a committee should be appointed which would select such papers as they thought proper to be printed, and should supervise the editing and publication of them. This, the so-called Committee of Papers, was approved and the officers and the vice-presidents were made permanent members of it. The quorum for the committee was at first fixed at five, but by the statutes of 1776 this was increased to seven. The committee was authorized by the statutes to call in the assistance of any member of the Society who was especially skilled in the branch of science which was dealt with in any paper, and as early as 25 May 1780 a paper by Mr Ludlow was referred to Mr H. Cavendish and Dr Hutton; the next case was on 21 March 1831 when a paper by Sir Humphry Davy was referred to Mr Michael Faraday; after this the employment of specialist referees was frequent.

The number of members on this Committee of Papers was probably about twelve to fourteen so that the quorum of seven was small when it is remembered that the papers to be reported upon would relate to many branches of science, and to do this properly would involve a considerable amount of work; at this time the scientific councillors varied between seven and nine even when the officers were included, and they in the latter part of the eighteenth century were often selected from the non-scientific Fellows though the papers referred to the committee must have been in almost all cases of a purely scientific character.

In 1830 Dr A. B. Granville, in his book entitled *Science without a Head*, criticized severely the shortcomings of the Society during the opening years of the nineteenth century. Living in London he had ready access to the Society's archives and records, and with the exception of a few trifling mistakes he has pointed out the shortcomings of the administration and inadequate control which the officers exercised at the beginning of the nineteenth century. He quotes numerous instances in which the selection or the rejection of papers by the Committee of Papers was ill-judged, and gives reasons for his conclusion. Sometimes the paper had not been read by any Fellow expert in the subject with which it dealt; in other cases none of the members of the committee who were present could have had any expert knowledge of the subject of the paper, and so on. Papers by astronomers of distinction were declined by a quorum on which no astronomer was present; mathematicians and physicists formed the majority on a committee which considered the merits of physiological papers. The definite impression which one gets from the various cases which he quotes is that at the beginning of the nineteenth century and probably for some years earlier the Committee of Papers had dealt negligently with the papers submitted to it; well-qualified referees were seldom employed, members of the committee voted on papers which were beyond the scope of their scientific experience, and with the aim of clearing off the business before them they seem to have given arbitrary decisions on unsufficient grounds. The cause is to be sought in the scientific weakness which had characterized the Councils throughout the eighteenth century; it had led to slack administration and encouraged the officers to decide matters which lay beyond their own experience, for in many cases they came from the literary group of Fellows. It was high time that the Society was completely reorganized and that a sufficient number of the councillors were selected from the Fellows who had adequate scientific qualifications; many of the Fellows were of this opinion although it was resolutely opposed by a number of others.

Another practice which had been formerly considered to be important was later being seriously neglected. In the early years of the Society matters which were brought before its meetings were keenly discussed, and in the statutes of 1663 (Chapter IV, para. v) this was included as part of its regular business; but in the course of the eighteenth century such investigation of scientific matters and communications was discouraged. By the early part of the nineteenth many of the Fellows insisted that all such debate or discussion was contrary to the statutes in force. In 1836 the reformers strongly opposed this view and discussions were therefore allowed but not encouraged. The statutes of 1847 lay down that the regulation of the debates of the Society is one of the president's duties; and also that at Ordinary meetings the business of the Society shall be to read and hear letters, reports and other papers concerning philosophical matters. From this time onwards discussion was allowed in accordance with standing orders determined from time to time by the President and Council so long as no matter relating to the statutes or management of the Society shall be discussed at them.

It will be seen therefore that many of those whose interest in the advancement of science was slight persisted during a series of years in hindering all scientific debate at the meetings of the Society. By stifling such discussion of papers, and by allowing the work of the Committee of Papers to be negligently performed the Councils had been rendering the Society's meetings less efficient for many years before 1820.

On 12 April 1832 the Council had ordered that a list of the candidates, who were to be considered at the forthcoming election, should be sent to all Fellows at least a fortnight before the ballot took place; but this was rescinded in June 1835, when the President was unable to be present on account of illness, on the plea that with four elections in each year the scheme caused too much work, but this could easily have been remedied. The motive would seem to have been the desire to obstruct any attempt to reform the election procedure. The Council in this year consisted of nineteen scientific members, the President and Davies Gilbert being the only ones who did not fall within this category; in the next year the President was the only non-scientific member of Council.

In 1831 the Treasurer drew the attention of the Council to the Croonian Lecture Fund since in his opinion it was difficult to justify the payment of the fee under the trust for some of the lectures which had been delivered in recent years, especially eleven lectures given between 1817 and 1829 by Sir Everard Home, the surgeon. No more lectures were therefore given until 1857 (except by Richard Owen in 1851), but the fees for the six years 1831 to 1836 were paid to the parish of St Mary-le-Strand for the poor in compliance with what was understood to be the conditions of Lady Sadleir's will. In 1856 the Treasurer reported that the Society held the sum of £59. 16s. 6d. which was due by the Society to the Croonian Lecture Fund. The Council ordered that a copy of Lady Sadleir's will should be procured and that the Treasurer should deal with the balance which he held for the fund in accordance with the terms of it.

In 1831 the Council ordered that the minutes of the Council meetings should be printed and copies circulated to members of Council, which was another useful step towards making the decisions and actions of the Council better known to the Fellows than they had been hitherto.

Since 1800 those whose contributions had been accepted for publication in the *Philosophical Transactions* had been required to supply also an abstract of their papers, and these were now printed and bound, to form the first two volumes of a new publication, *The Proceedings of the Royal Society*. In the third volume the abstracts were arranged under the meetings and following the order in which the papers had been read; the report of each meeting was headed by a brief account of the business which preceded the reading of the papers. The Anniversary meeting and the reports made to it were also included so that the Treasurer's statement became generally available to the Fellows, and could be referred to at any subsequent time. By 1904 seventy-five of these volumes had been published. After that date they were divided into two series: (A) Mathematical and Physical papers, and (B) Biological papers, in which form they are still appearing.

So far the President seems to have appreciated the need for a thorough revision of the Society's policy and to have carried out his duties with conspicuous success; his knowledge of science was slight, but he recognized in what respects the Society's administration had fallen behind what it should have been, and he realized that there was no justification for keeping the Fellows in ignorance of the Council's decisions or of the state of the Society's finances. By the latter part of 1833 his health began to fail; he only attended five of the eighteen Council meetings which were held in 1834, in 1835 he attended none, in 1836 he attended four out of eighteen for a part of the time only, and in 1837 one only, his place being usually taken by the Treasurer, J. W. Lubbock.

In June 1835 the Council rescinded their resolution of April 1832 laying down that a fortnight's notice of the election of candidates should be given to all Fellows; this was another retrograde step and showed that opposition to reform was still active.

On 6 July 1838 Mr D. Gilbert, then member for Falkirk, who had been responsible for bringing forward the name of the Duke of Sussex as a candidate for the presidency in 1830, introduced a motion in the House of Commons for an increase in the amount of the annual grant made from the Civil List to the duke on the grounds that he had not hitherto benefited so much as his brothers, and also on account of the expenses which he was accustomed to bear in entertaining literary and scientific visitors from other countries as President of the Royal Society. The government opposed the motion which was lost, greatly to the duke's annoyance.[1] It is evidently to this that the duke refers in his letter to the Council dated 17 August 1838 in which he resigned the presidency: 'Circumstances over which I have no control, and which I did hope to have seen remedied when once fairly represented and properly explained to those whose duty it was to have noticed and to have considered them, force me to absent myself for a while from London. This naturally must prevent my regular attendance and appearance amongst you and them, as you and they have a right to expect from the President of the Royal Society.'

If the scientific activity of the Society had not advanced during the past two decades as much as many of its Fellows had hoped, a definite step was taken in that direction in May 1838 by appointing eight committees to which matters relating to various branches of science were to be referred

[1] *Chronicles of Holland House*, p. 22, by the Earl of Ilchester. London, 1937.

in order that they should report on them to the Council. It will be remembered that so long ago as 1665 similar committees had been appointed to assist the Council. These had worked very successfully for some years but later they were allowed to lapse, only two or three of them having been reappointed for a further period. Now the experiment was to be tried again. Committees were appointed for Astronomy, Chemistry, Geology and Mineralogy, Meteorology, Mathematics, Physics, Botany and Vegetable Physiology, and Zoology and Animal Physiology; each was to consist of seven members, with power to add to their numbers, but the whole number of members in any committee was not to exceed fifteen. In December these committees were reappointed as well as the Committees of the Donation Fund, the Library, and the Catalogue. The Meteorology Committee was combined with that for Physics so the number of the scientific committees was reduced to seven. They were reappointed annually until 1849 when they were allowed to lapse. No reason for the change of policy is given so we are left wondering whether there was not yet enough work for them to do; or whether their activities were found to trench upon the privileges of the Council; or whether the non-scientific group had still enough influence to hinder their further employment. In 1851 and again in 1875 J. P. Gassiot, a physicist who had been an active member of Council in 1848 and 1849, raised the question of reappointing them, but on both occasions the decision of the Council was against doing so. It was not until 1896 that committees of a similar character, which are now called Sectional Committees, were appointed, and have been reappointed annually ever since.

The Council of 1838 had now to select and nominate a new president but the situation was by no means clear. It seemed, when Sir Humphry Davy had been elected, that opinion had turned in favour of having a scientific president, but the two elections which had taken place since his resignation had indicated a decided preference of the majority for men or political or social eminence rather than for those of scientific merit and ability. This time the choice fell upon the Marquis of Northampton. He was an antiquary of considerable distinction and was actively interested in geology and mineralogy; he was one of the early presidents of the Geological Society (1820–22) and had written on the geology of Mull, besides making large collections of fossils and minerals. He sat in the House of Commons for eight years, and then for several years resided in Italy. He succeeded his father in 1828 and returned to England in 1830 when he was elected a Fellow of the Royal Society; he had therefore little knowledge of the internal troubles which had disturbed it during the past twenty years, and when he assumed the presidency he found himself involved in the later stages of a movement which had aroused much conflict of opinion among the Fellows. His policy seems to have been to move cautiously and to avoid being carried away by the extremists of either

group. In this he was successful and though he was at one time appre-
hensive of what might be the effect of the proposals, which were strongly
supported in the Councils of 1846 and 1847, he was reconciled to their
adoption by the assurance that, so long as no change was made in the
Charters the Fellows of the Society would still possess the power to elect
whom they pleased without any restrictions. There was little doubt that
the scientific group of Fellows would soon return to the attack and press
again for some method of restricting the number of Fellows, as well as for
some form of selection to ensure that the scientific character and aims of
the Society would be fully maintained. Many of the Fellows realized the
urgent need for reorganization but no opportunity occurred until Lord
Northampton had succeeded the Duke of Sussex as President at the end
of 1838.

This question of reorganization was doubtless widely discussed among
the scientific Fellows from 1839 onwards, and informally by the members
of Council during the years 1839 to 1845 when from eighteen to twenty of
the Councillors were men of science, but no mention of it appears in the
minutes of Council during these years.

During the twelvemonths 30 November 1845 to 30 November 1846
the Council included:

Lord Northampton	President
*George Rennie, C.E.	Treasurer
Dr P. M. Roget, M.D. S. H. Christie, Astronomer }	Secretaries
Lieut.-Col. E. Sabine, Magnetician, **Foreign Secretary**	
*Dean of Ely (G. Peacock) T. Galloway W. R. Grove *Sir John Lubbock *Lord Wrottesley	Mathematicians and Astronomers
Bryan Donkin Charles Wheatstone }	Physicists
Capt. W. H. Smyth, R.N.	Hydrographer
John Bostock Sir W. Burnett John F. Royle William Sharpey	Doctors of Medicine
Charles Daubeny	Botanist
*Leonard Horne	Geologist
John Taylor	
Rev. R. Willis	

* Vice-Presidents

so that nineteen out of the twenty-one members were men of science.

A valuable recruit in the person of W. R. Grove joined the reformers in 1840 when he was elected a Fellow; at the end of 1845 he became a member of Council for 1846 and was renominated for 1847; in 1848–9 he was one of the Secretaries. Grove went to Brasenose College, Oxford, and took his B.A. degree in 1832; he selected the law as his profession and was called to the Bar at Lincoln's Inn in 1835. His legal work lay mainly in patent cases in which his mathematical and scientific training enabled him to build up a large practice. He was made a Queen's Counsel in 1853, a Judge of the Court of Common Pleas in 1871 and of the High Court in 1875. As early as 1839 he communicated a description of the 'Grove Cell' to the British Association; and this was followed by a long series of memoirs chiefly on electrical subjects. In 1842 he set forth in a lecture the idea of the interrelation of natural powers, which he elaborated later in a book published in 1846 under the title of *The Correlation of Physical Forces*. This and the independent study in 1847 by L. von Helmholtz, *Ueber die Erhaltung der Kraft*, contained the earliest general account of the principle now known as the 'conservation of energy'. When he was elected a member of Council in November 1845 Grove co-operated actively with those who were in favour of reform and were supported by the great majority of the scientific Fellows.

The first indication that new and far-reaching views were being considered by the Council was given by a resolution to the following effect:

That it is expedient to revise the Charters of the Royal Society with a view to obtaining a Supplementary Charter from the Crown.

This was adopted on 7 May 1846 as well as another:

That the President, the Vice-Presidents, the Secretaries and the Foreign Secretary be appointed a Committee to consider and report what alterations it would be desirable to introduce into a Supplementary Charter.

On 28 May 1846 the Council resolved:

That it be an instruction to the Council Committee appointed at the last Meeting to examine and report upon the several enactments in the Charters and Statutes, and how far they are adapted to the present state of the Society; whether any and what alterations are required, and whether such alterations would be most advantageously made by Charter or Statute; and that the Committee have power to add to their membership.

Also: Mr W. R. Grove was appointed a member of the Charters Committee.

Also: That at the next meeting of Council it be considered whether, in the event of the Society obtaining a New Charter, it would be expedient to introduce a clause limiting the number of Fellows to be elected in any one year, and another clause to alter the mode of election.

On 4 June 1846 the Council adopted another resolution:

That an instruction be given to the Charter Committee to consider whether any, and what improvements can be made in the mode of electing Fellows.

It is certain that the various points which were involved in the alterations which it was proposed to introduce in the Charters or the statutes had already been very fully discussed at informal meetings, since on 16 June the Charter Committee found themselves in a position to report very fully on the matters which had been referred to them. Their recommendations were as follows:

That the election of ordinary Fellows, not included in the privileged classes of Peers and Privy Councillors, shall take place on one day only in each year, and on the third Thursday of June.

This was a great improvement on the limit of four elections annually which had been laid down by the Council in 1831 only to be rescinded in 1835 by the Fellows who were opposed to the change.

That the number to be elected in any one year shall not exceed fifteen, exclusive of the privileged classes to be referred to hereafter.

This would at once bring about a reduction in the number of Fellows, and introduce a form of selection.

That at the first Ordinary meeting of the Society in March the names of all candidates proposed subsequently to the first meeting in March of the previous year be announced by the Secretary and their certificates suspended until the last meeting in April.
That in the first week in May a list of the candidates shall be printed with the names of their proposers and seconders.
That the Council shall have power to select by ballot from such printed list fifteen persons to be recommended to the Society for election.
That at the first Ordinary meeting in June the President shall read from the chair the names of the fifteen candidates whom the Council have selected; and that immediately afterwards a circular letter shall be sent to every Fellow naming the day and hour of election and enclosing a copy of the list of selected candidates.

This would ensure that every Fellow should be informed that an election would be held and know the names of the candidates who had been selected by the Council.

Every candidate who shall not be elected shall, if his proposers desire it, be included in the list of candidates of the following March.
The Committee after considering the statements of income and expenditure for the past five years were of opinion that the surplus of income over expenditure therein shown would be sufficient to offset any reduction of income caused by the reduction in the number of Fellows which their recommendations entailed.

They recommended also that, if Council were in favour of their proposals generally, the opinion of Counsel should be obtained on the legal point whether such alterations in the statutes would conflict with the Charter.

The Charters Committee also submitted the following estimates drawn up on the basis of the last five years:

INCOME

	£ s. d.	£ s. d.
Rent from estates	235 7 7	
Dividends on investments (excluding Trust Funds)	602 7 4	837 14 11
Sale of *Philosophical Transactions* (average of five years)		244 1 2
235 Contributing Fellows on 30 Nov. 1845 at £4	940 0 0	
19 do. do. at £2. 12s. 0d.	49 8 0	
		989 8 0
		2071 4 1
If 15 new Fellows elected: Admission fees	150 0 0	
If 8 of these compounded, say 6 at £60	360 0 0	
2 at £40	80 0 0	
		590 0 0
Total Income		£2661 4 1

EXPENDITURE

	£ s. d.	£ s. d.
Salaries	635 5 0	
Transactions and *Proceedings*	749 0 10	
Library	232 16 0	
Fuel, Light, etc.	62 9 3	
Furniture	38 1 11	
House repairs, etc.	53 16 9	
Stationery and Miscellaneous	60 1 9	
Petty expenses	83 2 0	
Taxes, rates, insurance	71 14 10	
		1986 8 4
Excess of Income over Expenditure		674 15 9
Balance in hands of Treasurer after investing £1000 on 30 November 1845		1076 11 10
Total available balance		£1751 7 7

In November the Council authorized the Charters Committee to obtain the opinion of the Attorney-General and the Solicitor-General whether the Council could pass a statute enacting that the number of Ordinary Fellows to be elected in any one year shall not exceed fifteen. The Law Officers of the Crown on 24 November 1846 gave it as their opinion that the Council could not pass a statute limiting the number of Fellows to be elected in any one year.

The Committee therefore recommended to the Council that the objects approved by Council might be legally attained by adopting statutes as follows:

(i) That the election should take place on one day.

(ii) That the Council should recommend to the Society the most eligible candidates; such selected candidates not to exceed fifteen in any one year.

This gave to the Society the Council's considered opinion which of the candidates were the most suitable and recommended that a specified group of them, fifteen in number, should be considered by the Society. The Fellows of the Society were thus able to accept the names recommended, or to exchange any of them for others, or to increase the number to be elected as they saw fit, thus retaining their rights as laid down in the Charters. The statutes as they stood in 1840 were revised to incorporate these and a few other minor alterations, and having been approved by the Council on 10 February 1847 were printed and circulated to the Fellows. The objects of the reformers had at last been achieved.

It seldom happens that the names of those who initiate or who take a leading part in reforms that change the whole future of an institution are recorded, but fortunately there exists in the library of the Society a pamphlet by J. P. Gassiot, F.R.S., who was a member of the Council from November 1847 to November 1849; this he printed and circulated privately in 1870, thereby throwing light on the part which Grove played in 1846–8. He wrote it in order to explain the reasons for Sir Edward Sabine's action in declining renomination for the presidency in November 1871. When recounting the services which Sir E. Sabine had rendered to the Society he says that Sabine 'as a member of the Charters Committee in 1846, jointly with Mr W. Grove and the late Mr L. Horner, had assisted in improving the then existing practice in the election of Fellows'. Later he goes on to say, 'Sir E. Sabine was Foreign Secretary in 1846 when Mr Grove *succeeded* in his proposition for altering the then existing practice in the election of Fellows'.

Very few Fellows of the Society have rendered to it such valuable services as did Sir William Grove. Sir Isaac Newton reorganized its Councils, and introduced the regular attendance at them by the President and officers which has characterized these meetings for two centuries past;

Sir John Lubbock in 1831 and the following years improved and modern-ized the financial procedure of the Society and provided the Fellows with the first annual statements which showed how their funds were being utilized. Sir William Grove, twenty years later, showed the Council how the proportion of non-scientific Fellows could be reduced and the scientific standard of the Society raised without any modification of the Charters. These results were obtained with little difficulty and, so far. as we can see to-day, without any serious opposition being aroused.

At the end of 1940 through the generosity of Mrs Grove-Hills, the wife of the late Colonel E. H. Grove-Hills, F.R.S., the Society received as a permanent loan a portrait of Sir William Grove by J. Edgell Collins as a record of the great services which he rendered to the Society as an ad-ministrator and a member of its Council.

There were a good many in the Society who did not like the new statutes and the changes which they were designed to bring about; this was fully recognized by those who had been most active in advocating their adoption; they felt that there was a danger lest the enthusiasm which had been aroused might wane, and that the old abuses might in time creep back. Some of the Fellows therefore, of whom Grove was the most active, determined to form a dining club which, instead of being almost wholly social in its aims as the Royal Society Club had at this time become, should aim at checking any retrograde influences which might make their appear-ance in the policy of the Councils, stimulating the intellectual activity of the members and strengthening the influence of science in Great Britain. The club, which was named the 'Philosophical Club', was founded on 12 April 1847 and consisted of forty-seven members, all Fellows of the Society. No members' guests were allowed but foreigners of distinction might be invited as guests of the club.

For a good many years the club flourished and its dinners were well attended; many proposals which in time reached the Council meetings and were there adopted owed their origin to the discussions which took place at the meetings of the club. The number of its members, which was limited to forty-seven, was meant to keep in mind the year in which the reformers achieved their aim. The history of this club has been compiled by the Rev. T. G. Bonney, F.R.S., and in it are recorded the names of the original members who were the most active and zealous of the reformers (see table on p. 264). Half a century later when it was felt that its work had been accomplished, it was amalgamated with the Royal Society Club.

One of the earliest of the suggestions which were discussed at the club was that the Society should endeavour to secure from the government Burlington House instead of Somerset House where the accommodation was inadequate and was likely to be required before long for government offices. In 1856 an official letter proposed this.

*The Original Members of the Philosophical Club, 12 April,
6 May, or 3 June 1847*

Name	Profession	Name	Profession
D. T. Ansted	Geologist	C. Lyell	Geologist
Admiral F. Beaufort	Meteorologist	*Prof. J. McCullagh	Nat. Philosopher
*Sir H. de la Beche	Geologist	Dr W. A. Miller	Chemist
T. Bell	Zoologist	Dr W. H. Miller	Mineralogist
W. Bowman	Surgeon Oculist	Sir R. Murchison	Geologist
W. J. Broderick	Zoologist	G. Newport	Entomologist
R. Brown	Botanist	R. Owen	Palaeontologist
Major P. T. Cautley	Palaeontologist	R. Partridge	Surgeon
*S. H. Christie	Mathematician	R. Pereira	Pharmacist
Sir P. Egerton	Geologist	Dr J. Phillips	Geologist
Dr W. Falconer	Botanist	G. Rennie	Civil Engineer
M. Faraday	Physicist	*Sir J. Richardson	Surgeon
*Prof. E. Forbes	Zoologist	Dr J. F. Royle	Botanist
J. P. Gassiot	Physicist	*Col. E. Sabine	Magnetician
J. Goodwin	Anatomist	Rev. A. Sedgwick	Geologist
Prof. J. Graham	Chemist	Dr W. Sharpey	Physiologist
J. T. Graves	Mathematician	*Capt. W. H. Smyth	Hydrographer
J. H. Green	Surgeon	E. Solly	Chemist
*W. R. Grove	Mathematician	W. Spence	Statistician
Sir W. Harris	Electrical Engineer	*Col. W. H. Sykes	Zoologist
Sir J. F. Herschel	Astronomer	W. H. Fox Talbot	Chemist
Dr J. D. Hooker	Botanist	Dr W. Wallich	Botanist
*W. Hopkins	Mathematician	*C. Wheatstone	Physicist
*L. Horner	Geologist		

* Members of Council, Nov. 1846–Nov. 1847.

Another and a very useful proposal was made by J. P. Gassiot that a
Scientific Relief Fund should be established. The revision of the election
statutes was another and the limitation of the President's tenure of office
was also discussed. Some of them were later brought before the Council by
those who advocated them and perhaps adopted; others were to be heard
of in later years, but in either case the club fulfilled a very useful part in
discussing improvements which might be of real use even if the time was
not suitable for their introduction. It was characteristic of the new spirit,
to try all things and hold by that which was good and likely to be beneficial, and recalls the early days of the Society when the same eagerness was
manifest generally until the growth of formal administration at first
slackened the effort and later almost brought it to a standstill.

In May 1848 the first selection of fifteen candidates was made; they were
recommended to the meeting of the Fellows for election, and were then
formally elected. The new statutes were now in operation, and the un-
wearied efforts of those who desired to see the Society working as a
scientific institution and devoting itself wholly to the 'improvement of

Natural Knowledge', had been crowned with success. Twenty-five years had passed since Sir Humphry Davy had appointed a committee to consider how the Society's organization and management could be improved, and only twenty since Dr Wollaston's committee pointed out that its most urgent needs were to select the most suitable candidates for election, to restrict the number of those elected, to control expenditure more effectively and to exercise a more critical selection of the papers which were published by the Society. In spite of several attempts to frustrate the reformers' efforts success had been achieved, and it remained only for the Councils to show that the right method had been selected in order to give the Society the distinguished position in the world of science which was due to it.

The practice of inviting guests to visit the rooms of the Society in order to meet the President and Fellows and to inspect collections of scientific instruments and other objects, which had been brought together for the purpose of illustrating the most recent advances in scientific research, seems to have been introduced in 1849. At this time the Council were very active in promoting the scientific influence of the Society, and in bringing it to the notice of as many as possible of the public; in March 1848 the Finance Committee, to which the Council had referred the question of obtaining additional accommodation in Somerset House for the meetings of the Society, recommended:

1. That the President of the Royal Society be requested to communicate with the President of the Society of Antiquaries respecting the use of a room on the north of the present Meeting-room of the Royal Society, permission to occupy which on certain Thursdays in the year is given with the sanction of the Government to the Society of Antiquaries;

2. That the President and the officers of the Royal Society be requested to communicate with the officers of the University of London respecting permission to occupy the room mentioned above, and also if thought expedient, other rooms in the possession of the University of London, at such times as such rooms may not be required by the University of London or by the Society of Antiquaries.

The University of London and the Commissioners of Woods and Forests, as the government department responsible for Somerset House, gave their consent in June and September 1848 respectively; and in 1849 the Society of Antiquaries was requested to allow the Royal Society the use of their meeting-room for the soirées of the President and consented.

At these soirées or conversaziones the President was the host and visitors were invited as his guests; he also defrayed the expenses and it was not until May 1871 that the Council resolved that the cost should be met from the funds of the Society. As these entertainments were at first given by the President, no mention of them occurs in the Council minutes between

1849 and 1871; but after this for some years the dates of the conversaziones are given and in some cases the number of guests attending them, usually about 500, was recorded. A set of the programmes, which is nearly complete, is preserved in the library, the earliest being that for 1862. Up to 1873, when the Society was transferred to its present rooms, these programmes were printed on large sheets of the 'poster' type, which were hung up in the rooms in which the exhibits were shown. The original sheet measured 20 inches by 30 inches. Programmes printed in page form were introduced in 1873.

In 1884 the Soirée Committee of that year discussed whether it was to the advantage of the Society and its aims that these annual entertainments, at which the Society's guests could see objects and apparatus related to scientific research, should be given. In the report which was rendered to the Council the Committee recommended that they should be continued and to this the Council agreed; they have therefore been held annually up to the present time except during the war. This committee, which is charged with the selection of the exhibits and with making the necessary arrangements for the two annual conversaziones or soirées and for the Anniversary dinner, was appointed in 1874, and has been reappointed at the beginning of each session since. The notable improvement of the constitution of the Society and the increase in its scientific reputation which had been successfully attained by the Fellows during the second quarter of the nineteenth century had not passed unnoticed by the government of the day. On 16 November 1849 a letter which Lord John Russell, the First Lord of the Treasury, had addressed to the Earl of Rosse, the President of the Society, informed him that Her Majesty's government proposed to make a grant for the promotion of scientific enquiries, and suggested that the Royal Society should administer the grant. At their following meeting the Council requested the President to inform Lord John Russell that the President and Council unanimously accepted the government's liberal offer, and would without delay consider the best means for giving effect to his lordship's object.

On 20 December a committee consisting of Sir Jonathan Frederick Pollock, the Lord Chief Baron, Professor Richard Owen, Sir Roderick Murchison, Dr W. A. Miller and the officers was appointed to consider and to report to the Council respecting the application of the grant which Her Majesty's government proposed to make for the promotion of scientific enquiries in this country. This committee presented their report to the Council on 7 March 1850, and in it recommended that:

Firstly, the grant be awarded in aid of private individual scientific investigations;

Secondly, in aid of the calculation and scientific reduction of masses of accumulated observations;

Thirdly, in aid of astronomical, meteorological and other observations, which may be assisted by the purchase and employment of new instruments;

Fourthly, and subordinately to the purposes above named, in aid of such other scientific objects as may, from time to time, appear to be of sufficient interest, although not coming under any of the foregoing heads.

The report was adopted by the Council and a Committee of Recommendations was appointed. The sum which the government proposed to grant was £1000, and it was to be administered by this committee.

This was quite a new departure on the part of the government, for while funds for special specified purposes had frequently been provided, making an annual grant for scientific investigations which was to be administered under the supervision of the Council of the Society was unprecedented. Some of the Fellows disliked the scheme and feared lest it might lead to the government exercising some form of control over the Society which had hitherto enjoyed complete freedom to act as seemed best to the body of the Fellows. Others again argued that if funds were available they would be better employed in increasing the salaries of the staffs employed at the National Museums of Science. Nevertheless the offer was accepted and, with occasional modifications to the regulations under which it was operated, it has proved to be of the greatest value.

A misunderstanding, which arose early and which exists even to-day in the minds of some, was due to the proposal being described as a grant to the Royal Society, whereas the Society derived no benefit from the grant but undertook to administer it under a set of regulations, which were drafted and accepted by the Treasury, and with the aid of a committee on which representatives of several scientific societies served together with those of the Royal Society.

On 6 January 1851 Lord John Russell wrote to the President informing him that 'one thousand pounds would be set apart from the fund for Special Service to be applied by the Council of the Royal Society in the same manner as the Grant made for scientific purposes last year'. After three years the fund from which the grant had been made had been expended and the Treasury informed the Society that a parliamentary vote for the amount of the grant would be asked for in the estimates for the following year. The Council accepted the proposal but took the opportunity to point out that the grant was not one made to the Royal Society for its own needs, but was a contribution on the part of the nation towards the promotion of science generally in the United Kingdom.

In 1876 a letter was received from the Lord President of the Council proposing that further aid should be given to research by raising the amount of the grant to £5000.[1]

[1] See the *Record of the Royal Society of London*, 4th ed., Chapter IX.

In 1848 the Assistant Secretary reported that thirteen sealed documents had been committed to the custody of the Society and placed in the 'iron chest used for this purpose'. The earliest had been received in 1829 and the latest in the year of report. In April of this year Sir Henry de la Beche moved in Council that 'it is inexpedient that the name of the same Fellow of the Royal Society should be inserted as President in the House-list of the Society for more than four successive years'. This the Council adopted, but the next President, Lord Rosse, held the post for seven years, Lord Wrottesley for four, Sir Benjamin Brodie for three and Sir E. Sabine for ten. No statute has ever been passed dealing with the tenure of posts by the officers but so far as that of President is concerned it has been held for not more than five years by the same Fellow since 1870.

Even under the reorganized administration which had been already brought into operation during several years there was still a good deal to be done, for the legacy of the past had been an unexpectedly heavy one. In 1851 one of the Fellows complained to the Council that on asking the Assistant Secretary for the minutes of Council for 1845 none were forthcoming of later date than 1832. It appeared that the rough manuscript copies of these minutes as confirmed by the Council were in the library but no fair copies had been made. It was then ordered by the Council that a fair manuscript copy of the minutes of Council should in future be entered in a book kept for the purpose, and that the minutes entered in this book should be signed by the chairman on being confirmed; and that these should be regarded as the authentic record of the proceedings of the Council.

In 1855 Sir James South called attention to the absence of Leuwenhoek's microscopes from the rooms of the Society; they had been bequeathed at his death in 1723. Careful enquiries were made but without any result; it seems most probable that they left the Society's rooms between 1800 and 1830. In the days when the Society changed its quarters from time to time, and when no store-ledgers in which its property was recorded were kept, instruments and other objects which were lent to Fellows temporarily were easily overlooked and later forgotten. Flamsteed's clock by Tompion and a clock by Fromantel given by Dr Seth Ward in memory of Lawrence Rooke provide instances of such mishaps, and there were others. In more recent years store-ledgers have been opened, and the condition and whereabouts of all objects on loan are checked and reported on yearly.

In 1856 a letter was received from the Treasury in reply to a petition from the Chartered Societies that accommodation should be provided for them in juxtaposition in a convenient and central locality. It was explained that while the government were not yet in a position to state their definite views they were prepared to concede the temporary location of the Royal Society, the Linnean Society and the Chemical Society in Burlington House, and the offer was accepted.

In 1851 the Great Exhibition was held in Hyde Park and proved to be unexpectedly successful in directing the attention of the nation to the importance of employing science in the advancement of both Art and Industry. So successful was it that when it was closed the Royal Commissioners had at their disposal the sum of £180,000 which, at the suggestion of the prince consort, then Chairman, was expended in purchasing a considerable area of land in South Kensington consisting mostly of fields and gardens. In the second report of the Commissioners a suggestion was made that when this estate came to be developed it might provide a suitable place in which to group the most important scientific societies. This did not at all accord with the views of the Chartered Societies and after a long discussion they requested the Royal Society to approach the government for accommodation in a more central position. By the end of 1856 the Royal Society was able to move from Somerset House to Burlington House where rooms were set apart for the use of the Linnean and Chemical Societies as well. The Geological Society remained for the time being at Somerset House.

In the year 1854 the Earl of Rosse intimated that he did not desire to be renominated for the office of President, which he had held for the past seven years. The Council therefore decided to approach the Earl of Wrottesley, who was well known as an active amateur astronomer, and had built a small observatory for his own use at Blackheath.

In 1855 the Society received notice that a railway company desired to acquire a part of the Society's property at Acton. This land, which consisted of many scattered plots, was increasing rapidly in value as the houses of London extended westwards, so the Council appointed a committee to consider the whole question of the future development of the property and especially whether the various separate holdings could not be united into a single estate under the powers of the Inclosure Commissioners. This very prudent and timely action was carried out in co-operation with the other neighbouring landowners and by the end of 1858 an agreement had been reached under which the area of the Society's land was fixed at 34 acres 3 roods 11 poles. The wisdom of having this complicated matter settled became evident about twenty years later when the Society accepted an offer to sell the estate.

Early in 1859 J. P. Gassiot explained to the Philosophical Club his proposal that a fund should be raised the income of which should be devoted to the relief of scientific men and their families. The members of the club warmly supported it, and in May the Council were informed of what had been done. The fund, it was suggested, should be invested in the names of the President of the Royal Society and the Council. At the next meeting the definite proposal was submitted to the Council together with the information that seventy-seven Fellows had already given their support and that sixteen of them had given £100 each to it. On 3 November the report of a committee which the Council had appointed to consider the proposal reported in favour of the scheme and it was approved by the Council. By the end of the year £3200 had been received and had been invested. This sum has grown steadily by gifts and bequests until the value of its invested funds now exceeds £19,000 and the annual income which is at the disposal of the Scientific Relief Committee is not less than £800.

The Society's business was now very considerable and the increasing amount of scientific communications added greatly to the work of the two Secretaries. Whereas between 1790 and 1799 only 319 papers were communicated to the Society, the number in the ten years 1850–59 was 672. The editing of the *Proceedings of the Royal Society* as well as of the *Philosophical Transactions*, both of which publications were becoming more voluminous, also added materially to their labours. The Council therefore at the beginning of 1860 resolved that the gratuity paid annually to each of the Society's Secretaries should be increased by £95, thus bringing the total amount of each gratuity to £200. That paid to the Foreign Secretary had hitherto been £20, which represented the income of the Keck bequest, but as the relations with foreign countries had increased and were likely to be largely extended the Council decided that it should be raised to £100.

By 1860 the reformers had succeeded in completely transforming the Society; for the first time since its foundation the scientific Fellows were more numerous than their colleagues, and as the councillors were now almost all men of science it was not to be expected that they would re-commend for election those who had no knowledge of natural science or any interest in it; the average number of Ordinary Fellows during the decade 1871–80 was about five hundred and after that it decreased slowly. Between 1830 and 1860 the scientific Fellows increased by 117, and their colleagues decreased in number by 146. At this period the Council con-sisted almost wholly of scientific men who actively promoted the aims of the Society, and the greatly improved attendance at the meetings led to a very considerable amount of useful work being done. As has been said the number of scientific communications had greatly increased, adding largely to the work of the officers and of the councillors, as well as to the costs of printing and publication. All this had come about as the result of raising the number of scientific members of Council which Dr Wollaston and those who were associated with him had advocated in 1820 and suc-ceeded in introducing a few years later; once the Council had become a body of scientific men holding similar views for the advancement of the Society further improvements could not be long delayed. By 1848 the new statutes, which restricted the election of candidates to one date in the year, and the recommendation which the Council might make to the Society to elect fifteen annually, put an end to the unrestricted admission of candidates whose qualifications were known to few. The Society at a general meeting could elect whomever it pleased, but in practice there was no desire to go beyond the recommendations of the Council whose selection of candidates in these times was accepted as having been impartially and dispassionately carried out.

It is impossible to form any reliable opinion on what the probable effect would have been had the scientific composition of the Council been increased early in the eighteenth century, and if scientific opinion in the Society been influential enough to have started such an arrangement in Sir Isaac Newton's or Sir Hans Sloane's presidencies, and to have continued in subsequent years the gradual increase in the number of councillors of this type. A change in the policy and character of the Society would probably have been brought about before the end of the century, but the opposition to any such scheme was doubtless too powerful. By 1820 the position was easier for there were in the Society double the number of scientific men, and the industrial revolution in this country had made many realize how much larger and more important a part science and the application of it to industry was to play everywhere in the immediate future. Among these may be mentioned: J. C. Adams [1849], mathematician; Sir J. Herschel [1813], astronomer; M. Faraday [1824], physicist; Sir J. Hooker [1847], botanist; Sir Ch. Lyell [1826], geologist; C. R. Darwin [1839], naturalist.

¹ The dates in square brackets indicate the year of election to the Royal Society.

REFERENCES

BABBAGE, Professor C. *The Decline of Science in England.* London, 1830.
BONNEY, Rev. T. G., F.R.S. *The Annals of the Philosophical Club.* London, 1919.
GEIKIE, Sir ARCHIBALD, F.R.S. *The Annals of the Royal Society Club.* London, 1917.
GRANVILLE, A. B., M.D., F.R.S. *History of the Royal Society in the Nineteenth Century.* London, 1836.
Royal Society, Record of. 4th ed. 1940.
WELD, J. C. *History of the Royal Society.* Vol. II. London, 1848.

CHAPTER VIII

A SCIENTIFIC SOCIETY: 1861–1900

THE important changes in the constitution and administration of the Society which have been described in the last chapter were not long in producing their effect. Though only thirteen years had passed since the new statutes governing the selection and election of candidates had come into force, and the total number of Fellows had already fallen from 764 to 630, only one hundred and ninety-five candidates had been elected, in addition to a few who were admissible as privileged candidates; this was a much smaller number than would have been elected in former years when the average number admitted annually had been from twenty-five to thirty and occasionally as many as fifty. From the names of the candidates which are printed in the *Record* (pp. 478–98) as having been selected between 1861 and 1900, it will be seen that selection under the new procedure was already increasing the admissions of men of science. Most of the new Fellows preferred to pay their subscriptions annually so that the number of those who had compounded for their subscriptions had fallen from 480 to 375; this practice was certain to continue, so that the Society's income would be reduced as the Council had anticipated. Now for the first time in the history of the Society the scientific Fellows exceeded their colleagues in number, and their majority was to increase steadily for some years to come until by 1900 or even somewhat earlier all but those who had been admitted as privileged candidates, some twenty in number, were men of science. The number of scientific men on the Councils was now rarely less than eighteen. By the end of the nineteenth century the Society, after overcoming much opposition and indifference, had realized the aims of its founders, and at last had become an institution for promoting natural science, having an organization which was adequate for effecting this.

THE OFFICERS

At the Anniversary meeting in 1860 Sir Benjamin Brodie, a distinguished surgeon, was re-elected President for the third time. In the previous October he had written to say that his eyesight was much impaired and that a recent operation had not improved it; he thought therefore that he should not stand for re-election. The Council however pressed him to continue in office and he consented to do so, but in June 1861 he decided that his resignation had become unavoidable. At the Anniversary meeting of that year therefore the Society elected Major-General Edward Sabine as his successor. The other officers were: Dr W. A. Miller, Treasurer, who took the place of General Sabine; Dr W. Sharpey, M.D., and Professor George Gabriel Stokes, a physicist, the Secretaries; and W. H. Miller was the Foreign Secretary.

General Sabine had been elected a Fellow in 1818 and had occupied the posts of Secretary 1827–30, Foreign Secretary 1845–50, and Treasurer 1850–61. Dr Sharpey and Professor G. G. Stokes had already held the post of Secretary for eight and seven years respectively and therefore were well acquainted with the administration of the Society under the new conditions.

In 1836 Dr A. B. Granville in describing the procedure at a meeting of the Society at that time says that the papers to be communicated were read in abstract by the senior Secretary, these abstracts of the mathematical and physical papers having been prepared by one of the Secretaries and those relating to biological science by the other; this required that one of them should be familiar with mathematical and physical science while the other should be well qualified in biological science. Previously the Council had never laid down that the responsibility for matters relating to physical and biological science should be shared between the two Secretaries of the Society since the posts were not always held by men of science; from 1778 to 1784, for example, both of them were literary men, being members of the staff of the British Museum. It was not until 1827 that the practice of electing a physical and a biological Secretary began, and Lieut.-Colonel E. Sabine was then elected as the physical, or as it is now called the 'A' Secretary, and Dr P. M. Roget, M.D., was elected as the natural history or 'B' Secretary; and this division of the Secretaries' work has been customary ever since. In November 1848 W. R. Grove, the mathematician, and S. H. Christie, the astronomer, were elected Secretaries though both of them were mathematicians, but this was doubtless due to the convenience of having Grove, who had been very closely connected with the revision of the statutes relating to elections, as one of the officers during the first year during which they were in operation. He was replaced by T. Bell, a surgeon, in the following year.

Under the conditions which now prevailed the regularity with which the officers and councillors attended the meetings of Council is of less interest than it had been a century earlier because the officers and the other councillors were now present at most of the meetings. On the other hand, presidents were being changed more frequently; they were eminent scientific men who were fully occupied with their other duties and with their own researches, so that it was difficult for them to continue in office for long periods. The Council had recently expressed the opinion that four years was a suitable period for which a president should remain in office, but no statute laying this down as a definite rule had been adopted. General Sabine held office for ten years, but all the other presidents after his tenure of office came to an end have held office for not more than five years. His successor, Sir Joseph Hooker, told his colleagues before he was elected President in 1873 that he did not wish to hold the appointment for more than five years as his work at Kew was exacting and was likely to increase; this has been the usual practice ever since, though in a few cases a president has declined renomination on the ground of ill-health after a shorter period. Eight presidents held office between 1860 and 1900, namely Sir Benjamin Brodie, Sir Edward Sabine, Sir George Airy, the Astronomer Royal, Sir Joseph Hooker, William Spottiswoode, Professor T. H. Huxley, Sir George Stokes, and Lord Lister.

No restrictions had been suggested for the other officers and the tenure of their posts varied considerably at this time. Four physical Secretaries, Sir George Stokes, Lord Rayleigh, Sir Arthur Rucker and Sir Joseph Larmor, held office in this period, and so did three biological Secretaries, Sir William Sharpey, Professor T. H. Huxley and Sir Michael Foster. The Treasurers up to 1897 were W. A. Miller, W. Spottiswoode and Sir John Evans; the last of these was succeeded by Sir Alfred Kempe, who then held the treasurership until 1919.

How long the different officers should hold office was discussed in Council from time to time but further consideration of the matter may conveniently be left until the next chapter since in 1933 the matter again came before the Council, and in 1934 a memorandum on the subject which was signed by a number of Fellows was presented to the Council and was later remitted to a General meeting of the Fellows (see p. 323).

Throughout the second half of the nineteenth century the work of the Society increased greatly and expanded in many new directions. There may therefore have been good reasons for some of the officers continuing to act for exceptionally long periods; Professor G. G. Stokes was physical Secretary for thirty-one years; W. Sharpey was biological Secretary for seventeen years, and Professor Michael Foster for twenty-two years. Sir John Evans was Treasurer for twenty years, and was followed by Sir Alfred Kempe, who acted for twenty-two years. These exceptionally long tenures of office may then have been justifiable, but it is now generally regarded as preferable that they should not exceed ten years. At any rate it was a great advantage to the Society that these posts were held by scientific men of eminence, who could form reliable opinions on the questions which now came before Council and could deal efficiently with the many difficulties which the selection of papers communicated to the Society for reading and publication often presented at a period when it was developing its constitution as a modern scientific institution. Various old-time practices had to be discarded, and new ones introduced.

COUNCILS

When in 1821 the number of scientific Fellows selected to serve on the Councils was increased, the attendance of the councillors at the meetings at once improved considerably, and in the decade ending 1860 the average number of scientific members of Council was 19·2 out of twenty-one, the average attendance at a meeting being 13·9 which was three higher than in the ten years 1811–20. In the last forty years of the nineteenth century the number of scientific members of Council in any year varied from eighteen to twenty-one and the average attendance at a Council meeting rose from fourteen to nineteen. When it is remembered that the councillors were all busy men it is hardly to be expected that such regularity could be much improved. The administration of the Society and its contributions to the improvement of natural science had been now brought up to the standard which many of the Fellows had long desired to see attained; shortcomings which had been transmitted from former times were being rectified, and matters which had been overlooked in the past were considered and dealt with so that the Society's efficiency rapidly improved. Government departments were now referring various matters to the Society for its advice more frequently than in the past and some of these enquiries, such as meteorology in the service of the State, reached the magnitude of a lengthy enquiry which extended over several years.

Towards the end of the century international co-operation in scientific research attained considerable importance and occupied much of the Council's time. Intimate relations had been maintained with the scientific men of other countries from the earliest days of the Society, but international co-operation in scientific research began much later. Apart from the comparison of standards of length and weight by the Royal Society and the Académie des Sciences in 1742 and the geodetic operations undertaken by English and French geodesists in 1787 and 1821 to determine the difference in longitude between the observatories of Greenwich and Paris, it may be said to have been started by von Humboldt in the early part of the nineteenth century. In 1836 he wrote to the Duke of Sussex urging that magnetic observing stations should be established in the British possessions in North America, Australia, the Cape of Good Hope and between the tropics for obtaining records of periodical and secular changes.

THE FELLOWSHIP

At the end of 1860 the scientific Fellows in the Society and their non-scientific colleagues were for the first time nearly equal in numbers; the former had increased by 117 since 1830 and the latter had fallen off by 146; the total number of Fellows was now 630 besides 42 foreign members. Under the new statutes governing the elections the number of Fellows continued to decrease and by 1900 had fallen by nearly two hundred. The average number of Fellows for the period 1891–5 was 457 and it has only varied from this figure to 475 since then; it may be said therefore that by the end of the nineteenth century the Society's Fellowship had reached a stable figure of from 455 to 475, which varied as the yearly losses by death were greater or less than usual.

At the end of 1847 the number of Fellows who had compounded for their annual subscriptions was 480, and nine years later it was 376, 164 having died and only 60 of those newly elected having paid compounders' fees. Within the same period those paying annual subscriptions had only increased by one, deaths and resignations having accounted for 99, and 100 new subscribers having been added to the list. The decrease of 102 in the Fellowship during the period was therefore mainly due to the retirement of many who had compounded.

In 1874 a new statute was enacted which limited the privileged class of candidates to Princes of the Blood Royal, and Members of the Privy Council, so that the class of those 'having the title and place of a Baron or having any higher title and place' could no longer be proposed and elected on the same day, a privilege which had been theirs since the foundation of the Society;[1] they are now subject to the ordinary procedure of election. Privy Councillors continued to be qualified for privileged election until 1917. Except therefore for the few Fellows who entered the Society by privileged election, before the end of the century the whole body of the Fellows consisted of those who had been accepted for their scientific qualifications; those who had been admitted earlier without such knowledge had by this time died or had resigned their fellowships.

One result of restricting the number of candidates recommended in each year for election to the Fellowship must have been foreseen by the Charters Committee of 1846 though no mention of it is made in their report to the Council. Previously nearly all the candidates whose names were submitted to the Society were elected, and in some years, in 1834 for example, the number so added to the Fellowship had been as high as fifty. In 1848 and for eighty-two years after that year, however many candidates might have been proposed only the names of a definite number, fifteen, might be recommended to the Society for election; these the Council, now composed almost wholly of men of science, would select for their scientific knowledge, and for the quality of such research work as they might have carried out. This elimination of candidates whose qualifications were less noteworthy soon produced an effect, and the standard of the Fellowship rose as had been the intention of the reformers. Those who were not successful candidates at the first election might remain on the list and be considered with any new candidates in the one which was prepared in the month of March following. Thus in any year the list of candidates would now include those who had been proposed in a previous year but had not been elected, and also any candidates who had been proposed since the last election. From time to time some names might be withdrawn, but the general tendency would be for the number of candidates on each year's list to increase as the standard of the Fellowship rose, and the distinction of election to it became more widely recognized.

[1] Statutes of 1663; Chapter VI, § 1.

In 1848 and again in 1849 there were twenty-two candidates, in 1850 there were twenty-seven, so seven in each of the two earlier years and twelve in the last year were not recommended for election. The average number of candidates who had been proposed for election in each year of the following decades has been:

Period	Average no.	Period	Average no.
1851–60	37	1901–10	87
1861–70	50	1911–20	111
1871–80	54	1921–30	118
1881–90	61	1931–40	145
1891–1900	72		

Thus the large number of candidates whose names come before the Council each year is made up of those whose names are on the list for the first time, together with a larger number of those who have already been candidates for one or more years, and so far have not been selected.

In 1930 the number of candidates to be recommended to Council was raised to seventeen and in 1938 to twenty in order to meet the cases of those whose research work lay in fields between two main groups of science and thus were closely related to both; such candidates were now becoming more numerous; but it would be a mistake to assume that the large number of names which now came before the Society each year as candidates is a true indication of those who ought to receive its recognition, for the names of some may have been left on the candidates' list by their proposers after it had become clear that there was but a small chance of their ever being elected. The whole question is a very difficult one and has for many years been discussed from time to time by successive Councils with the object of carrying out the task of selection as fairly as possible, and of only altering the regulations when it was clear that a real need existed which could be met in this way without lowering the standard of the Fellowship.

Between 1901 and 1940 the number of candidates from whom a selection had to be made had risen from 97 in 1901 to 152 in 1938; but before attempting to draw any conclusion from these figures it will be well to see how many candidates have been proposed for the first time in each of these years, for presumably these are the men who are in the opinion of their proposers likely to become leaders in their own line of research and therefore worthy of early admission to the Society. In the first decade the number varied between fourteen and thirty, in the second between fifteen and thirty-seven, in the third between twenty and thirty-six, and in the fourth between twenty-one and forty-four. When the number to be selected by the Council was increased, at first to seventeen and then later to twenty, the number of candidates put forward rose somewhat. The average number of those who were proposed for the first time in each year of the four decades was: between 1901 and 1910—22·7; between 1911 and 1920—24·5; between 1921 and 1930—26·7; and between 1931 and 1940—29·3. It cannot be said that these figures justify us in drawing any very definite conclusions from them, but they may serve as a warning against accepting the large numbers on the present annual list of candidates as proof that the Fellowship is unreasonably restricted; the present number of Fellows should only be increased after lengthy and very careful consideration.

FINANCE

By the middle of the nineteenth century the financial position of the Society was satisfactory; the financial reports which had been prepared for the Council in 1847 had shown that the receipts were in excess of the expenditure, and with care and economy should be sufficient, although the number of Fellows was decreasing and composition fees were now to be invested instead of being treated as income. In 1833 the invested capital amounted to £14,000, the capital of the trust funds not being included; and by 1857 this had risen to £30,000. The expenses due to the transfer of the Society and its furniture and library to Burlington House was estimated at £1500, which was to be met by selling some investments, thus reducing the reserve by this amount. While the scientific efficiency of the Society was growing the number of its Fellows was decreasing as had been foreseen; by 1878 the Ordinary Fellows numbered 505 against 764 in 1847, 253 being annual subscribers; by 1900 there were only 449 of whom 327 were annual subscribers. This reduction did not however introduce any serious financial difficulty; the estimates of the Finance Committee in 1847 and of the Treasurer in 1857 had shown that the Society should be able to count upon having an excess of receipts over expenditure amounting to a few hundreds of pounds annually, and this proved to be the case. The total sum derived from Fellows' contributions, admission fees and composition fees in 1850 amounted to about £1500, and these items brought in substantially the same amount until 1900. The principal increase was in dividends on investments which rose from £753 in 1850 to £1511 in 1875 and to £2650 in 1900. From time to time Treasurers suggested that the Council might use a part of the annual surplus to assist scientific research but the Council were disinclined to do this, at any rate until a stable figure for the Fellowship had been reached. The cost of publishing the *Philosophical Transactions* and other scientific publications was considerable, and now that the Fellowship was wholly composed of men of science not only were the papers which were accepted for publication in the *Philosophical Transactions* and *Proceedings* more numerous but the copies which were issued to Fellows free of charge were many more than they had been formerly. Fortunately before the century had come to an end the situation had been radically modified by the sale of the Acton estate. At the end of 1872 the firm of estate agents who had dealt with the Society's Acton estate for several years past wrote to say that there had been a great increase in building in that neighbourhood during recent years, and suggested that the Society might be well advised to advertise the land for sale. The Council agreed in principle and instructed that a detailed plan for laying out the land in building sites should be prepared. This was received by the end of the year and the Council authorized the Committee of the Acton estate to let the land on building leases. A few years later, in 1880, an offer was received to buy the whole of the land owned by the Society at Acton at a price which the Council regarded as being quite satisfactory, and the contract for the sale was approved and sealed in October. The purchaser however then raised the question whether the Society had legal power to dispose of the land in question under the provisions of the Vendor and Purchaser's Act. The case was argued in Court before Vice-Chancellor Hall on 5 March 1881, when the Society was able to show that it had purchased the land with funds which were at its free disposal. Judgment was given that, it having been proved to the satisfaction of the Court that the estate at Acton had been purchased or acquired by the Royal Society out of property which might legally be applied by the Society as income and did not form an endowment within the meaning of the Charitable Trusts Acts of 1853 and 1855, the Court accordingly declared that notwithstanding the 29th Section of the Charitable Trusts Acts, 1855, the Society had power to sell their estate at Acton without obtaining the consent of the Charity Commissioners. It will be remembered that this land had been bought in 1732 for £1600 with the money provided by the payments made by Fellows who had been in arrear in paying the fees and subscriptions as they had given an undertaking to do regularly in a Bond which they had signed at the time of their election (p. 127). The Council of that day, 1728, after taking legal advice, advised all defaulters that they would be sued for their outstanding debts to the Society. The sale was completed in May 1882 for the sum of £32,207. 13s. 7d. £15,000 of this was invested in government stock and £6940 was invested in twenty-three houses in Wharton Road, Kensington; the ground rent of 27 Basinghall Street, E.C., was also acquired for £9500. This last brought in a rent of £387, and is still in the possession of the Society. By this sale the financial position of the Society was greatly improved; careful administration had contributed to this satisfactory result; and by the end of 1900 the value of the securities in the General Fund as reported by the chartered accountants to the Society amounted to £81,076. There were twelve other special funds which were held in trust by the Society and administered in accordance with the expressed wishes of the donors or founders. These provided the Bakerian and Croonian lecture fees, the Copley, Rumford, Buchanan, Davy, Darwin and Sylvester Medals; also the Brady Fund for the Library, the Keck Fund, the income of which is paid annually to the Foreign Secretary under the terms of the bequest, as well as the Publication Fund and the Scientific Relief Fund. The total value of the securities belonging to these funds in 1900 was about £46,000. By the end of the nineteenth century there were six medals besides the two royal medals to be awarded annually or at longer periods by the Society for distinguished work in scientific research.

Every year the Council have to award several medals, and have been led by experience to the conclusion that it is neither to the advantage of the Society nor in the interests of the advancement of Natural Knowledge that this already long list of medals should be added to, and that, therefore, no further bequests to be awarded as prizes for past achievements should be accepted by the Society. They desire, however, to remind would-be benefactors that the funds belonging absolutely to the Society, funds tied down by no special directions as to their applications, funds which the Society are free to use for general purposes, are very few indeed. The President and Council have again and again had the experience that the usefulness of the Society for the advancement of Natural Knowledge has been greatly hampered by the lack of funds which they could freely use according to their own judgment.

The President and Council are confident that it would not be difficult, wherever desirable, to associate in some conspicuous manner with any gift to the Society the name of the benefactor, and indeed they would wish to do so. They have therefore made it known that, while they will willingly receive gifts to be applied to special objects or for the benefit of particular sciences indicated by the donors, they consider that, in view of the varying necessities of science, the most useful benefactions are those which are given to the Society in general terms for the advancement of Natural Knowledge.

This information is now printed in each issue of the *Year-book* of the Society.

There were also some funds, the income of which was according to the testator's or donor's wishes to be used for some specified branch of scientific research; these were: the Donation Fund, the Jodrell Fund, the Gassiot Fund for Kew Observatory, the Handley Fund, the Joule Fund, the Gunning Fund. The total value of the securities belonging to these research funds in 1900 was about £40,000. Full details of all these funds and the purposes to which they may be applied are given in the *Record of the Royal Society*, Chapter IV, and in the *Year-book* of the Society. The Society was therefore not only in a sound financial position but possessed an income of about £1375 a year derived from the research funds which was to be devoted to the furtherance of certain branches of scientific research.

The time had now arrived, to which Bishop Sprat had confidently looked forward, when the Society would receive gifts and bequests from those who were well-to-do and were acquainted with the value of the Society's work. He had not anticipated that more than two centuries would elapse before the greatly needed financial assistance from benefactors would be enough to afford material support to the Society, but so it was. The almost complete absence of any financial assistance being given to the Society during the eighteenth century by the non-scientific Fellows, from whom much had been expected, has been mentioned already. During the first half of the nineteenth there was no increase in benefactions though greater activity in the administration was to be seen in many directions. Between 1801 and 1850 the President, D. Gilbert, subscribed £1000 to the Donation Fund which Dr W. H. Wollaston founded, his brother-in-law, J. Guillemard, contributed £100, and one other Fellow, T. Botfield, who was not a man of science, gave £60; and in 1874 £500 was received from Sir Francis Ronalds. There were at this time 376 Fellows in the Society, many of whom were well off, but they had no knowledge of natural philosophy nor apparently did they take any interest in it. Not until the second half of the nineteenth century, when the Society was being rapidly converted into a scientific institution, did it begin to receive the generous support which it deserved, and then the greater portion of it came from various men of science who were Fellows of the Society itself. This financial support took the form of thirty-one gifts made by various benefactors whose names are given in the *Historical Record of the Society* (4th ed., pp. 142–4). The total value of them amounted to about £50,000 and included bequests, gifts, etc., some of which were in the form of trusts of which the income only was for annual use, while others were contributions to projects which the Society were engaged in at the time. The majority of these benefactions came from Fellows of the Society but six of them were given by members of the public in order to assist the Society in its work.

This summary of the administrative organization of the Society will serve to show that it passed through a rather difficult period between the success of the scientific Fellows in obtaining the control of the selection of candidates, and the creation of a properly organized scientific Society capable of carrying out research or of selecting and financing scientific workers whom it considered to be competent to carry out such work. Sir John Lubbock's work as Treasurer had clearly demonstrated that with care and prudent management the Society's resources should suffice for its needs; it was now housed by the State and the salaries of a moderate staff was not a heavy charge. On the other hand, the number of Fellows who were engaged in scientific research was increasing and the specialized societies recently organized were now promoting more actively than hitherto work of this character. The Council had to look forward to a considerable increase in the expenditure on publication which would only be met in part by the sale of the *Philosophical Transactions* and the *Proceedings* of the Society.

During the last half of the nineteenth century several proposals came before the Council for extending the Society's activity in international as well as in national fields of scientific research, and some of these obliged the Society to incur not only considerable expenditure from its private funds but also added materially to the work of its officers and of a number of its Fellows who generously offered to assist.

The reorganization which had been carried out since 1840 enabled the Society as its resources increased to undertake work on a larger scale and over a wider field than it had previously attempted. On numerous occasions in the seventeenth and eighteenth centuries the Society had petitioned the government to make grants to the Society in order that it might carry out researches of scientific importance for which its own re-sources were wholly inadequate; these included such projects as geodetic operations, the observation at places abroad of solar eclipses and transits of Venus across the sun's disc, voyages of exploration and much else. These applications usually received sympathetic consideration and were often granted; ships to carry out the explorations, or to transport observers to the most suitable places for their work, were provided which necessitated an expenditure far in excess of what the Society could have provided. At first however it had to proceed with caution as composition fees were now to be invested, and the increase in annual subscriptions only made good the difference gradually; for some years therefore strict economy had to be practised.

One small change, which was introduced at the beginning of the period under consideration, is worth mentioning since it modified the hospitable practice of entertaining scientific guests which Sir Joseph Banks had in-troduced about eighty years before; it did not entail any considerable expenditure. In 1860 a charge for teas served after the Ordinary meetings appears in the accounts for the first time. The cost of this had for some years past been borne by the Presidents, and the custom had its origin in the Sunday-evening receptions which Sir Joseph Banks had been in the habit of giving at his house in Soho Square. Sir Humphry Davy con-tinued the practice so long as he lived in Grosvenor Place, but when he moved to Park Lane he arranged that tea should be provided after the evening meetings at Somerset House, and he defrayed the expense of this. Davies Gilbert, who was the next President, entertained parties of the Fellows at dinner at the Thatched House Tavern in St James' Street, but there is no record of his having provided the teas as well, though he may have done so. The Duke of Sussex gave several receptions at Kensington Palace when he became President but these were discontinued during his illness. In 1840 Lord Northampton resumed Sir Humphry Davy's practice of providing tea for those who attended the evening meetings and of defraying the cost; his successors seem to have continued the practice since it was not until 1860 during Sir Benjamin Brodie's presidency that the cost was first charged to the Society. In 1880 the meetings were held at 4.30 p.m. instead of in the evening and the teas were then provided before the meetings instead of after them. Now that the Presidents were no longer selected as being men of means as well as of scientific eminence, it was but reasonable that they should be relieved of any expenses which were incurred on behalf of the Society as a whole, and in 1871 the Council decided that the cost of the soirées, which had been borne by the Presidents since they were instituted in 1848, should in future be met from the funds of the Society, as the teas after the weekly meetings had been since 1860. The soirées were in fact a very useful extension of the Society's activities for they provided pleasant and welcome opportunities for the Fellows to meet one another and to discuss matters of common interest as well as showing to their guests recent advances in science. In these days of constant occupation for those engaged in scientific work such occasions as are afforded by the meetings in the tea room before and during meetings, and the more formal evening gatherings at the soirées as well as the dinners at the Royal Society Club, provide opportunities for scientific men to discuss with their colleagues matters of common interest.

In 1861, C. R. Weld, the Assistant Secretary, retired after eighteen years' faithful and valuable service to the Society, and was succeeded by W. White. For about a century Weld's *History of the Royal Society* has been the only full account of the Society's activities up to 1830 that has existed. The scientific movement was then only beginning, and the full importance of its results could not be estimated until towards the end of the century.

The business which came before the Council was now increasing year by year, so in order that important matters might be adequately discussed, special committees were appointed which rendered their reports to the Council on such matters as had been referred to them. In this way the Council could base their decisions on the advice given to them by those who were fully acquainted with all the circumstances and could judge how any action that the committee recommended would affect the policy of the Society. The members of these committees were reappointed annually in December, when any changes that were desirable were made. Some of these committees, such as the Library Committee, the Finance Committee, the Scientific Relief Committee, and one or two more, have been re-appointed annually ever since they were first appointed as the matters with which they deal are constantly coming up for consideration; others, like that which dealt with the arrangements for the sale of the Society's land at Acton, were discharged as soon as the matters with which they were concerned were settled.

The Library Committee, which had been first appointed before 1831, was the earliest of these committees and in 1860 was still the only one in operation; in 1873 the House Committee was added and also one for dealing with the Society's property at Acton and its disposal. Later a number of others were added in order to deal with matters relating to the annual soirées, the Society's estates, the *Challenger* Expedition, the volcanic explosion at Krakatoa, Borings in the delta of the Nile, the Catalogue of Scientific Papers, and other matters. Towards the end of the century there were twenty-one of these committees in operation. They relieved the Council of a great burden of detailed business which was dealt with by men who had a full knowledge of the subject and had by some years' service on a committee gained special familiarity with all the factors involved. These special committees enabled the Council to save much time as well as ensuring a notable increase of efficiency in dealing with the business which came before it. By the end of the century they had become a normal part of the Society's administrative organization.

In 1861 the Society of Apothecaries had for many years held the Botanic Garden at Chelsea under a deed of gift from Sir Hans Sloane subject to certain conditions failing the due performance of which the garden might be held by his heirs in trust for the President and Fellows of the Royal Society. As the Master and Wardens of the Society now contemplated obtaining a release from the responsibility of maintaining the garden they informed the Royal Society; but in the following year the Society of Apothecaries carefully reconsidered the action which they had proposed taking and rescinded their former resolution, having determined to retain the garden and to continue their responsibility for its maintenance. Thus the long-standing association of the two institutions and their interest in the maintenance of Sir Hans Sloane's Physic Garden was not interrupted.

The Second Centenary of the sealing of the First Charter occurred in 1862 but there is no record of any special arrangement having been made to celebrate it, nor do the minutes of Council contain any record of the First Centenary which occurred during the presidency of Lord Macclesfield.

As custodians of one of the sets of the legal standards of length and weight and as those responsible for many objects of value belonging to the Society, the Council decided in 1862 to provide a fireproof safe in which the national standards, the Charter-book, the Mace, the Charters and other muniments of the Society could be preserved in safety. Two years later the Treasury invited the President (Major-General E. Sabine) to form with the Astronomer Royal and the Master of the Mint a commission to examine the national standards of length and weight at the Exchequer Office and to bring with them the standards which had been deposited with the Royal Society in October 1853. These were a gun-metal Standard Yard No. 3 and a platinum avoirdupois Pound No. 2.

At the meeting of the British Association which was held at Glasgow in 1855 Professor Henry of Washington communicated a proposal for the publication of a Catalogue of the Philosophical Memoirs which were to be found in the *Transactions* of Societies in Europe and America. The catalogue was to include the mathematical and physical sciences only, and subject and author catalogues were to be included.

In March 1857 General Sabine enquired, on behalf of the British Association, whether the Royal Society would co-operate in the undertaking. This the Society agreed to do and it was decided to include all the sciences and not the mathematical and physical only; also the indexing was to be extended to all libraries except that of the Royal Society. The whole work was undertaken by the Society at its own charge.[1] The catalogue was originally intended to be in manuscript but by 1864 it was thought to be desirable to print it. By then the progress which had been made with the Catalogue of Scientific Papers since its commencement in 1857 was reported, and it was decided to propose to the government that the expenses of its publication might be borne by public funds since the catalogue would be for the use and convenience of scientific men everywhere and not for the Fellows of the Royal Society only; they had already carried out the work by the voluntary co-operation of many of the Fellows and by providing for the clerical and other assistance which had been necessary from the private funds of the Society. Official approval was given to the request and the printing of the first series covering the scientific serials from 1800 to 1863 was commenced by H.M. Stationery Office in 1866, the last volume being issued in 1872. Two more volumes to include the years 1864-73 were completed in 1876 and were published by the Stationery Office in 1879. Three more volumes were still needed to cover the next decade and the cost of these was met by a parliamentary grant, a handsome donation from Dr Ludwig Mond, and from the Society's own funds. These volumes were published by 1896.

In 1866 the Council received notice that the government had decided to allocate Burlington House to the Royal Academy and that other accommodation would be provided for the scientific societies which were then occupying it. It was proposed to construct new buildings on either side of the entrance court at Burlington House between it and Piccadilly in order to provide the necessary accommodation. After some discussion between the Office of Works, the contractors and the two institutions concerned, an agreement was reached whereby temporary accommodation would be provided in the courtyard during the construction of the new buildings, which would have to be constructed on either side of the courtyard of Burlington House. Ultimately the transfer of the Royal Society, its library, furniture and other property to the rooms that it now occupies on the east side of the courtyard was completed by 30 November 1873 and the Anniversary meeting of that year was held there. In the evening a soirée was given in the new rooms in order that the Fellows might see their new quarters; five hundred guests were present.

[1] Cf. *The Royal Society Record*, 4th ed., pp. 180-2.

Under the terms of the will of Dr Davy, to whom Sir Humphry Davy had bequeathed the service of silver plate which had been presented to him for his invention of the Miners' Safety Lamp, it now reverted to the Society that it might be sold to provide a medal to be awarded annually for the most important discovery in Chemistry made in Europe or Anglo-America. It was handed over in 1869 by the executors of the will and a sum of £736 was realized, of which £700 was invested in the purchase of £660 Madras Railway Stock to form the capital of the Davy Medal Fund.

Kew Observatory had been for some years administered by the British Association, but this institution now desired to be relieved of the responsibility. Fortunately Mr J. P. Gassiot, who had long been interested in the future of the observatory, offered to deposit securities of a value sufficient to secure an income of £250 per annum if the Royal Society would undertake the responsibility of the future maintenance of a Central Magnetical and Physical Observatory at Kew; to this the Council agreed.

Sir Edward Sabine having indicated that he would not seek renomination for the office of President at the end of 1871, the Council invited Mr G. B. Airy, the Astronomer Royal, to stand. He consented and was duly elected, but he only held office for two years, being succeeded by Sir Joseph Hooker after the Anniversary meeting of 1873, and on the following day the Society formally took over the building which it still occupies on the east side of the courtyard of Burlington House.

At the end of 1871 the Council wrote to the Admiralty to represent to the Lords Commissioners of the Admiralty that the experience of recent scientific investigations of the deep sea carried on in European waters by the Admiralty at the instance of the Royal Society has led to the conviction that advantages of great importance to science and to navigation would accrue from the extension of such investigations to the great oceanic regions of the globe. The President and Council therefore submitted a proposal for fitting out an expedition commensurate with the objects in view which are briefly as follows:

1. The physical conditions of the deep sea in all the great ocean basins;
2. The chemical constitution of the water at various depths from the surface to the bottom;
3. The physical and chemical characters of the deposit;
4. The distribution of organic life throughout the area explored.

For effectively carrying out these researches there would be required a ship of sufficient size for a probable absence of about four years, a staff of scientific men fully qualified, and a supply of all necessary materials and instruments. The proposal was approved and H.M.S. *Challenger* was provided for the service of the expedition, which was wholly successful and resulted in valuable and notable additions to our knowledge of oceanography. The results were published in a series of volumes after the return of the expedition.

When the Society moved into its new quarters, the Office of Works wrote requiring the Society to insure them for £35,000. This was an innovation, for no claim of the kind had been made when the Society accepted the accommodation at Somerset House, which was described as being 'for the purpose of holding the meetings of the said Society therein during the pleasure of the Crown without payment of any rent or other pecuniary consideration whatsoever'. The Society therefore claimed that its tenancy of Burlington House should be on similar terms and their protest was accepted by the Treasury, the claim being withdrawn in May 1874.

On 30 November 1875 the report of a committee, which had been appointed to consider whether it was desirable or not to make any alterations in the statutes relating to the election of Fellows, was presented to the Council. These statutes had been in operation for about thirty years and the number of ordinary Fellows had fallen from 764 to 503, of whom 253 had compounded for their annual subscriptions: thus there had been a decrease of about 230 in the number of those who had compounded as compared with 30 of annual subscribers. In order to investigate the position thoroughly, the committee sent to all the Fellows a questionary requesting each of them to furnish information as regards age, date of election and on other matters in which the records of the Society were incomplete. These were carefully considered by the members of the committee who came to the conclusion that the average number of Fellows, exclusive of foreign members, which would be reached under the rules for election then in force would be about 430; actually the number for the next fifty years (1876–1925) was 461. At the conclusion of their discussions the committee recommended to the Council that:

1. The duty of selecting the candidates to be recommended to the Society for election should be left in the hands of the Council as was then the custom;
2. The number of candidates to be selected and recommended for election by the Society should be fifteen as at present.

The Catalogue of Scientific Papers, which had been printed in the *Transactions* and *Proceedings* of scientific societies and scientific journals from 1800 had been undertaken by the Society in 1858 and had cost £3720. It had proved to be of great value to scientific men both in England and abroad, and the government's approval given in 1864 that it should be printed at the national expense had been fully justified. The Society now asked that two supplementary volumes, bringing the catalogue up to the end of 1873, be printed at the national expense; this the Treasury sanctioned.

In 1876 the Parliamentary Grant for Scientific Investigations was increased to £5000. In the same year the Oriental Manuscripts which Sir William and Lady Jones had presented to the Society in 1792 and 1797 were lent to the India Office and placed in the library there where they would be more accessible to students desiring to consult them than at the Society. The India Office agreed to bind any that required it and to compile and print a catalogue of the whole collection, which was done shortly after.

In April 1877 the Foreign Secretary, Professor A. W. Williamson, wrote to the President to say that he understood the financial position of the Society was causing some anxiety to the Council; he proposed therefore that the foreign secretaryship, which he had held since 1873, should be transformed from a paid office to an honorary office with charge of the Foreign Correspondence of the Society, adding that he, if the Society desired it, would be quite willing to retain the office on the new footing. The honorarium of the Foreign Secretary at this time was £100 per annum, at which sum it had been fixed in 1860. The Foreign Secretary's offer was accepted, but it was afterwards realized that of the hundred pounds twenty were provided by the Keck bequest of £500 which had been left to the Society in 1719 by Dr Keck in order that the income of the bequest should be paid to the Fellow who, for the time being, was charged with the Foreign Correspondence of the Society. Four years later, in 1881, the Treasurer called the Council's attention to the fact that up to then the securities representing the Keck bequest had always been included in the general funds of the Society. He recommended that the sum of £666. 13s. 4d. Consols should be sold and reinvested in Midland Railway 4 per cent Debenture Stock in order that the account should be in future kept as a separate fund in the same manner as those of other trust funds. This was approved and the change was carried out. The Keck Fund is now included in the Special Funds under the classification of the funds which was made in 1934; the income received from the securities, about £24, is paid annually to the Foreign Secretary as required by the testator's will.

The Society's temporary financial difficulty at this time seems to have been caused by the increasing costs of publication, and the outlay on the preparation of the Catalogue of Scientific Papers and similar scientific objects; at the same time the annual subscriptions had fallen off to some extent as the number of Fellows had decreased, and composition fees were fewer. Gifts and bequests were being received from time to time but most of them were given on the condition that the income derived from them should be used for scientific research. However, a few years later the situation was greatly eased by the sale of the Acton estate for a large sum which was at the free disposal of the Society without any restrictions, and could therefore be placed to the credit of the General Fund.

The original statutes of 1663 required that the Treasurer's accounts should be audited before the Anniversary meeting by five Fellows nominated by the Council from their members, and another five Fellows who were to be appointed by the Fellows of the Society by ballot at one of the three weekly meetings preceding the Anniversary meeting. This had been the regular practice for more than two centuries, but for some years past the business of the Society had increased so much and the sums which were dealt with were so much larger than they had been in the past that a few hours' examination of them by the auditing committee, whose members were not specially familiar with this class of work, was not likely to provide an adequate control; accounting errors or unauthorized operations might and occasionally did continue for years unnoticed. The number of vouchers alone must have taken a considerable time to check efficiently. It was time therefore that expert knowledge should be utilized, and in 1878 for the first time the Council authorized the payment of ten guineas to a firm of accountants for examining the accounts of the past year, and the expenditure of fifteen guineas for a similar purpose in the following year. This now became the regular practice, and in 1886 Messrs W. B. Keen and Co., Chartered Accountants, were appointed to check the Society's accounts; this firm have continued to discharge this duty up to the present time to the complete satisfaction of the Society. In 1891 the number of auditors nominated by the Council and the Society, which had hitherto been five from each body, was reduced to three. At first the examination demanded from the professional auditors was not very detailed as the modest amount of their fee indicates; but as time went by the Society received large bequests for research, and accepted the responsibility for the administration of additional government grants so that the supervision of its financial work became increasingly important. The system which had been introduced with such excellent results in 1831 now needed extending and modernizing, and this was undertaken under the advice and guidance of the Society's accountants between 1930 and 1940.

In 1881 an addition was made to one of the Society's benefactions, that of £100 which had been bequeathed by Sir Godfrey Copley, Bt., in 1709. Sir Joseph Copley, Bt., a descendant of Sir Godfrey, gave to the Society £1666. 13s. 4d. in 3 per cent Consols 'to provide in perpetuity a yearly bonus of £50 to be given to the recipient of the Copley Medal'. As the fund had at this time a balance in hand there was no difficulty for many years in paying the bonus, but later when the 3 per cent Consols had been converted to 2½ per cent and when the price of gold for the medal had increased the income of the fund did not suffice for both the medal and the bonus. The Council therefore in 1937 decided to reduce the bonus to £35.

In June 1887 a loyal Address, which was to be presented to H.M. Queen Victoria, the Patron of the Society, on the completion of the fiftieth year of Her Majesty's prosperous reign, was prepared; it was transmitted to the Home Secretary according to the usual procedure. At the meeting of

Council on 7 July the President reported that, in conformity with the Home Secretary's instructions, he and the Senior Secretary had proceeded to Windsor Castle on Monday, 27 June, where in company with many representatives of bodies of very varied character they were admitted to the Royal Presence, and the Address had been graciously received by Her Majesty from the President's hands. This procedure was not in accord with the custom followed on such occasions in the past, and the Council resolved that a letter should be addressed to Her Majesty's Secretary of State for Home Affairs representing that on previous occasions, when the Society had had the honour of presenting Addresses of a like character to the Sovereign, the representatives of the Society had enjoyed the privilege of being admitted to the Royal Presence in a different manner. This was supported by various extracts from the Society's Journals and Minute-books. The letter ended by expressing the hope that the procedure which had been followed in June 1887 might not be considered as forming a precedent, but that on future occasions their old privilege of direct approach to the Royal Presence might be allowed to them. The Home Secretary replied to the President on 10 December to the effect that he had submitted the claim of the Royal Society to be admitted to the number of those public bodies to which is conceded by prescription or otherwise the privilege of presenting their Addresses to the Sovereign on the Throne. He added that Her Majesty had favourably considered such claim and had been graciously pleased to command that the Royal Society be allowed to enjoy the privilege, on all fit and proper occasions, of presenting their Addresses to the Sovereign on the Throne.

Under the Public Schools Acts of 1868 the Society was desired to appoint a representative on the governing bodies of Winchester, Charterhouse, Rugby and Shrewsbury: other schools were added later, namely Dulwich, Eton, Harrow and Westminster. Vacancies which occur from time to time are filled by the Council who select so far as possible Fellows who are connected in some way with the institution to the governing body of which they are to be nominated.

The instruments which were purchased with the funds of the Scientific Investigations Grant-in-Aid had been, when no longer required in the investigation for which they were provided, returned to the Society for custody. It was decided in 1869 that, when they were not likely to be required in the future, they should be transferred to the Science Museum at South Kensington for safe custody where they would be included in the collections there and if suitable be placed on exhibition. At first the number lent or transferred in this way was small, but from 1890 it increased and in 1900 about fifty instruments, mostly astronomical, were disposed of in this way. All the instruments which have been received at the Science Museum, whether by gift or loan, and whether they were the Society's own property or purchased from the Scientific Investigations

Grant-in-Aid, are now entered in the store-ledgers of both institutions and verified annually so that any of them are readily traceable.

In 1879 the sum of ten guineas was placed at the disposal of the President in order that distinguished guests might be invited to the Anniversary dinner. Hitherto any guests that there may have been at this dinner had been invited by individual Fellows at their own expense, and this was certainly the case at the end of the eighteenth century, when they were very few. It was now considered very desirable that more guests of distinction should be invited to the Anniversary dinner, in order that they might become acquainted with the Fellows and learn something of the aims and activities of the Society. At first they principally included ministers and other high officials, but as years passed and the number of the Society's guests had increased it was seen to be highly desirable that the leading men in industry as well as those of the official classes should have an opportunity of becoming acquainted with the Society's work. The importance of having such guests is now fully realized and they usually number about sixty.

In 1890 the Council decided that sealed packets should no longer be accepted by the Society for safe custody, as at this time there existed many places such as banks, safe-deposits, etc., in which valuables or confidential documents could be lodged more suitably and in quite as great security as in the rooms of the Society. The Society had first accepted this responsibility in 1667 when Robert Boyle made considerable use of the privilege, but no reference to it is made during the eighteenth century. About 1834 the question was reconsidered and it was decided that under certain circumstances such documents, etc. should be accepted with the approval of the Council; these were to be placed in the 'iron chest', the key of which was kept by the Treasurer who reported to the Council from time to time on the number of such documents which had been placed in custody. In December 1890 the question was further considered and the Council decided that no more documents should be accepted for custody by the Society, and this continued to be the practice for fifty years. A request was then received that a file of documents relating to a Fellow of the Society should be accepted for custody; after discussion it was resolved that they might be accepted if the Council thought it desirable.

In 1895 the Society represented to the government the heavy expense which the Royal Society was incurring in printing and publishing its scientific publications which were of value and benefited not only its own Fellows but also scientific workers in other countries. The government agreed to make a grant-in-aid amounting to £1000 per annum to the Society on the understanding that it would by means of the grant assist not only the Society's own publications but also the adequate publication of scientific matter provided through other channels and in other ways. This grant was increased to £2500 in 1925, and is still being made.

The increase in the amount of scientific business with which the Council now had to deal led in 1896 to a decision to appoint eight special scientific committees to which the Council could refer scientific matters for consideration and report. It will be remembered that such committees were first appointed in 1664 but after a few years these were allowed to lapse; others of similar character were appointed in 1838, which were reappointed annually until 1849 when they were discontinued. No reason was given for their suspension but it may have been found that there was not enough work for them without encroaching on the responsibilities of the Council. J. P. Gassiot, who from the first had been a strong advocate of them, proposed their reappointment in 1851 and again in 1871, but the Councils of those years would not consent. By 1896, however, it was agreed that such bodies would be useful and the Scientific, or as they are now termed Sectional, Committees were appointed and still form an important part of the Society's organization, dealing with the following branches of science: Mathematics, Physics, Chemistry, Engineering, Geology, Botany, Zoology and Physiology, and Medical Science. This division of subjects is nearly the same as that employed in 1838 except that Astronomy and Meteorology, for which separate committees were then appointed, have been replaced by Engineering and Physiology and Medical Sciences. These Sectional Committees were only concerned with the scientific matters which were referred to them; there still remained others which had for a good many years past dealt with such matters as finance, the library, and much besides; these standing committees number about twenty-five, besides those which include among their membership representatives of other institutions. Complete lists of all these committees and their membership are published annually in the *Year-book* of the Royal Society.

The Society has always maintained close relations with the scientific workers in other countries; the philosophers of the seventeenth century corresponded freely with their colleagues in France, and about 1640 several of them were frequently communicating with Père Marin Mersenne, Renaudot, and later with de Montmor, Sorbière and others. Some, like Sorbière, Huygens and Malpighi, were admitted to the Fellowship of the Society when it was formed. They exchanged information, sent scientific books and in many ways helped one another in times when communications were slow and difficult.

In 1742, as has already been related, the Society had a yard measure and a pound weight constructed with all the accuracy which was possible at the time, and sent them to the Académie des Sciences at Paris in exchange for copies of the corresponding French measures. In 1787 the Society co-operated with French geodesists in carrying out a triangulation, with all the accuracy then attainable, between the observatories of Greenwich and Paris; in 1821 this work was repeated. In 1836 von Humboldt advanced such international co-operation considerably when he wrote to the Duke of Sussex, who was then President of the Society, to seek his support for the establishment of stations for the study of terrestrial magnetism in England and her overseas possessions; in this aim he was successful. This seems to have been one of the earliest international organizations which have been established in increasing numbers in order to initiate and foster international co-operation in many branches of science. With many of these the Society has been directly concerned, and has been represented at their conferences by its delegates. Some of them have had for their object the establishment of uniformity in standards of measurements, others to advance science by the co-operative effort of several countries, and others again sought to encourage the personal exchange of opinions. The underlying motive was that the effective promotion of science would only be possible by utilizing observations from all parts of the world, and for this uniformity of standard both in methods and in observations was essential. The earliest of these organizations was the Bureau Internationale des Poids et Mesures, which was established at Sèvres near Paris in 1873 on the recommendation of an international commission appointed in 1869 to consider the construction of metric standards; Great Britain became a member of the convention in 1884.

An International Mid-European Geodetic Association had been established at Berlin even earlier, in 1861, to co-ordinate the geodetic operations which were being carried out in Germany and the neighbouring countries since they had to be connected at each frontier, and for this standards of accuracy had to be laid down. Great Britain joined this Association in 1884, but did not contribute to it financially until twelve years later when a Geodetic Convention was concluded to which twenty-three states adhered, Great Britain being one; each of these paid an annual subscription calculated on the basis of its population. The British delegate was recommended by the Council of the Royal Society for appointment by the government.

In July 1894 the Royal Society of Göttingen invited the Royal Society to send a representative to a conference at Innsbruck where international action in relation to gravitational observations would be discussed; the Council accepted the invitation. This led to another proposal being made from the same quarter four years later. For some years before 1898 the Academies of Munich and Vienna, as well as the Royal Societies of Göttingen and Leipzig, had met annually at one or other of these places as an Association in order to discuss scientific matters of common interest to them, and in 1898 their meeting was to be held at Göttingen. In May of this year the Council considered a letter which had been received from the Royal Scientific Society of Göttingen inviting the Royal Society to nominate delegates to attend a cartel meeting of the Academies and Scientific Societies of Göttingen, Leipzig, Vienna and Munich, at the end of the month. This invitation the Council accepted, and instructed their delegates that, in case the question of the Royal Society adhering to the Association of Academies and Scientific Societies was brought up at the meeting, it should be made clear that such adhesion must be contingent on the Association being rendered truly international by the adhesion of other nations, and more particularly of France. The principal subjects discussed were the organization of seismological observations, and the determination of the gravitational acceleration by pendulum observation. The scheme of an international catalogue of scientific literature was favourably received, and support was given to the Royal Society's proposals on the subject. The delegates of the cartel agreed to propose to their respective Academies the international association of important scientific societies, and requested that a similar proposal should be conveyed to the Council of the Royal Society by its delegates. If the principle of these resolutions was accepted it was suggested that the Royal Society should ascertain the views of the Academies of Science at Paris and St Petersburg, and of the Academia dei Lincei at Rome, while the cartel would communicate with the Academy of Science at Berlin. The suggestion was accepted by the Society, whose decisions as an independent institution did not require the approval of any Minister of State, and did not necessitate the use of diplomatic channels during the preliminary discussions.

In response to the enquiries which were sent by the Council of the Society to the Academies at Paris, Washington and St Petersburg, and to the Academia dei Lincei at Rome, these institutions declared themselves to be in favour of accepting the proposal; the Academy of Sciences at Berlin also concurred. A preliminary conference was therefore held at Weisbaden in October 1899, at which draft regulations were prepared which were adopted by the General Assembly of the Association at its first meeting which was held at Paris on 31 July 1900.

Under the regulations which had been adopted an academy might be admitted to the literary section or to the scientific section of the Association, but not to both unless its constitution provided that the sphere of its activities included both literary and scientific subjects as was the case with most, if not all, the other academies. The Royal Society, however, by its aims and its constitution could only be a member of, and vote in, the scientific section. At that time there was no society existing in England which was deemed competent to represent the United Kingdom in the section dealing with historical, philosophical and philological studies in such a manner as would satisfy the conditions; the Royal Society could represent it in the scientific section but it had no vote in the literary section.

The Council therefore consulted several distinguished men of letters in order to see how this disability could be removed, who suggested that the Royal Society might enlarge its own scope so as to include a section corresponding to the philosophical, historical and philological divisions of some foreign academies. The matter was one of great importance to the advancement of two great branches of human knowledge in this country and the Council discussed it at several meetings, with the result that they recommended that a memorandum, which the Council had prepared dealing with the whole question, should be laid before a General meeting of the Society in order that all aspects of the question might be thoroughly examined. Finally it was agreed that while fully sympathizing with the desire of those interested in the study of history, philosophy, philology and archaeology, for a representative institution dealing with their branches of knowledge, the Society found much difficulty in extending its own field of activity to include the humanities as well as science. The whole question was further considered in 1901 (see Chapter IX, pp. 307–10).

As early as 1684 the Council had given instructions that a detailed index of the information which was to be found in its Minute-books and in the Journals of the Society was to be prepared; with this was included another requiring that a second copy of the registers and Journals should be made to guard against loss or damage. This latter was carried out but it involved much additional work; of the index nothing more is recorded. From time to time the need for arranging and cataloguing the archives and correspondence of the Society was emphasized and orders were given by the Council but the efforts at improvement were spasmodic and soon died away. In 1823 a resolution to this effect was adopted by the Council, but five years later Dr Granville records that little had been done beyond tying up papers in bundles, though the library had been put in order. These intermittent displays of energy usually aimed at achieving too much at once which resulted in some improvements being carried out more or less thoroughly while others involving an increase in the office routine were ignored. In 1844 Weld, the Assistant Secretary, found the lack of information about the Society's past history so inconvenient that he compiled his *History of the Society*, and this improved the situation for a time. By the end of the nineteenth century however the administrative work of the Society had increased so much that the Council decided on the compilation of a Record of the History of the Society which was published as a small octavo volume in 1897. It was then found that some of the material which had been collected, such as the membership of the Council, and of the various committees, changed at the end of each year; the list of Fellows and foreign members, and the annual reports on the Society's administration and its finances could not be held up for the quinquennial issue of the Record which was contemplated. The issue of a Year-book to contain such information was therefore authorized to appear as early as possible in January so as to contain the names of the members of Council who had been elected at the Anniversary meeting and of the members of committees who had been appointed by the Council at its first meeting in December. The issue of these two publications was not only useful to the officers and administrative staff but they mark a new attitude of the Council towards the Fellows of the Society. The circulation of the Treasurer's annual report on the finances of the Society for 1830 and all later years to every Fellow before the Anniversary meeting had been the first step; now the *Year-book* gave them a full and detailed account of the way in which their Society was being administered from year to year.

During the forty years which have been described very considerable changes in the Society and its administration had taken place. No sooner had the revised statutes dealing with the selection of candidates been adopted than their effect became noticeable. Since fifteen candidates only were admitted to the Fellowship in each year, and the majority of these were scientific men, this class of Fellows rapidly outnumbered their non-scientific colleagues and before the end of the century had replaced all of them except a few specially elected ones.

The officers continued to attend both Council meetings and the Ordinary meetings regularly as had been the custom for many years past, but the councillors now were much more regular in attendance, the average number at a meeting being sixteen in the first decade and between eighteen and nineteen in the last. Council meetings were held twelve or thirteen times annually and this number was found to be sufficient to deal with the business promptly by the well-attended meetings with the assistance of the special committees. There were signs too that gifts and bequests were becoming more frequent though these did not materially improve the position of the Society itself; the income of such gifts had to be used for such purposes as the donor or testator had laid down and these usually included the provision of a medal, or the delivery of a lecture or research work of some specified character. It was only occasionally that it might be used for the general purposes of the Society. For several years past composition fees had been invested instead of being used as income, so the Society's income at this time had decreased as compared with what it had been at the beginning of the century. These restrictions made the financial position somewhat difficult for the Council until the sale of the Acton estate had been successfully negotiated in 1882 since the income available for administrative expenses was provided by the Society's investments, subscriptions, sale of publications, etc.

During the latter half of the nineteenth century the scientific reputation of the Society had increased greatly for it was known to be administered and its policy determined by a Council of men of science. It had been throughout this period a scientific Society in a true sense and was no longer an institution the majority of the members of which had little if any scientific knowledge. It had taken two centuries for the successors of its founders to arrive at this satisfactory result. Mention has already been made of the indifference shown by the non-scientific Fellows during the eighteenth century to the needs of the Society. It was well known that they had been elected because most of them were in a position to assist it financially though not by their scientific work; Sir Joseph Banks laid it down for his own guidance that men who had done good scientific work should be admitted whatever their social or financial position might be, but that the rest might be expected to assist it from their private resources; Sir Humphry Davy followed the same line but no benefaction of any importance resulted in either case. Between 1820 and 1850 Dr W. H. Wollaston and some others had given or bequeathed about £10,000 (see *Royal Society Record*, p. 142) for scientific research. This at any rate was an improvement.

In the second half of the nineteenth century there was a notable change, and evidently this was due to the control and direction of the Society having passed from Councils of the former type to those which were wholly composed of men of science.

Between 1851 and 1900, rather more than £50,000 was received by gifts, bequests and contributions to schemes which the Society had undertaken. Most of these were provided by Fellows of the Society, but six of the benefactors were from the general public. Seven of the larger bequests were in the form of trusts, the income only being used for scientific research. Other considerable sums were given by various Fellows to such schemes as the Fee Reduction Fund, the Scientific Relief Fund and the preparation of the Catalogue of Scientific Papers.

By such contributions the Society was being enabled to finance research on selected subjects by men who were specially fitted for the task, and also to press forward schemes of its own which were of special importance. Bishop Sprat's hopes had been at last realized by the voluntary help of the scientific Fellows, but not by that of the 'gentlemen free and unconfined' to whom he had looked to give to the Society that support which it was certain to need for many years to come.

The Society might now reasonably look forward to even greater successes than it had yet achieved.

REFERENCE

Royal Society, Record of. 4th ed. London, 1940.

CHAPTER IX

RESEARCH: 1901–1940

B Y THE middle of the nineteenth century the principal reforms, which the scientific Fellows of the Society had been advocating for many years past, had been adopted; it remained to be seen how successful they would prove to be in practice. During the latter half of the century the Councils had worked hard to reorganize the administration in many directions, the most important being the economical use of the Society's resources, and the advancement of science by all available means. Their work was thereby greatly increased as is shown by the number of committees which it was found necessary to appoint if the Councils were to keep pace with the numerous demands made upon them; every branch of the administration was overhauled and many new activities were introduced during these years.

By the beginning of the twentieth century the Society was in possession of a well-organized administration; its Councils were taking their duties seriously, and had shown an efficiency which had been sadly lacking in the eighteenth century. They had begun work on several large problems which had to be left by former Councils for their successors to complete, such as the International Association of Academies, International Science, the International Catalogue of Scientific Literature and the National Physical Laboratory. On all these a start had been made, but there was much work still to be done before their purposes had been achieved; a number of smaller schemes which might or might not develop into activities of wide scope and great importance had been started, and would need attention for some time to come.

This body of about four hundred and fifty men of science, all of whom had been selected for their zeal and ability, had before them a wide field in which to carry out research of all kinds, while the Society was now being directed by their colleagues who were in full sympathy with the aims of all who were interested in research; its resources were now being organized so as to assist research workers in every possible way. Such a state of affairs was a new experience for the Society; the rapid advances which were being made in every branch of natural science, and the large endowments of research which many generous benefactors have given to the Society in recent years, have made it possible for the Councils to initiate research on a much larger scale than had been possible at first. It is too early to forecast the direction in which new advances will be made in the future, and to-day all that can be done is to describe the situation as it stands. The preceding chapters have provided an opportunity of taking stock of what the Society had accomplished in the two centuries and a half which have passed since it was founded; in the present one we are looking forward towards new fields of endeavour still unexplored. This chapter is in truth not so much the close of the first volume of the Society's history as the introduction to a second volume which must be left for another hand to write a century or two hence. What can be done now is to record what has taken place in the last forty years.

THE OFFICERS

In 1873 when Sir Joseph Hooker had been nominated for the presidency he expressed the hope that he would not be expected to serve for more than five years since his duties as Director of the Royal Gardens at Kew occupied his time very fully. The Councils of those years respected his wishes in this matter but without making any statute limiting the tenure of the presidency or recording any formal decision in the matter. Sir William Huggins, the astronomer, was elected President by the Society at the Anniversary meeting in 1900 and was re-elected at the four succeeding Anniversary meetings. His successors were Lord Rayleigh, Sir Archibald Geikie, Sir William Crookes, Sir Joseph Thomson, Sir Charles Sherrington, Lord Rutherford, Sir Frederick Gowland Hopkins and Sir William Bragg; all of these served for five years except Lord Rayleigh (1905–8) and Sir William Crookes (1913–15). The five-year period for a President's tenure of office having become customary was adopted by the Council on 7 March 1935 as a by-law; the Presidents have all been men of exceptional ability and wide experience who were still actively engaged in scientific work when they were elected, and could not reasonably be expected to undertake the onerous duties of President for a longer period.

Out of the forty-four eminent men who have held the office of President of the Royal Society the portraits of thirty-three are in its possession and are hung in its rooms. So long as its Presidents were chosen from men of title, or from those who were well off, it was not unreasonable to expect that they would present the Society with their portraits painted by well-known artists as an acknowledgment of their election to the highest scientific position that the Fellows had to offer; portraits have sometimes been given by relatives or friends, and in these ways the Society has acquired pictures of three-fourths of its Past Presidents. In the latter half of the nineteenth century it had become the practice to elect Presidents for their eminence in the scientific world rather than for their social position or the administrative posts which they had occupied; such men could not be expected to contribute their portraits to the Society's collection. It therefore became usual for a number of the Fellows, who desired to show their friendship and high esteem for a retiring President as a colleague and as the titular head of British science for the time being, to contribute to a fund so that a portrait painter of distinction might be commissioned to execute his portrait for presentation to the Society. In this way the Society has been enabled to add to its collection portraits of the last fourteen of its Presidents. This series of some of the most eminent scientific men of this country is of great and increasing historical importance.[1] To add to it as the years roll by and to preserve it from damage or deterioration might well be accepted by the Society as a graceful act of recognition which it owes to its Past Presidents.

The Treasurer, Sir Alfred Kempe, who was in office at the end of the previous century, continued to carry out his duties with great benefit to the Society until 1919; besides being a man of science and a very competent administrator his wide legal experience was of great value to the Society on many occasions. Each of the next two Treasurers who succeeded him held office for ten years only so that the principle that this office should not be held for more than ten years had by then been tacitly accepted. At a meeting of Council on 7 November 1901 it had been proposed that the Society's Secretaries should not be elected for more than ten successive years; after some discussion this was adopted by the Council and has since been the recognized practice. The Secretaries in office at this time were Sir Arthur Rücker, the physical Secretary, who had been elected in 1896, and Sir Michael Foster, the biological Secretary, who had been first elected in 1881; he had thus served for twenty years. In the nineteenth century Dr P. M. Roget had served as a Secretary for twenty-one years, S. H. Christie for twenty-seven and Sir George Stokes for thirty-one years (1854–85). It has already been suggested that the interests of the Society were not as a rule best served when the officers held their posts for very long periods or for a few years only. The last half of the nineteenth century was perhaps an exceptional period in the Society; efficient and orderly administration had been planned but many features of it had still to be modified so as to meet the growing needs of the time; work on the Society's scientific activities and on its executive reforms was increasing rapidly. At first the Councils were able to deal with this adequately, but before the end of the century the assistance of twenty standing and occasional committees to examine and report upon questions on which the Council had to take their final decisions, had become necessary. The talent and long experience of such men as Sir George Stokes and Sir Michael Foster may well have been of exceptional value to the Society at such a time although the long tenure of the secretaryships aroused some little criticism. An accurate appreciation of special circumstances is seldom possible until some time has elapsed.

[1] *Royal Society Record*, ed. 1940, pp. 154-61.

COUNCILS

There is not much to be said about the Councils and their work in this period for it had become the normal custom for some years past that the officers and the councillors should attend the meetings regularly unless they were unavoidably prevented from doing so; consequently matters which came before them were dealt with more promptly and efficiently than had been the case when only about half the twenty-one councillors attended the meetings.

Though the Council had never relinquished the power conferred upon it by the Charter to control the administration of the Society and to direct its policy, early in the nineteenth century signs were not wanting of a growing tendency to take the Fellows more into the Council's confidence by informing them of what had been done, or by advising them of the changes that the Council proposed to carry out. The first step in this direction had been to circulate a copy of the Treasurer's report to Council to every Fellow before the Anniversary meeting, which was first done at the end of 1830 and has been carried out regularly ever since under a statute of 1831. In the next year the Council decided to circulate to all Fellows before the Anniversary meeting, the list of officers and councillors who were recommended for election. This was followed in 1848 by sending a list of all the candidates for election to each Fellow two months before the date on which the elections were to be held, and later by the publication of the *Year-book* and the *Record* in 1897, both of which gave to the Fellows much information which previously had not been readily accessible. Another important innovation was introduced in 1915 whereby the Annual Report of the Council to the Society, which is circulated to all the Fellows, is now brought before a Special General meeting of the Society about the middle of November, in order that any additional information which the Fellows may desire to have on the matters dealt with in the report can be given to them and explained more fully than is possible at the Anniversary meeting when the report is formally presented. In 1916 the Council decided that the list of candidates for the Fellowship should be laid before the Sectional Committees at a Special meeting, so that the qualifications of the various candidates might be discussed by them. The final selection of those to be recommended for election to the Fellowship still remains with the Council, but by this arrangement a wider range of opinion drawn from those working in the same fields of scientific research as a candidate and likely to be well acquainted with his qualifications would be at the disposal of the Council. In 1918 the Council resolved that before any new statute, or any modification of an existing statute, was approved and adopted it should be brought before a Special meeting of the Society at which the Fellows could discuss it and ask for additional information. Such a meeting can neither approve nor reject the statute, but any representations made by the Fellows present for its modification is given the serious consideration of Council at its next meeting. By this means the rather arbitrary practice dating from 1663, whereby a new statute had only to be approved at one meeting of Council and adopted at another one for it to become operative, though the general body of the Fellows had no knowledge of its terms or even of its existence, was amended so as to allow the Fellows to discuss and criticize the Council's proposals, a procedure which was most desirable. In 1935 the Council decided that the annual reports of the Research Professors, Fellows and Students who had been appointed by the Council, and whose expenses were met from some of the various research funds administered by the Society, should be added to the Council's report as an appendix so that the Fellows might know what researches were being carried out under these trusts, and where the work was being done.

Another move in the same direction was made in 1936 when it was suggested to the Council that their decisions and other matters bearing on the Society's policy and administration should be published and circulated to the Fellows in order that they might be kept informed of what was taking place without having to consult the Minute-books at the Society's rooms as every Fellow has the right of doing. The Council adopted the suggestion and gave instructions for *Occasional Notes* to be published once or twice yearly; the first number appeared in April 1937. In the following year these *Notes* were replaced by the *Notes and Records of the Royal Society of London* in which the Councils' instructions were interpreted more widely. A considerable amount of current information relating to the Society's administration, its work, its relations with other scientific bodies both at home and abroad and other activities, as well as short historical articles, are now included, many of which are of considerable interest; but scientific subjects are not discussed since these lie within the province of the *Philosophical Transactions* and the *Proceedings* of the Society. In all these ways the Fellows are now kept informed of many things relating to the Society and its work of which they would formerly have been left in ignorance.

THE FELLOWSHIP

The number of Ordinary Fellows at the beginning of the century was 456 and has since varied from 442 to 480 between 1901 and 1940; it has indeed altered but little since 1890. The number of foreign members has ranged from thirty-six to fifty, but as they are usually well past middle age when they are elected the losses by death among them are in some years considerable. There was now greater uniformity in the Fellowship than in former years since all the Fellows had been selected and recommended for election under the same procedure, and by councillors all of whom were now men of science. The statute authorizing the privileged election of peers had been annulled in 1874, and in 1902 a similar privilege for Privy Councillors was withdrawn. It was however thought to be of advantage to the Society that it should be able to include in its Fellowship a few men of exceptional ability or of wide experience whom the Council might consult or who might be asked to serve on special committees. A special statute was therefore adopted to the following effect:

> In cases in which the Council is of opinion that, in the interests of the advancement of Natural Knowledge, it is desirable that persons be elected Fellows of the Society otherwise than as provided by Statutes 1 to 11 of this Chapter they may recommend to the Society for election persons who, in their opinion, either have rendered conspicuous service to the cause of science or are such that their election would be of signal benefit to the Society; Provided that not more than two persons shall be so elected in any one calendar year; and if two persons be elected in any one year there shall be no election in the following year.

In the year 1940 there were ten Fellows in the Society who had been elected under this statute; the number admitted in this way will always be small and the Society can thereby avail itself of the knowledge and experience of distinguished men who are interested in the advancement of the Society's aims.

Analyses of the number of scientific Fellows who were working on various subjects in several years are given in Appendix II B, and these show how rapidly research in many fields extended among the Fellows as soon as the Society was being administered as a scientific institution. By 1860 330 Fellows out of 630 came within this category. In 1940 the number of ordinary Fellows, including ten special elections, was 456, all of whom except those specially elected were men of science; the table on p. 304 shows the number of Fellows who were working in various branches of science according to the classification now in use, and as many of them are active in more than one field of research the number so working is 640 or about two-fifths greater than the number of ordinary Fellows. The number of Fellows who are classified as working on subjects in the Mathematical and Physical (A) group are 296 as compared with 344 on those of the Biological (B) group or in the ratio of about five to six. The distribution of the Fellowship among the various departments of science is given on p. 304.

In 1875 the statistics of the Fellowship were examined by a committee appointed to report on the working of the new election statutes which had been approved in 1847. They had before them a large number of returns prepared by the Fellows themselves for the purpose, and when these had been analysed and studied the committee reported that the average age of election of the Fellows then alive was 40·9 years. In an analysis of 410 Fellows who had been admitted to the Society up to 1812 the age of election is given by Dr Thomson in his *History of the Society* as 40·8. In 1925 Sir Arthur Schuster published an interesting study of the life statistics of Fellows (*Proc. Roy. Soc.* vol. 107), and from the tables which he prepared for the period 1848–73 the median age of election was about 43; from 1873 to 1897 it was about 44½; Professor A. V. Hill[1] has shown that the median age of election for recent years, 1920–38, is about 47 years, so it has been rising for a century at least. He also gives a table

Subject	Number of Fellows working in 1940	Percentage
(A) *Physics, Mathematics, etc.*		
Astronomy	15	2·3
Chemistry	79	12·3
Engineering	26	4·1
Geodesy and Surveying	10	1·6
Mathematics	49	7·7
Metallurgy	6	0·9
Meteorology	7	1·1
Physical Chemistry	11	1·7
Physics	86	13·5
Statistics	7	1·1
	—— 296	
(B) *Biology, etc.*		
Agriculture and Botany	73	11·4
Anatomy and Anthropology	13	2·0
Archaeology	6	0·9
Biochemistry	32	5·0
Geology and Palaeontology	36	5·6
Neurology and Psychology	11	1·7
Pathology and Bacteriology	33	5·2
Physiology and Pharmacology	50	7·8
Medicine	28	4·4
Zoology	62	9·7
	—— 344	
Total	640	100·0

showing the different ages of Fellows in various scientific categories, and the results are interesting; the lowest figures are for mathematics and physics where they are 42 and 39; the highest, 62, is naturally found for the special elections, that is, for those who have been selected for their personal ability and distinction and not solely for eminence in scientific research. The next highest figures, 53 and 55, are found in geology and engineering, as though there had been a tendency to select men who had a long record of distinguished work to their credit rather than those of younger age and of high promise. The 'median' age of election is that age below which as many are elected as above it.

[1] *Royal Society Notes and Records*, 1938.

In January 1902 the Secretaries having reported that a certificate of candidature for election in favour of Mrs B. Ayrton had been received, it was resolved that counsel's opinion should be obtained as to whether women were eligible as candidates for the Fellowship under the Charter. The legal opinion given by two eminent counsel in the month following was to the effect that women were not eligible as Fellows of the Royal Society. By 1923, however, in consequence of legislation which had been enacted since the war of 1914–18, this disability had been removed, and both men and women are now eligible for election.

FINANCE

As the result of the sale of the Acton estate in 1882, and the receipt of a number of gifts and bequests, the Society's financial position at the end of the nineteenth century was satisfactory even though expenditure, especially on the publication of the *Transactions* and the *Proceedings*, was increasing. The form in which the accounts had been presented since 1833 when there were but three or four funds no longer met the needs of the Society a century later when they numbered about fifty. In 1934 therefore the Council approved a new classification of the Society's various funds which it will be convenient to use here in describing the rapid growth of the Society's resources which took place between 1901 and 1940.[1] Under the scheme which was then introduced the funds over which the Society has full control for any purposes that may commend themselves to it were included in the 'General Purposes Fund' which is the direct successor of the Society's 'General Fund' established in 1660. In the early years all sums received were paid into this fund for whatever purpose they might be intended, but gradually those which had been bequeathed for use under specific conditions were removed from it and treated separately as special or trust funds; this was done with the Keck bequest of 1719, the Copley bequest of 1736, and the Rumford bequest of 1796. At the beginning of the twentieth century the value of the securities belonging to the General Purposes Fund amounted to £81,076; in 1833 their value had been £14,000 and by 1875 it was about £50,000; by the end of 1939 this had risen to £116,556.

The Special Funds are those of which the income is only available for the objects specified by the donors and testators, and include those for the library, pensions, scientific relief, medals and lectures, but not for scientific research.

The other funds which, by the terms of the gift or bequest, are to be devoted to the improving of Natural Knowledge (research) in such special fields as may have been prescribed by deed, will or other document, are classified as Research Funds. These began with the Donation Fund which was founded by Dr W. H. Wollaston and some of his friends in 1828, and

[1] *Year-book*, 1940, pp. 222–7.

have been increased by numerous other gifts from time to time. In the annual statement of accounts presented to the Society by the Treasurer the market value of the capital of each fund is now stated, and since 1936 the amount of the unexpended income belonging to each fund which has been accumulated is also reported. At the end of 1940 there were twenty-four Special Funds and twenty-seven Research Funds, besides five grants voted by Parliament which are administered by the Society not for its own purposes but for specified scientific objects in accordance with regulations drawn up by the Treasury. The Warren Research Fund was constituted in 1936 by an agreement between the late Mr H. B. Gordon Warren and the Society. The capital which must be kept intact amounts to about £200,000; the income is to be applied from time to time to research as specified by the testator.

Funds of the Royal Society, 1940

A. GENERAL PURPOSES FUND

B. SPECIAL FUNDS

Arundel Library
Brady
Church
Embossed Scientific Books
Federal Council of Chemistry
Keck
Parsons' Memorial
Pension
Petavel
Publication
Scientific Relief
Travelling Expenses
Wintringham

Yarrow Educational
Medals and Lectures:
  Bakerian and Copley
  Buchanan
  Croonian
  Darwin
  Davy
  Ferrier
  Hughes
  Pilgrim Trust
  Rumford
  Sylvester

C. RESEARCH FUNDS

Browne
Caird and Scott
Darwin
Dewrance
Donation and Jodrell
Foulerton
Foulerton Gift
Gassiot
General Research Fund
Gore
Gunning
Handley
E. Alan Johnston
Joule

Lawrence
Mackinnon
Medical Research
Messel
Mond
Moseley
Pedler
Rosse
Smithson
Sorby
Tomes
Tyndall
Yarrow

D. PARLIAMENTARY GRANTS

Scientific Publications

International Research Associations
and Scientific Congresses

International Polar Year

Bermuda Oceanographical Scientific
Investigations

E. WARREN RESEARCH FUND

The representation of literature, history and other humanistic studies of this country at the International Association of Academies was still awaiting a decision when the century began; early in 1901 the Council had before it the report of the committee which had been appointed in January 1900 to consider a letter from the President of the Society of Antiquaries, and a plan from Professor H. Sedgwick for the institution of a new Academy or Section in the Society to deal with departments of study other than Mathematics and Natural Science. The Council resolved to refer both of these for report to a committee consisting of the President, Lord Lister; the Treasurer, Sir A. Kempe; the Secretaries, Sir Michael Foster and Lord Rayleigh and Professor H. E. Armstrong, Rt Hon. J. Bryce, Sir J. Evans, Professor A. R. Forsyth, Professor E. Ray Lankester, Sir Norman Lockyer, Sir W. C. Roberts-Austen, Professor A. Schuster and Professor E. B. Tylor. When their report came before the Council on 21 February 1901 it was decided that a special meeting of the Fellows should be summoned in order that the President and Council should have an opportunity of hearing the views of the Fellows on the questions raised in the report. This Special meeting was held on 9 May 1901 when a very full and interesting discussion of the whole subject took place. At this meeting the Fellows had before them the report of the Council's committee in which the various aspects of the proposals were very fully set out. After discussing the possible organization of Philosophical-Historical research it went on to describe what the relation of the Royal Society might be to such an organization, and gave arguments in favour of action by the Society; the constitutional powers of the Society and its past practice were described; the reception of papers on subjects not hitherto regarded as properly within its scope would not be unlawful, it was suggested, if the Society deliberately came to the conclusion that in view of the scientific method of their treatment those subjects ought not to be excluded. Four possible solutions had been laid before the committee:

1. The creation of an organization independent of the Royal Society though possibly in connection with it in somewhat similar a manner to that in which the French Academies form parts of the Institute of France.

2. The creation of two 'Academies' within the Royal Society, one of Mathematics and the Natural Sciences and the other of Philosophy-History, each Academy having its own Council and Officers.

3. The creation of two or of three sections of the Royal Society either A and B corresponding to the Academies just named, or A, B and C Sections, the last being that of the Philosophico-Historical Sciences.

4. Representatives of the various Philosophico-Historical Sciences whose views had been obtained were not generally in favour of attempting to establish an independent body, and appeared to prefer that any corporate representation of those sciences should be associated in some way with the Royal Society.

There was also the fundamental issue to be considered: Would the Royal Society be more useful if the scope of its interests were to be so much enlarged? The important question to decide was therefore whether the gains which might result if the Royal Society represented History, Economics and Philosophy, as it already represents Physics and Biology, would compensate for the disadvantages which might arise from the loss of singleness of purpose and the ultimate complication of organization. The Society had only just succeeded in establishing itself as an exclusively scientific institution after its development had been hindered for two centuries during which period the majority of its members had but a limited knowledge of natural science, and was not in sympathy with its aims. Safeguards might be provided, but past experience must have made many disinclined to run the risk of any repetition of the Society's earlier difficulties. The feeling of the Special General meeting was on the whole not in favour of any form of amalgamation.

The committee's report came before the Council at their meeting in July 1901, and after full consideration of all the views which had been expressed at the Special General meeting of the Fellows it was resolved that

The Council, while sympathizing with the desire to secure corporate organization for the exact literary studies considered in the Report, are of opinion that it is undesirable that the Royal Society should itself initiate the establishment of a British Academy.

The outcome of all these discussions was that at the end of 1901 a number of those who had taken part in them formed a new body, 'The British Academy for the Promotion of Historical, Philosophical and Philological Studies'. This body drew up a petition to His Majesty in Council for the Grant of a Royal Charter; and on 8 August 1902 a British Academy having the title of the petitioning body was constituted and incorporated by Royal Charter. Lord Reay was its first President.

The institution of the new Academy was warmly welcomed by the Royal Society who had petitioned His Majesty's Privy Council in favour of the Charter being granted. Several years were to elapse before the British Academy was able to obtain suitable accommodation, but in February 1902 the Council had authorized the officers to place the rooms of the Royal Society at the disposal of the Academy whenever they were required by it. In December 1902 the Academy requested the Royal Society to propose the admission of the new Academy to the International Association of Academies, a request to which the Society gladly acceded. The Academy continued to hold its meetings at the rooms of the Royal Society until 1928, when accommodation in Burlington Gardens was placed at its disposal by the government.

Towards the end of the previous century when the Royal Society's important 'Catalogue of Scientific Papers, 1800–1900' was approaching completion, the Council realized that a continuation of so large and costly a work would far exceed what the resources of the Society or of any single body could bear. It was clear that the only means of ensuring the continuation of the work would be by utilizing international co-operation for collecting the material and providing the funds. Accordingly the Royal Society consulted a large number of representative bodies and individuals, from almost all of whom they received replies in favour of the work being undertaken internationally; steps were therefore taken to summon a conference of delegates appointed by various governments. This conference was held in London in July 1896, and was followed by two others in October 1898 and June 1900. At the last of these a scheme for the publication of an International Catalogue of Scientific Literature was agreed upon.

The supreme control of the Catalogue was to be invested in an International Convention; but in the intervals between the meetings of the Convention the administration of the Catalogue was entrusted to an International Council, the editing and publication being carried on by the staff of a central office. In October 1900 the Society undertook to act as the publishers of the Catalogue on behalf of the International Council thereby giving the necessary legal status to the undertaking; it also agreed to advance the capital needed to start the enterprise on the understanding that the sums so advanced should be repaid during the next five years. The remaining difficulties were thus removed, but the Society had accepted the heavy responsibility of recovering from various international bodies the capital which had been advanced, and this caused considerable trouble later on.

The first meeting of the International Council was held in London in December 1900 when it was decided that the Catalogue should begin in January 1901, that each issue should consist of seventeen volumes, and that the price of a set should be seventeen pounds.

At the second meeting of the Council, which was held in May 1904, it was decided that: 'In view of the success already achieved by the International Catalogue of Scientific Literature, and of its great importance to scientific workers, it is imperative to continue the publication of the Catalogue beyond the first five years.'

The first meeting of the International Convention was held in London in July 1905 and accredited delegates from the following countries were present: Austria, Belgium, France, Germany, Greece, Holland, India, Italy, Japan, Mexico, Russia, South Africa, the United Kingdom and the United States of America. The Convention adopted the resolution of the Council which recommended that the work should be continued for a second period of five years; they requested the Society to continue to act as the publishing body and to conclude a contract for printing and publishing the volumes of the Catalogue which would include the scientific literature of the years 1906-10. The Convention also asked the Society to provide the capital which would be required.

The second meeting of the Convention was held in July 1910 when it was resolved to continue the publication of the Catalogue for a third period, 1911-15, and, on the recommendation of the Council, for a fourth period 1916-20 also. By 1910 the Society had advanced sums amounting to £7500 for working expenses on which interest at the rate of 4 per cent was still due. The printing and publishing contracts were renewed until 1915. The outbreak of war in 1914 seriously hindered the work on the Catalogue and in 1916 it was decided to suspend all work on it until the international position became more stable. When peace was declared, the financial position of the Catalogue was carefully examined and was found to be far from satisfactory. Little had been done towards repaying the loans which had been advanced by the Royal Society, or the interest upon them; moreover considerable arrears of subscriptions were due from several countries and organizations for the copies of the Catalogues which had been supplied; this made the resumption of publication very difficult. In 1922 the Executive Committee of the International Council of the Catalogue discussed the position in detail and resolved that work on the Catalogue should be brought to an end forthwith. At their request the Council of the Royal Society agreed that the Treasurer of the Society should be appointed Receiver to close the accounts, and to wind up the undertaking.

During the next few years efforts were made to collect the unpaid subscriptions which were outstanding, but by 1930 about £3000 were still owing. One or two countries which were in default with their contributions and a large number of organizations were approached with requests that the outstanding debts, representing the cost of the published volumes of the Catalogue which had already been supplied to them at their request, should be paid forthwith in order that the accounts might be closed; but only about £600 were received, and there was little prospect of any more being obtained. In 1935 therefore the Council of the Society, on the recommendation of the Treasurer, in which the Society's accountants concurred, decided to write off as irrecoverable the sum of £12,725 as representing what the Society had lent together with accrued interest which had not been paid, besides a loan of £1892 which had been provided for staff salaries, from the Donation Fund.

Even if the war had not brought the publication of the Catalogue to an end, it seems very doubtful if such a project would have succeeded. A scientific institution is not suitably organized to carry on such commercial operations as printing, publishing and selling publications on a large scale in all parts of the world; moreover, since the Royal Society is a national academy it seemed to be impossible to dispel the firm belief which existed in many quarters that sooner or later the British Government would come to its aid and make good any deficit which might have been incurred. Foreign States and international organizations appeared to find it incredible that the funds used to produce this Catalogue had been advanced from the Society's *private* funds and were not a part of some State subvention.

In this case the Council of the Society had expressed themselves as willing in 1919 to renew the attempt within any amount that the signatories to the Convention were willing to place at its disposal, but they were not willing to advance any more capital unless it was fully guaranteed against loss. No such guarantees were forthcoming so the production of the Catalogue was not resumed.

For some years past the amalgamation of the two dining clubs which were connected with the Society had been discussed from time to time. The Philosophers' Club of forty-seven members had, it will be remembered, been established by the 'reformers' of 1847 when the new statutes for the election of candidates had been adopted by the Council; its purpose was to keep later Councils up to the mark and to resist any tendency to relax the standard of candidates recommended for election. When the Philosophers' Club was formed, the Royal Society Club, which was the older body by a century at least, was considered to be too interested in the social duties of a dining club to watch over the policy which the reformers had succeeded in introducing and to maintain it. The members of the new club discussed new ways of promoting the Society's scientific aims and of improving its methods and its organization; hospitality was limited to foreign men of science. During the years which followed conditions changed rapidly and before the end of the century the Fellowship of the Society consisted almost wholly of men of science and the membership of both clubs was of the same character; the Philosophers' Club was therefore no longer needed. The attendance at its meetings had been decreasing for some years past and it was agreed in 1901 that the membership of the two clubs should be combined to form a single dining club; in May 1942, by the death of Sir Joseph Larmor, the last of the members of the Philosophers' Club passed away. The Treasurer of the younger club joined the officers of the other, which explains why the Royal Society Club has now three Treasurers.

Since the first edition of the *Record* had been published about four years before, several improvements had been suggested and it was decided to prepare a second edition which appeared near the end of 1901. This included two lists of all the Fellows of the Society, one being arranged alphabetically and the other chronologically; considerable difficulty was experienced in verifying the names of some of the earlier Fellows, and it was realized that numerous corrections might have to be made in a later edition. The volume contained rather more than four hundred pages and provided a very useful book of reference for those who had to administer the affairs of the Society. It had been intended originally that the *Record* should be published at intervals of five years, but the experience gained in the preparation of the first two editions had proved that this was too short a period in view of the labour and expense involved in the preparation of a new edition. The third edition was published in 1912, on the occasion of the two hundred and fiftieth anniversary of the sealing of the First Charter, in the form of a handsome quarto volume of nearly five hundred pages. The introductory chapter had been re-written and expanded from eighteen pages to forty-eight, but even then it did not carry the history of the Society beyond the early years of the eighteenth century. The statutes which were adopted by the Council in 1663, 1847 and 1905 were printed in full and all the sections of the second edition were revised and brought up to date. This edition of the Society's *Record* continued to be a most useful book of reference on all matters relating to the Society and its administration for about twenty-five years, but by that time much of the information which it contained was out of date; the Society had been relieved of many of its responsibilities and had assumed others; some five hundred names had been added to the lists of its Fellowship and its foreign membership. In July 1935 therefore the Council appointed a committee to consider the preparation and publication of a fourth edition and early in 1936 the recommendations of the committee were approved. The first chapter of the third edition was re-written and enlarged to ninety pages in order to describe, in a very condensed form, the history of the Society since its foundation. The texts of the Charters and the Statutes as well as the lists of the Fellows, of the Patrons and Officers of the Society and of the Medallists and Lecturers were very carefully checked, and have been printed as appendices, which made it possible to bring the more descriptive part of the historical material together in nine chapters at the beginning of the book.

It had been customary to include in each edition portraits of some of the more distinguished Fellows of the Society; in the fourth edition portraits of those who had played an important part in improving the organization and development of the Society were selected, including those of Sir John Lubbock who reformed its finances in 1831, and Sir William Grove who played a leading part in the drafting and adoption of the new statutes for the election of candidates in 1846-7. Photographs of the Society's meeting room at Crane Court, Somerset House, Old Burlington House and at its present rooms were also included.

Among the many important projects which were brought forward in the latter part of the nineteenth century was one supported by Sir Douglas Galton who, in November 1896, asked whether the President and Council would join the Council of the British Association in urging upon the Government the pressing need for establishing a National Physical Laboratory. After an influential deputation in favour of the proposal had been received by the Prime Minister the government appointed a committee to consider and report on the subject; and their recommendations also being strongly in favour of the proposal the government agreed to place on the estimates for 1899 a grant-in-aid for the expenses of the laboratory and another for the erection of the necessary additional buildings. The hope was also expressed that the Royal Society would accept the responsibility of directing the new institution. A scheme of organization was drawn up early in 1899 which was adopted by the Treasury and by the Royal Society. The ultimate control of the laboratory was invested in the President and Council of the Society; and a Governing Body was formed consisting of the President and the officers, the Permanent Secretary of the Board of Trade and thirty-six additional members. Of these last twenty-four were nominated by the President and Council and the other twelve by the six leading institutions in technical industry. There was also an executive committee consisting of the President, Treasurer and one Secretary of the Society, the Permanent Secretary of the Board of Trade and twelve ordinary members. A director was appointed in 1899 and work was commenced at Kew Observatory, but it was found that the plan of enlarging this observatory to meet the requirements of the laboratory presented many difficulties; in December 1900 however H.M. Queen Victoria made a grant of Bushy House, an old royal residence at Teddington, together with its twenty-three acres of ground, to the Commissioners of Works for the use of the National Physical Laboratory; the grounds have since been increased by the inclusion of an additional twenty-five acres of land.

The alterations and additions at Bushy House were completed in 1901 and the laboratory was formally opened by the Prince and Princess of Wales in March 1902. At the end of that year the staff consisted of twenty-six persons, but since then the laboratory and its staff have grown continuously until by January 1938 they numbered 740, of whom 163 were scientific officers.

Government departments were still in the habit of asking the Society to carry out scientific investigations from time to time, the cost being met usually from national funds; such was an investigation into recent volcanic activity in the West Indies desired by the Colonial Office; a similar investigation in Monserrat was carried out in 1937–8, and to this the Society contributed part of the cost. In 1903 reports of investigations by Captain James on malaria and by Drs Low, Castellani and Christy into sleeping sickness were received. In 1904 the Malta fever was studied and the results of a careful investigation were reported.

In 1904 the *Discovery* under the command of Captain Scott, R.N., returned from an expedition to the Antarctic which had been jointly arranged by the Society and the Royal Geographical Society assisted by a grant from national funds for a scientific expedition of exploration. A scheme had been prepared in 1901 and had been successfully carried out. The Society undertook the reduction and publication of the magnetic and meteorological observations.

By 1906 a very considerable stock of back numbers of the *Philosophical Transactions* and the *Proceedings* had accumulated, which was surplus to the Society's requirements, copies of past issues being only occasionally asked for. The Council therefore decided that a large quantity of this stock should be struck off charge and sold to be pulped. There was also a quantity of old papers unregistered and unindexed which had been accumulating for many years and it seems probable that much of these were also disposed of since the quantity which now remains is not excessive. No list of such papers was made nor does it appear that they were viewed by any Board of Survey for this had never been the custom in the past; with the rapid increase of the Council's work this accumulation of documents, many of them of quite temporary interest, was becoming a serious problem.

In 1909 the Council gave to Cambridge University a telescope which had been on loan to Sir William Huggins since 1870 but which he no longer required. The instrument had been bought from funds bequeathed by B. Oliveira in 1865 to the Society, and was a very useful one of its kind.

In February 1911 the President reminded the Council that the two hundred and fiftieth anniversary of the sealing of the Society's First Charter would take place in the next year. It was decided that a suitable celebration of it should be arranged to include a reception of the delegates of foreign institutions who should attend; a banquet was also to be held in the Guildhall by the permission of the Lord Mayor and the Corporation of the City of London. H.M. the King signified his pleasure to give a Garden Party at Windsor, to which the Delegates and Fellows of the Society were invited. Neither in 1812 nor in 1862, the years of the 150th and 200th Anniversaries, had any special notice been taken of them, though the Society's finances could have supported a moderate expenditure on either occasion. There was all the more reason to celebrate the 250th in 1912. The second edition of the *Record* was now eleven years old and needed revision, so the Council decided to publish a third edition considerably enlarged and in quarto form; the Council also decided to reproduce by photography the signatures of all the Fellows as they appear in the Charter-book, and this formed a most interesting and appropriate record of the occasion. It is not known from whom the suggestion came, but to make available in facsimile the signatures of all those who had been admitted to the Fellowship of the Society and had signed the Obligation in the Charter-book over a period of two hundred and fifty years was a most fitting and original idea. Fifteen years later fourteen plates containing the signatures of those who had been admitted since 1912 were added in order to bring this publication up to date, and doubtless similar additional plates will be provided from time to time since each year between twenty and twenty-five names are added.

During the past fifty years the Society had received a number of bequests from various benefactors, and by 1912 the number of Research Funds had increased to nine while the income received from them amounted to about £2400. The war of 1914–18 interrupted the flow of benefactions but it was resumed in 1919 on an even more generous scale. In this year Miss L. A. Foulerton, in accordance with her father's wish, transferred £20,000 to the Society for original research in medicine, and later bequeathed the residue of her own estate, about £70,000, for the same purpose, namely, research for the discovery of the causes of disease and the relief of human suffering. In 1920 Dr R. Messel left four-fifths of his residuary estate to the Society for scientific research. In 1923 Dr Ludwig Mond bequeathed £50,000 to the Society for research in Natural Science. In the same year Sir Alfred Yarrow handed over to the Society securities to the value of £100,000 for 'the purposes of the Society as the Council may think fit'. Under the terms of a trust deed executed by Mr H. B. Gordon Warren shortly before his death in 1933 the Society receives the income from a sum of about £200,000 to be devoted to scientific research in specified branches of physical science. Fuller details will be found in the Royal Society's *Record* and in the *Year-book*. When the accounts for the year 1939 were closed the Society's accountants reported that the value of the invested capital of the General Purposes Fund was £116,556, that of the twenty-four Special Funds was £109,041, and that of the twenty-seven Research Funds was £555,828. There was also the sum of £94,180 which had been received as income from various research funds but had not been expended; this had been reinvested and was available for use in the research for which it had been originally received. Thus the income which was available for the promotion of scientific research had increased in twenty years by about tenfold, and with its aid from fifteen to eighteen research Professors, Fellows and Students have been enabled to devote the greater part of their time to some branch of research approved by the Council of the Society. Reports are rendered by them annually and these are printed as an appendix to the Report of the Council so that the Society may know how the income from the Research Funds is being utilized.

In 1915 a freehold in the City of London which Lady Sadleir, the widow of Dr W. Croone, had constituted a trust in 1709 in order to remunerate the Croonian lecturer, was sold for £6500; one-fifth of this, being the Society's share, is now held by the Official Trustee of the Charity Commission for the Society, to whom the income is paid.

In 1916 the magnetic survey of the British Isles, which had been begun by Dr G. W. Walker in March 1914 at the charge of the Society, was continued despite the outbreak of war, and was completed by him in 1916 for the mainland area: that of the outlying islands was observed later. This was a repetition of a similar survey which had been carried out by Sir T. Thorpe and Sir A. Rücker in 1886–8.

The International Association of Academies, like many other international organizations, was brought to an end by the war of 1914–18, but the Neutral countries attempted to carry on such of its work as was already in hand. This was also done in the case of the International Geodetic Association, Holland and Switzerland being especially active. In the early part of 1918 a few men in France, the United States and Great Britain were considering what arrangements should be made to resume international scientific co-operation as soon as hostilities should cease, for it was manifestly undesirable that in these and other international organizations separate groups representing scientific work in the Allied countries, the Central Powers and the Neutral countries should be established permanently.

In 1918 therefore the Royal Society, the Académie des Sciences of Paris and the National Academy of Sciences at Washington discussed informally the future organization of the scientific undertakings which had previously been carried on by international co-operation. The outcome of this was that the academies of all the Allied countries, which alone were at the time accessible, were invited by the Royal Society to send representatives to a conference which took place in London at the rooms of the Society on 9, 10 and 11 October 1918.

Though the Association of Academies had proved very successful in establishing closer and more frequent contact between the academies of various countries, it had not been able to do very much to advance international scientific co-operation. It had assisted from time to time the Solar Union, the Marey Institute and the International Committee for the annual publication of Tables of physical and chemical Constants, and had appointed a permanent committee to deal with the Functions of the Brain. One inconvenience which had come to light during the Association's existence was that, as one country might have more academies than another, it would therefore be entitled to more votes than another irrespective of their relative scientific activities. Germany had then four academies, to which Heidelberg had since been added, and their delegates, who held preliminary meetings to discuss the agenda that were to be dealt with, were able thereby to speak and vote as a German group, thus introducing a political element into the meetings which was neither anticipated nor desired.

The meeting of October 1918 in London was followed by another in Paris which met from 26 to 29 November 1918, and at this, after much discussion, it was decided that another meeting of delegates from the academies of the Allied countries should be held in Brussels in the course of the summer of 1919. In the meantime a small committee was appointed to prepare a draft set of Statutes defining the constitution of the new organization which it was proposed to form. This only concerned the Allied countries who were represented at the London and Paris meetings; the admission of the Central Powers was not considered until later, and the inclusion of the Neutral States was postponed until the meeting at Brussels had taken place.

In July 1919 the first General Assembly of the International Research Council, as it was to be called, met in Brussels to discuss the draft Statutes. The Assembly was welcomed by the King of the Belgians who addressed the delegates and invited them to hold the periodical meetings of the General Assembly at Brussels. This act of well-meant hospitality, which was quite unexpected, was warmly greeted and gratefully accepted; but later it was quoted by the nationals of the Central Powers as indicating the definitely political character of the whole organization, and the subordination of its scientific aims to it; it was made an excuse for declining the invitation to join which was sent to each of them a few years later. The Statutes which the committee had drafted were adopted with certain modifications, and it was decided that they should come up for consideration, and if desirable for revision, in twelve years time, i.e. in 1931.

At the First General Assembly, International Unions for Astronomy, Geodesy and Geophysics, Chemistry and Mathematics were formed and their Statutes were approved, as well as those of the Research Council. Other Unions for Physics, Scientific Radio, Geography and the Biological Sciences were proposed but the consideration of their Statutes was postponed until the next General Assembly which met at Brussels in the summer of 1922.

The unit of subscription which each country was to pay annually on becoming a member of the Research Council or of an International Union, was laid down in the Statutes of the Council or Union concerned, and the number of units to be paid by any country was dependent on its population and varied from one to eight units. These Statutes having been accepted by the representatives of the Union concerned and approved by the International Research Council, it was the duty of each nation's delegates to obtain their acceptance by their Government, Academy or other Institution which was to be responsible for the regular payment of the subscriptions to the Research Council, or to a Union. This system worked satisfactorily except that the financial years of different countries did not always coincide and consequently subscriptions were not always received as regularly as was desirable.

The Unions were left completely free to manage their own affairs in accordance with their Statutes when these had been approved by the General Assembly of the Council; they admitted new members who were qualified, fixed the dates of their meetings, and collected the subscriptions of the countries adhering to them as prescribed in their Statutes, and defrayed their expenses.

By its Statutes each International Union required that each adhering Academy, Society, Research Council or Government should appoint a committee to keep in touch with the Bureau of the Union and to advise its appointing authority of all that the Union was doing.

At the next General Assembly of the Council in 1922 the Neutral countries were invited to join the Research Council and the Unions, and accepted. The Statutes of the Unions for Physics, Scientific Radio, Geography and the Biological Sciences were also approved.

In 1925 Great Britain, the United States and the Scandinavian countries associated themselves with a proposal made by Norway that the Central Powers should now be invited to join also, in order to form a truly international organization. This involved a modification of the Statutes for which a majority of the votes held by all the adhering countries was necessary. The number of such votes held by any country varied from one to five according to its population, and therefore at any discussion it was difficult to say until the final voting by countries' delegates had been carried out what the result was likely to be. From the discussions which took place on the Norwegian proposal in 1925 it was fairly certain that the necessary majority would be obtained, but during the adjournment for luncheon a rumour was current that this was not the case. Consequently several delegates who had intended to support the proposal voted against it and others abstained. The motion was consequently lost. This, of course, was taken in some quarters as indisputable proof of the political character of the whole organization and greatly increased the difficulty of the situation. The Royal Society was much dissatisfied with the result, which was wholly unexpected, and demanded that a Special General Assembly should meet in 1926 and reconsider the proposal. This was agreed to and the proposal was then adopted by a large majority. The mischief had however been done.

On the whole the eight International Unions which were appointed in 1919 and 1922 have worked reasonably well, and in some cases much better than might have been expected.

The Astronomical Union utilized the numerous international astronomical committees which were already in existence, discharged some and arranged that the others should report to the Union at its triennial General Assembly. This has apparently worked very successfully.

The Union of Geodesy and Geophysics probably undertook too large a field of action; it has been suggested that there should have been separate

Unions for (a) Geodesy, and (b) Geophysics, but this has never gained any wide support. There seems always to have been a fear lest Geodesy should retain the larger share of the income from subscriptions, and that the Geophysical units would then be financially worse off than they were. It is always difficult to obtain increased grants from the finance ministries of a number of countries unless they can include them in the estimates of the technical departments most directly concerned, and these object to being responsible for grants which they do not themselves administer. At first this Union consisted of seven Sections for Geodesy, Meteorology, Seismology, Atmospheric Electricity and Magnetism, Physical Oceanography, Volcanology, and Scientific Hydrology; but of recent years these have all called themselves Associations and are almost independent organizations.

At the time that the new organization was formed in 1918–19, it was agreed that there should be no permanent bureaux established in any country; that for Geodesy which had existed for many years at Potsdam had made the International Geodetic Association practically a German organization; nevertheless there has been a bureau of Geodesy at Paris since 1920, and also one for Seismology at Strasbourg; and for some years past one for Volcanology has been maintained at Naples. In these and in other ways the organization soon developed a somewhat political character which by some was thought to be prejudicial to the scientific aims which it had been formed to develop.

The Union of Chemistry had the advantage that in every country there existed already a well-organized body of chemists who had been in the habit of meeting periodically to discuss matters of common interest. Their Union therefore was quickly formed and has worked very satisfactorily. For a good many years its National Committee in this country was not appointed by the Royal Society but by the Federal Council of Chemistry, which accepted responsibility for the expenses incurred by its appointed delegates. In 1936 however it was transferred to the Royal Society's list and is now administered in the same way as the other National Committees.

The Union of Mathematics which was appointed in 1919 is no longer in existence; its General Assembly met three or four times but in 1932 at Zürich the liquidation of the Union was decided upon by a majority of the countries represented at that General Assembly.

The Union of Scientific Radio which was founded in 1922 under the presidency of the late General Ferrier has had a most successful career, and under his direction and later under that of Sir Edward Appleton has done excellent work.

The Union of Pure and Applied Physics was formed in 1922 but no meeting of its General Assembly was held until 1934 when a joint meeting of this Union and that for Scientific Radio was held in London. Since then there has been no meeting nor have any reports been issued.

The Union of Geography which was formed in 1922 followed the example of the Union of Chemistry and took over the arrangements for periodical meetings of the International Geographical Congress which had existed for several years previously. The Union at first met triennially but recently the interval between its General Assemblies has been changed to four years.

The Union of Biological Sciences was established in 1922, but for some years it was not very active. In late years however Professor F. C. Went of Holland and Sir Albert Seward reorganized the Botanical Section of it.

At the first General Assembly which was held at Brussels in 1919 proposals were made for the formation of an International Union of Geology and also another for the Medical Sciences. These have not however received much support and the proposals were dropped at the next General Assembly which was held in 1925.

The funds, from which the various National Committees paid the subscription of their country to the Research Council and the Unions, had to be obtained from their respective governments, and this inevitably suggested that political influence might override the scientific aims of some of the Unions. Dr George Hale and Sir Arthur Schuster, who had much to do with the planning of the organization and the drafting of its statutes, had given much attention to the need for avoiding any appearance of political influence, but, as it has turned out, not wholly successfully. When the time comes for planning anew some form of international scientific co-operation, it would be well to consider whether one based on the various sciences might not be preferable to the national representation of the past twenty years. The numerous Congresses such as those of Chemistry, Geology, Geography, Mathematics, and many others have accomplished a large amount of useful scientific work, have brought together many scientific workers, and have done these things without any interference from political aspirations. Their cost has been moderate, and was probably considerably less than that of the Council and Unions.

During the war of 1914–18 a new State organization, the Department of Scientific and Industrial Research, had been established and to this the National Physical Laboratory was transferred since the financial responsibilities for this institution had greatly increased and would certainly become even greater in the near future. The Royal Society however retained a large representation on the General Council and on the executive committee.

In 1930 the Council decided that an increase of two in the number of candidates to be elected annually would not be prejudicial to the Society's scientific standing but would enable the Council to include in their recommendation men of high scientific standing who, since their field of work lay in two closely related branches of sciences, might not otherwise secure election.   No objection to this very reasonable proposal having been made by the Fellows at a Special Ordinary meeting, the necessary statute was passed by the Council.

It has already been mentioned that in these years members of Council were very regular in their attendance at its meetings, and since the amount of business to be dealt with had greatly increased and scientific matters of all kinds were likely to come up for discussion a full attendance of the councillors was more important than ever. This imposed a heavy burden on those who lived at a distance and made it difficult to utilize the services of members of the staffs of the northern Universities whose advice and experience it was often most desirable to have. In 1930 therefore the Council removed the hardship by authorizing the repayment of travelling expenses to all Fellows who were required to attend meetings of Council or of the Society's committees in London.

The numerous gifts and bequests, some of them of considerable value, which had been recently received by the Society necessitated a reorganization of the method of accounts which had been introduced by Sir John Lubbock in 1831. In 1933-4 therefore Messrs W. B. Keen and Co., Chartered Accountants, who had audited the Society's accounts since 1886, were consulted. Hitherto they had only been required to report upon the correctness of the statements which were to be presented to the Council and to the Anniversary meeting, and to verify that payments made were duly supported by vouchers; but it was very desirable that the Society should have the benefit of their expert knowledge of accountancy in order that as efficient and economical a system as possible should be introduced. With their assistance the various books of accounts were brought into line with modern practice; a single bank account replaced the five or six that had been in use for several years; and they were invited to prepare a special report on the Society's finances at the end of each financial year in which they would draw attention to any matter where improvement was desirable, and to any case where the control could be improved. This report is laid before the Finance Committee by the Treasurer, and then, with any recommendation that the committee may wish to make, is presented to the Council. Another change to facilitate dealing with the increasing work of the Society was made in 1938 when the accounts were audited monthly instead of half-yearly; a year later the Society's financial year was made to run from 1 October to 30 September instead of from 1 November to 31 October in order to give sufficient time for the effective examination of the accounts of the past year.

It will be remembered that the first bequest which the Society had received was the sum of £400 which Dr J. Wilkins, one of the Society's first Secretaries, left to it in 1672. This was invested by the Council in certain fee-farm rents at Lewes and produced an annual rental of £24. At the end of the eighteenth century this was reduced to £22. 16s. 0d. by the imposition of a land tax but by some oversight £19. 4s. 0d. as the annual payment due was accepted by the Society from about 1835 onwards. In 1870 the Estate Office of Lord Abergavenny enquired where the lands were situated by which the annual charge of £19. 4s. 0d. was borne. The precise location and description of them were not to be found in the archives of the Society; a suggestion made by the Estate Office that the annual payment might be redeemed provided a convenient way of settling the matter but it was not then followed up. In 1937 the Estate Office renewed the enquiry and the suggestion which had been made in 1870; an examination was made of all relevant documents which were to be found and an arrangement was reached whereby the annual payment of £22. 16s. 0d. should be redeemed on the basis of twenty-five years' rent by the payment of £570. This, having been approved by the Council, was accepted by Lord Abergavenny and the confirming deed was signed and sealed on 2 March 1939.

From time to time the suggestion has been made that the Second Charter, by which the Society has been governed since 1663, needs revision in order that it should be applicable to more modern conditions; though this would only be when changes of a fundamental character were contemplated. In 1831 a committee of forty-two members was nominated by the Council to report on a number of proposed changes, which included: (i) that elections should be held four times in each year and not at any Ordinary meeting; (ii) that the annual subscription should be raised to £4 but no bond was to be demanded; (iii) that the list of officers and councillors recommended for election should be circulated *before* the Anniversary meeting; (iv) that the Treasurer's report and statement of accounts should be printed and circulated to the Fellows *before* the Anniversary meeting. On this occasion the adoption of new statutes was recommended but no change in the Charters was contemplated. On 7 May 1846 the Council appointed a 'Charters Committee', which was so termed since its instructions were 'to consider and report what alterations might with advantage be introduced into a supplementary Charter'. Three weeks later Sir William Grove was added to the Charters Committee, and on his advice a careful study of the whole question showed that the desired ends could be attained by revising the statutes relating to elections or adopting new ones without a new Charter being required. The committee recommended that this procedure should be adopted, and that no revision of the Charters should be undertaken nor any supplement to them applied for; in this the Council concurred and the action which was then taken has proved to be quite satisfactory.

For more than ninety years the annual elections have been carried out under the statutes as revised in 1847 which have proved to be quite adequate. It is not to be expected that the varying fortunes of the Society over a period of about three centuries can be known with accuracy by the majority of the Fellows whenever such proposals are brought forward, and yet the history of the past often provides a valuable guide to the probable effect of any new regulations which may be proposed. It is therefore important that when such are being considered the Society's experience in the past should be at hand for reference.

At the end of 1933 a memorial signed by ninety-one Fellows was sent to the Council for their consideration. The memorial asked that certain modifications should be made in the Charter of 1663 and in the statutes in order to give the general body of the Fellows a more active part in the administration of the Society, and to make fuller use of the experience and ability of many of the Fellows. This memorial recommended that the maximum period of office for the officers should not exceed five years, and that officers should be elected or re-elected by the new Council rather than, as now, nominated by the existing Council. With regard to the nomination, election and re-election of members of Council, the memorial submitted the following suggestions, as a basis of discussion, for the consideration of Council:

(a) Only five members should retire every year. If holding an office, but not otherwise, they should be eligible for consecutive re-election.

(b) Retirements should be determined first by insufficient attendance (suitable allowances being made for distance of residence) and then by seniority as reckoned from the date of election.

(c) Council should declare the number of vacancies and invite nominations from the Fellows (the candidate's consent and the support of a minimum number of Fellows being required) in Section A and Section B severally to preserve the due balance between the physical and biological sciences.

(d) After due consideration of the nominations so received, Council should make its own nominations in such manner that the total number of nominations, including those under (c) not adopted by them, shall exceed the number of vacancies by a minimum number, say one or two in each section.

(e) The election should be by postal ballot, each Fellow having as many votes as there are vacancies, the voting in the two sections to be separate and time being allowed for Fellows resident abroad to vote.

(f) A first meeting of the newly elected Council for the purpose of electing or re-electing officers should take place within the period of office of the retiring Council.

(g) The maximum consecutive period of office of each officer should be determined by statute, and officers should be subject to the conditions under (a) and (b) as regards retirement and re-election.

After carefully considering these proposals the Council unanimously approved the following decisions on February 28 and confirmed them on 7 March 1935:

(i) That Council re-affirm Minute 4 of October 26, 1933, which approved the custom by which the President holds office for a period of five years, confirmed the order restricting the tenure of office by the Treasurer and

Secretaries to a period of ten years, and established the maximum tenure of the Foreign Secretaryship at five years.

(ii) That the maximum periods of office given above be printed in the *Year-book* as Standing Orders.

(iii) That the officers should, as at present, be nominated by the outgoing Council to the Society for election, and not elected by the incoming Council.

(iv) That it is not desirable to introduce a system whereby the number of nominations for new members of Council is in excess of the number of vacancies.

(v) That it is not desirable to make any alterations to the procedure in the nomination of the Council, which require any alteration of the Charters.

These decisions are now printed in the *Year-book* at the end of the Standing Orders for election of officers and Council. There is no previous record of any criticism of the Charters or of any challenge to the constitution which they gave to the Society in 1663.

The Anniversary meeting of the Fellows, at which the election of the officers and councillors for the following twelvemonth took place on 30 November 1935, was largely attended. A memorandum setting forth the views of the memorialists had been sent to all the Fellows, who had also received the Council's memorandum dealing with the whole matter. When the ballot was taken to determine the names of the officers and councillors for the following twelvemonth each Fellow could either hand in the ballot paper prepared by the outgoing Council, or alter his copy to support those whom the memorialists had put forward as an alternative list. When the scrutineers of the ballot made their report to the meeting they announced an overwhelming majority for those whose names the outgoing Council had proposed and recommended.

In all institutions in which the members meet periodically to discuss the technical or scientific matters in which they are specially interested, a problem which is for ever presenting itself in one form or another is: How can the communications contributed by the members be most effectively dealt with? The Society has experienced every variety of this difficulty. Samuel Pepys was besought by John Evelyn to attend the meetings as contributions were few and unimportant; Lord Macclesfield on becoming President begged for the Fellows' assistance in restoring the standard of the Society's communications; Granville and Babbage early in the nineteenth century criticized the negligent way in which papers offered to the Society were accepted or rejected, and the inefficiency of some of the referees. Before the end of the nineteenth century communications had trebled in number and had greatly improved in their quality. Competent referees were forthcoming, and the number of papers which the Council considered worthy of publication often embarrassed the Treasurer.

With the increasing specialization of many branches of science a new difficulty presents itself: valuable communications treating of some new and important aspect of the subject may be so specialized in their treatment as to be fully appreciated only by those who have worked on this special aspect of the subject. The increasing number of such communications has led to the suggestion that discussion of a selected subject might usefully replace the reading of some of the more abstruse papers. The experiment has been tried with complete success, and the Society has not hesitated to invite as its guests scientific men of other countries who had distinguished themselves by the advances which they had made in the subject to be discussed. Instead of the two hours which are usually allotted to a meeting these discussions often begin in the forenoon and continue, with a short interval, until the evening, so as to give sufficient time for those who are expert in the subject to present their views.

One result of these lengthy meetings has been that the benches with which the meeting room was furnished in the eighteenth and nineteenth centuries (see *Royal Society Record*, Plates XVII and XVIII) have proved to be extremely uncomfortable when discussions began at 10 a.m. and might last until 6 p.m. In 1939 the benches which had been used by many generations of Fellows and had become very shabby and worn were replaced by others of modern type more suitable for meetings which lasted long and demanded the close attention of the audience to the addresses of the various speakers.

REFERENCE

Royal Society, Record of. 4th ed. 1940.

CONCLUSION

SINCE its foundation in 1662 the Royal Society has grown in size and influence until it is now widely recognized as an institute which is playing an important part in promoting research in all branches of natural science. This has been accomplished by the zeal and energy of those of its Fellows eminent in learning and distinguished for the accounts of their discoveries in Natural Knowledge which they have contributed to the Society.

Those who drafted the Charters performed their task with great skill and foresight when they placed very wide powers in the hands of the councillors, thereby making it possible for the Council to introduce improvements and to rectify mistakes without any alteration in the Charter of 1663 being needed. The Council could at any time revise, or revoke, any existing statute, or adopt a new one under the powers conferred on it by the Charter; no reference to the general body of the Fellows was necessary except when the admission of new Fellows, or the appointment of members of Council was concerned. It may seem strange that such far-reaching powers should have been placed in the hands of the Council without the body of the Fellows having any control of the final decision, but this was in accordance with the custom of the time.

Among the problems which the Council had to solve was the size of the Fellowship; this the Charters did not restrict in any way, but allowed candidates to be admitted to the Fellowship by the vote of the Fellows who could increase it to any number they pleased, and could if they saw fit demand specific qualifications, though in fact they did not do so. The result has been that the membership is much larger than that of most of the other national academies. For two centuries the men of science in the Society were in a minority, and not until after 1850 when they controlled the Councils were the aims of its founders fully realized.

Since the beginning of the present century the Fellows have all been men of science, except a very few who have been specially elected, and the Society's aims and policy have been directed wholly to the advance of scientific research.

The original founders planned that the Society should be a democratic institution of scientific men, but financial difficulties necessitated the admission of a small number of well-to-do candidates from 1660 onwards, and this category soon increased in number until they were a majority which hindered the activity of the Society for nearly two centuries. But when all the councillors were men of science, they wisely took the Fellows more fully into their confidence by explaining the policy which they had designed. The effect has been that the Fellows are to-day more fully acquainted with the policy of the Society than formerly. So long as men of prudence and foresight constitute its Councils and its benefactors are as generous as they have been in recent years, there seems to be no limit to the progress which may be anticipated. A new and even more propitious future may await the Society in the progress which it may achieve under the motto which John Evelyn suggested for us two hundred and eighty years ago: *Nullius in Verba.*

APPENDIX I

SECOND CHARTA: 22 APRIL 1663

———

CHARTA SECUNDA

Praesidi, Concilio, et Sodalibus REGALIS SOCIETATIS Londini,
a Rege CAROLO SECUNDO concessa, A.D. MDCLXIII

———

Translation of Second Charter, A.D. 1663

CHARLES THE SECOND, by the grace of God King of England, Scotland, France, and Ireland, Defender of the Faith, &c., to all to whom these our Letters Patent shall come, greeting.

We have long and fully resolved with Ourself to extend not only the boundaries of the Empire, but also the very arts and sciences. Therefore we look with favour upon all forms of learning, but with particular grace we encourage philosophical studies, especially those which by actual experiments attempt either to shape out a new philosophy or to perfect the old. In order, therefore, that such studies, which have not hitherto been sufficiently brilliant in any part of the world, may shine conspicuously amongst our people, and that at length the whole world of letters may always recognize us not only as the Defender of the Faith, but also as the universal lover and patron of every kind of truth:

Know ye that we, of our special grace and of our certain knowledge and mere motion, have ordained, established, and granted, and by these presents for us, our heirs, and successors do ordain, establish, and grant, that henceforth for ever there shall be a Society consisting of a President, Council, and Fellows, who shall be called and named The President, Council, and Fellows of the Royal Society of London for improving Natural Knowledge (of which same Society we by these presents declare Ourself Founder and Patron); And by these presents for us, our heirs, and successors we do make, ordain, create, and constitute the same Society, by the name of The President, Council, and Fellows of the Royal Society of London for promoting Natural Knowledge, one body corporate and politic, in fact, deed, and name, really and fully, and that by the same name they may have perpetual succession; And that they and their successors (whose studies are to be applied to further promoting by the authority of experiments the sciences of natural things and of useful arts, to the glory of God the Creator, and the advantage of the human race), by the same

name of The President, Council, and Fellows of the Royal Society of
London for promoting Natural Knowledge, may and shall be in all future
times persons able and capable in law to have, acquire, receive, and possess
lands [and] tenements, meadows, feedings, pastures, liberties, privileges,
franchises, jurisdictions, and hereditaments whatsoever to them and their
successors in fee and perpetuity, or for term of life, lives, or years, or otherwise in whatsoever manner, and also goods and chattels, and all other
things, of whatsoever kind, nature, sort, or quality they may be (the
Statute concerning alienation in mortmain notwithstanding); and also
to give, grant, [demise,] and assign the same lands, tenements, and hereditaments, goods and chattels, and to do and execute all acts and things
necessary of and concerning the same, by the name aforesaid; And that
by the name of The President, Council, and Fellows of the Royal Society
of London for promoting Natural Knowledge aforesaid, they may henceforth for ever be able and have power to plead and be impleaded, to
answer and be answered, to defend and be defended, in whatsoever Courts
and places, and before whatsoever Judges, Justices, and other persons and
officers of us, our heirs, and successors, in all and singular actions, both
real and personal, pleas, suits, plaints, causes, matters, things, and demands
whatsoever, of whatsoever kind, nature, or sort they may or shall be, in
the same manner and form as any of our lieges within this our Realm of
England, being persons able and capable in law, or as any body corporate
or politic within this our Realm of England, may be able and have power
to have, acquire, receive, possess, give, and grant, to plead and be impleaded, to answer and be answered, to defend or be defended; And that
the same President, Council, and Fellows of the Royal Society aforesaid
and their successors for ever may have a Common Seal, to serve for
transacting all causes and affairs whatsoever of them and their successors;
and that it may and shall be good and lawful to the same President, Council,
and Fellows of the Royal Society aforesaid, and to their successors for the
time being, to break, change, and make anew that Seal from time to time,
as it shall seem most expedient to them.

We give and grant moreover by these presents to the President, Council,
and Fellows of the Royal Society aforesaid, and to their successors for ever,
in testimony of our royal favour towards them, and of our peculiar esteem
for them, to the present and future ages, these following blazons of honour,
that is to say: in the dexter corner of a silver shield our three Lions of
England, and for Crest a helm adorned with a crown studded with florets,
surmounted by an eagle of proper colour holding in one foot a shield
charged with our lions: Supporters, two white hounds gorged with crowns;
to be borne, exhibited, and possessed for ever by the aforesaid President,
Council, and Fellows, and their successors, as occasion shall serve.

And that our royal intention may obtain the better effect, and for the
good rule and government of the aforesaid Royal Society from time to

time, we will, and by these presents for us, our heirs, and successors do grant to the same President, Council, and Fellows of the Royal Society aforesaid, and to their successors, that henceforth for ever the Council aforesaid shall be and consist of twenty-one persons (of whom we will the President for the time being, or his Deputy, to be always one); And that all and singular other persons who within two months next following after the date of these presents shall be received and admitted into the same Society as Members of the Royal Society aforesaid, by the President and Council, or by any eleven or more of them (of whom we will the President for the time being, or his Deputy, to be always one), or by two third parts or more of the aforesaid eleven or more, and in all time following by the President, Council, and Fellows, or by any twenty-one or more of them (of whom we will the President for the time being, or his Deputy, to be always one), or by two third parts or more of the aforesaid twenty-one or more, and shall have been noted in the Register by them to be kept, shall be, be called, and be named Fellows of the Royal Society aforesaid, as long as they shall live, unless it shall happen that any one of them be amoved for any reasonable cause, according to the Statutes of the Royal Society aforesaid, which are to be drawn up; whom, the more eminently they are distinguished for the study of every kind of learning and good letters, the more ardently they desire to promote the honour, studies, and advantage of this Society, the more they are noted for integrity of life, uprightness of character, and piety, and excel in fidelity and affection of mind towards us, our Crown, and dignity, the more we wish them to be especially deemed fitting and worthy of being admitted into the number of the Fellows of the same Society.

And for the better execution of our will and grant in this behalf, we have assigned, nominated, constituted, and made, and by these presents for us, our heirs, and successors do assign, nominate, constitute, and make, our very well-beloved and trusty William, Viscount Brouncker, Chancellor of our very dear consort Queen Catharine, to be the first and present President of the Royal Society aforesaid; willing that the aforesaid William, Viscount Brouncker, shall continue in the office of President of the Royal Society aforesaid from the date of these presents until the feast of St Andrew next following after the date of these presents, and until one other of the Council of the Royal Society aforesaid for the time being shall have been elected, appointed, and sworn to that office in due manner, according to the ordinance and provision below in these presents expressed and declared (if the aforesaid William, Viscount Brouncker, shall live so long); having first taken a corporal oath well and faithfully to execute [his office] in and by all things touching that office, according to the true intention of these presents, before our very well-beloved and very trusty Cousin and Councillor Edward, Earl of Clarendon, our Chancellor of England: to which same Edward, Earl of Clarendon, our Chancellor aforesaid, we give and grant full power and authority to administer the oath aforesaid in these words following, that is to say:

I, William, Viscount Brouncker, do promise to deal faithfully and honestly in all things belonging to the trust committed to me, as President of the Royal Society of London for improving Natural Knowledge, during my employment in that capacity. So help me God!

We have also assigned, constituted, and made, and by these presents for us, our heirs, and successors do make, our beloved and trusty Robert Moray, Knight, one of our Privy Council in our Realm of Scotland; Robert Boyle, Esquire; William Brereton, Esquire, eldest son of the Baron de Brereton; Kenelm Digby, Knight, Chancellor to our very dear mother, Queen Maria; Gilbert Talbot, Knight, Treasurer of our Jewels; Paul Neile, Knight, one of the Ushers of our Privy Chamber; Henry Slingesby, Esquire, one of the Gentlemen of our aforesaid Privy Chamber; William Petty, Knight; Timothy Clarke, Doctor in Medicine and one of our Physicians; John Wilkins, Doctor in Divinity; George Ent, Doctor in Medicine; William Aerskine, one of our Cup-bearers; Jonathan Goddard, Doctor in Medicine and Professor of Gresham College; William Balle, Esquire; Matthew Wren, Esquire; John Evelyn, Esquire; Thomas Henshaw, Esquire; Dudley Palmer, of Grey's Inn, in our County of Middlesex, Esquire; Abraham Hill, of London, Esquire; and Henry Oldenburg, Esquire, together with the President aforesaid, to be and become the first and present twenty-one of the Council and Fellows of the Royal Society aforesaid; to be continued in their offices of the Council aforesaid from the date of these presents until the aforesaid feast of St Andrew the Apostle next following, and thenceforth until other fitting and able and sufficient persons shall have been elected, appointed, and sworn into the offices aforesaid (if they shall live so long, or shall not have been amoved for any just and reasonable cause); first taking corporal oaths before the President for the time being of the aforesaid Royal Society, well and faithfully to execute their offices in and by all things touching those offices, according to the form and effect of the aforesaid oath, *mutatis mutandis*, to be administered to the President of the Royal Society aforesaid by our Chancellor of England; (to which same President for the time being, for us, our heirs, and successors, we give and grant by these presents full power and authority to administer the oaths aforesaid to the aforesaid persons, and to any others whomsoever hereafter from time to time to be elected into the Council aforesaid); And that the same persons, so as it is aforesaid elected, appointed, and sworn, and hereafter to be elected, appointed, and sworn from time to time, to the Council of the aforesaid Royal Society, shall be and become aiding, counselling, and assistant in all matters, business, and affairs touching or concerning the better regulation, government, and direction of the aforesaid Royal Society, and of every Member of the same.

We also grant to the President, Council, and Fellows of the aforesaid Society, and to their successors for ever, that they and their successors, or any nine or more of them (of whom we will the President for the time being, or his Deputy, to be always one), may be able lawfully to make and hold assemblies or meetings of themselves for the examination and investigation of experiments and of natural things, and for other affairs belonging to the Society aforesaid, as often as and whenever it shall be needful, in a College or Hall or other convenient place within our City of London, or in any other convenient place within ten miles of our same City.

And further we will, and by these presents for us, our heirs, and successors, do grant to the aforesaid President, Council, and Fellows of the Royal Society aforesaid, and to their successors, that the President, Council, and Fellows of the Royal Society aforesaid for the time being, or any thirty-one or more of them (of whom we will the President for the time being, or his Deputy, to be one), or the major part of the aforesaid thirty-one or more, may and shall have from time to time in all future times for ever power and authority to nominate and elect, and that they may be able and have power to elect and nominate, every year, on the aforesaid feast of St Andrew, one of the Council of the aforesaid Royal Society for the time being, who may and shall be President of the Royal Society aforesaid until the feast of St Andrew the Apostle thereafter next following (if he shall live so long, or shall not be amoved meanwhile for any just and reasonable cause), and thenceforth until another shall have been elected, appointed, and nominated to the office of President of the Royal Society aforesaid; and that he, after he shall so have been elected and nominated, as it is aforesaid, to the office of President of the Royal Society aforesaid, before he be admitted to that office, shall take a corporal oath before the Council of the same Royal Society, or any seven or more of them, rightly, well, and faithfully to execute that office in all things touching that office, according to the form and effect of the aforesaid oath, *mutatis mutandis* (to which same Council, or to any seven or more of them, we give and grant by these presents for us, our heirs, and successors full power and authority to administer the oath aforesaid from time to time, as often as it shall be needful to elect a President); and that after having so taken such oath, as it is aforesaid, he may be able and have power to execute the office of President of the Royal Society aforesaid until the feast of St Andrew the Apostle thereafter next following; And if it shall happen that the President of the Royal Society for the time being, at any time, so long as he shall be in the office of President of the same Royal Society, shall die, retire, or be amoved from his office, that then and so often it may and shall be good and lawful to the Council of the Royal Society aforesaid, and to their successors for ever, or to any eleven or more of them, to assemble or meet for the election of one of the aforesaid number of the Council aforesaid as

President of the Royal Society aforesaid; and that he who shall have been elected and sworn by the Council aforesaid, or by the aforesaid eleven or more, or by the major part of the aforesaid eleven or more, as it is aforesaid, may have and exercise that office during the residue of the same year, and until another shall have been in due manner elected and sworn to that office, first taking a corporal oath in the form above specified: and so as often as the case shall so happen.

And further we will, that whenever it shall happen that any one or any of the Council of the Royal Society aforesaid for the time being shall die, or be amoved from that office, or retire (which same [members] of the Council of the Royal Society aforesaid, and every one of them, we will to be amovable for misbehaviour or any other reasonable cause, at the good pleasure of the President and of the rest of the Council aforesaid, of whom we will the President for the time being, or his Deputy, to be one, or of the major part of the same), that then and so often it may and shall be good and lawful to the aforesaid President, Council, and Fellows of the Royal Society aforesaid, and to their successors for ever, or to any twenty-one or more of the same (of whom we will the President of the Royal Society aforesaid for the time being, or his Deputy, to be one), or to the major part of the aforesaid twenty-one or more, to nominate, elect, and appoint one other or several others of the Fellows of the Royal Society aforesaid, in the place or places of him or them so dead, retired, or amoved, to fill up the aforesaid number of twenty-one persons of the Council of the Royal Society aforesaid; and that he or they so elected and appointed in that office may have the same office until the feast of St Andrew the Apostle then next following, and thenceforth until one other or several others shall have been elected, appointed, and nominated; first taking a corporal oath before the President and Council of the Royal Society aforesaid, or any seven or more of them (of whom we will the President for the time being, or his Deputy, to be always one), well and faithfully to execute that office in and by all things touching that office, according to the true intention of these presents.

And further we will, and by these presents for us, our heirs, and successors do grant to the aforesaid President, Council, and Fellows of the aforesaid Royal Society, and to their successors [for ever], that they and their successors, or any thirty-one or more of them (of whom we will the President for the time being, or his Deputy, to be always one), or the major part of the aforesaid thirty-one or more, every year, on the aforesaid feast of St Andrew the Apostle, may and shall have full power and authority to elect, nominate, appoint, and change ten of the Fellows of the Royal Society aforesaid, to fill up the places and offices of ten of the aforesaid number of twenty-one of the Council of the Royal Society aforesaid; for we do declare it to be our royal pleasure, and by these presents for us, our heirs, and successors we do grant, that ten of the aforesaid Council, and no more, shall be annually changed and amoved by the President, Council, and Fellows of the Royal Society aforesaid.

We will also, and for us, our heirs, and successors do grant to the aforesaid President, Council, and Fellows of the aforesaid Royal Society, and to their successors for ever, that if it shall happen that the President of the same Royal Society for the time being is detained by sickness or infirmity, or is employed in the service of us, our heirs, or successors, or is otherwise occupied, so that he shall not be able to attend to the necessary affairs of the same Royal Society touching the office of President, that then and so often it may and shall be good and lawful to the same President so detained, employed or occupied, to nominate and appoint one of the Council of the aforesaid Royal Society for the time being to be and become the Deputy of the same President; which same Deputy, so to be made and appointed in the office of Deputy of the President aforesaid, may and shall be the Deputy of the same President from time to time, as often as the aforesaid President shall happen to be so absent, during the whole time in which the aforesaid President shall continue in the office of President; unless in the meanwhile the aforesaid President of the Royal Society aforesaid for the time being shall have made and appointed one other of the aforesaid Council his Deputy; And that every such Deputy of the aforesaid President so to be made and appointed, as it is aforesaid, may be able and have power to do and execute all and singular things which pertain or ought to pertain to the office of President of the aforesaid Royal Society, or which are limited and appointed to be done and executed by the aforesaid President, by virtue of these our Letters Patent, from time to time, as often as the aforesaid President shall happen to be so absent, during such time as he shall continue the Deputy of the aforesaid President, by force of these our Letters Patent, as fully, freely, and wholly, and in as ample manner and form, as the aforesaid President, if he were present, would be able and have power to do and execute those things; a corporal oath first to be taken by such Deputy upon the holy Gospels of God, in the form and effect above specified, well and faithfully to execute all and singular things which pertain to the office of President, before the aforesaid Council of the aforesaid Royal Society, or any seven or more of them; and so often as the case shall so happen: to which same Council, or to any seven or more of them, for the time being, we do give and grant by these presents, power and authority to administer the oath aforesaid, as often as the case shall so happen, without procuring or obtaining a writ, commission, or further warrant in that behalf from us, our heirs, or successors.

And further we will, and by these presents for us, our heirs, and successors do grant to the aforesaid President, Council, and Fellows of the Royal Society aforesaid, and to their successors, that they and their successors henceforth for ever may and shall have one Treasurer, two

Secretaries, two or more Curators of Experiments, one Clerk or more, and moreover two Serjeants-at-Mace, who may from time to time attend upon the President; and that the aforesaid Treasurer, Secretaries, Curators, Clerk or Clerks, and Serjeants-at-Mace, to be elected and nominated by the President, Council, and Fellows of the Royal Society aforesaid, or by any thirty-one or more of them (of whom we will the President for the time being, or his Deputy, to be one), or by the major part of the aforesaid thirty-one or more, before they be admitted to execute their special[1] and respective offices, shall take their corporal oaths in the form and effect above specified, before the President, or his Deputy, and the Council of the same Royal Society, or any seven or more of them, rightly, well, and faithfully to execute their several and respective offices in all things touching the same; and that after having so taken such oaths, as it is aforesaid, they may exercise and use their respective offices; to which same President and Council, or to any seven or more of them, we do give and grant by these presents full power and authority to administer the oaths aforesaid from time to time to the aforesaid several and respective officers and their successors: And we have assigned, nominated, chosen, created, appointed, and made, and by these presents for us, our heirs, and successors do assign, nominate, choose, create, appoint, and make, our beloved subjects the aforesaid William Balle, Esquire, to be and become the first and present Treasurer, and the aforesaid John Wilkins and Henry Oldenburg to be and become the first and present Secretaries, of the aforesaid Royal Society; to be continued in the same offices until the aforesaid feast of St Andrew the Apostle next following after the date of these presents: And that from time to time and at all times on the aforesaid feast of St Andrew the Apostle (unless it shall be Sunday, and if it be Sunday, then on the day next following), the President, Council, and Fellows of the aforesaid Royal Society for the time being, or any thirty-one or more of them (of whom we will the President for the time being, or his Deputy, to be one), or the major part of the aforesaid thirty-one or more, may be able and have power to elect, nominate, and appoint upright and discreet men, who are and shall be of the number of the Council of the Royal Society aforesaid, as Treasurer and Secretaries, from time to time; and that those who shall so have been elected, appointed, and sworn to the aforesaid several and respective offices, as it is aforesaid, may be able and have power to exercise and enjoy those respective offices until the aforesaid feast of St Andrew then next following, their aforesaid oaths, as it is aforesaid, first to be taken; and so as often as the case shall so happen. And if it shall happen that the aforesaid elections of President, Council, Treasurer, [and] Secretaries, or of any one or any of them, cannot conveniently be made or finished on the aforesaid feast of St Andrew, we give and grant to the aforesaid President, Council, and Fellows, and to their successors for ever, that they or any

[1] So in the original; *qu.* several.

thirty-one or more of them (of whom we will the President for the time being, or his Deputy, to be one), or the major part of the said thirty-one or more, may lawfully name and assign one other day, as near to the feast of St Andrew aforesaid as can conveniently be done, for making or finishing the aforesaid elections; and so from day to day, until the aforesaid elections be finished: And if it shall happen that any one or any of the aforesaid officers of the same Royal Society shall die, retire, or be amoved from their respective offices, that then and so often it may and shall be good and lawful to the President, Council, and Fellows of the aforesaid Royal Society, and to their successors for ever, or to any twenty-one or more of them (of whom we will the President for the time being, or his Deputy, to be one), or to the major part of the aforesaid twenty-one or more, to elect and appoint another or others to the office or offices of those persons so deceased, retired, or amoved; and that he or they so elected and appointed may have and exercise the respective offices aforesaid during the residue of the same year, and until another or others shall have been in due manner elected and sworn to those respective offices; and so as often as the case shall so happen.

And moreover we will, and of our special grace and of our certain knowledge and mere motion do grant to the aforesaid President, Council, and Fellows of the Royal Society aforesaid, and to their successors for ever, that the President and Council of the aforesaid Royal Society for the time being (due or lawful summons or citation being always first made of all the Members of the Council aforesaid to extraordinary meetings), or any nine or more of them (of whom we will the President for the time being, or his Deputy, to be one), may be able and have power both to meet together and assemble in a College or Hall or other convenient place within our City of London, or in any other convenient place within ten miles of our same City; and that they so met together and assembled, or the major part of them, shall and may have full authority, power, and faculty from time to time to draw up, constitute, ordain, make, and establish such laws, statutes, acts, ordinances, and constitutions as shall seem to them, or to the major part of them, to be good, wholesome, useful, honourable, and necessary, according to their sound discretions, for the better government, regulation, and direction of the Royal Society aforesaid, and of every Member of the same, and to do and perform all things belonging to the government, matters, goods, faculties, rents, lands, tenements, hereditaments, and affairs of the Royal Society aforesaid; all and singular which laws, statutes, acts, ordinances, and constitutions so to be made as it is aforesaid, we will, and by these presents for us, our heirs, and successors, firmly enjoining, do order and command, that they shall be inviolably observed from time to time, according to the tenor and effect of the same: so nevertheless, that the aforesaid laws, statutes, acts, ordinances, and constitutions so to be made as it is aforesaid, and every one of them, be reasonable, and not repugnant or contrary to the laws, customs, acts, or statutes of this our Realm of England.

And further, of our more ample special grace and of our certain knowledge and mere motion, we have given and granted, and by these presents for us, our heirs, and successors do give and grant to the aforesaid President, Council, and Fellows of the aforesaid Royal Society, and to their successors for ever, or to any twenty-one or more of them (of whom we will the President for the time being, or his Deputy, to be always one), or to the major part of the aforesaid twenty-one or more, full power and authority from time to time to elect, nominate, and appoint one or more Typographers or Printers, and Chalcographers or Engravers, and to grant to him or them, by a writing sealed with the Common Seal of the aforesaid Royal Society, and signed by the hand of the President for the time being, faculty to print such things, matters, and affairs touching or concerning the aforesaid Royal Society, as shall have been committed to the aforesaid Typographer or Printer, Chalcographer or Engraver, or Typographers or Printers, Chalcographers or Engravers, from time to time, by the President and Council of the aforesaid Royal Society, or any seven or more of them (of whom we will the President for the time being, or his Deputy, to be one), or by the major part of the aforesaid seven or more; their corporal oaths first to be taken, before they be admitted to exercise their offices, before the President and Council for the time being, or any seven or more of them, in the form and effect last specified; to which same President and Council, or to any seven or more of them, we do give and grant by these presents full power and authority to administer the oaths aforesaid.

And further, in order that the aforesaid President, Council, and Fellows of the aforesaid Royal Society may obtain the better success in their philosophical studies, of our more ample special grace and of our certain knowledge and mere motion, we had given and granted, and by these presents for us, our heirs, and successors do give and grant, to the aforesaid President, Council, and Fellows of the aforesaid Royal Society, and to their successors for ever, that they and their successors, or any nine or more of them (of whom we will the President for the time being, or his deputy, to be one), or the major part of the aforesaid nine or more, may and shall have from time to time full power and authority to require, take, and receive from time to time, and at such seasonable times, according to their discretion, by their assign or assigns the bodies of such persons as have suffered death by the hand of the executioner, and to anatomize them, in as ample manner and form, and to all intents and purposes, as the President of the College of Physicians and the Company of Surgeons of our City of London (by whatsoever names the two aforesaid corporations shall have been distinguished) have used or enjoyed, or may be able and have power to use and enjoy, the same bodies.

And further, for the improvement of the experiments, arts, and sciences of the aforesaid Royal Society, of our more abundant special grace and of our certain knowledge and mere motion, we have given and granted, and by these presents for us, our heirs, and successors do give and grant, to the aforesaid President, Council, and Fellows of the aforesaid Royal Society, and to their successors for ever, that they and their successors, or any nine or more of them (of whom we will the President for the time being, or his Deputy, to be one), or the major part of the aforesaid nine or more, may and shall have from time to time full power and authority, by letters or epistles under the hand of the aforesaid President or his Deputy, in the presence of the Council, or of any seven or more of them, and in the name of the Royal Society, to enjoy mutual intelligence and affairs with all and all manner of strangers and foreigners, whether private or collegiate, corporate or politic, without any molestation, interruption, or disturbance whatsoever: Provided nevertheless, that this our indulgence, so granted as it is aforesaid, be not extended to further use than the particular benefit and interest of the aforesaid Royal Society in matters or things philosophical, mathematical, or mechanical.

And further we have given and granted, and by these presents for us, our heirs, and successors do give and grant to the aforesaid President, Council, and Fellows of the Royal Society aforesaid, and to their successors for ever, or to the President and Council of the Royal Society aforesaid, or the major part of them, full power and authority to erect, build, and construct, or to make or cause to be erected, built, or constructed, within our City of London, or ten miles of the same, one or more College or Colleges, of whatsoever kind or quality, for the habitation, assembly, and meeting of the aforesaid President, Council, and Fellows of the aforesaid Royal Society, and of their successors, for the ordering and arranging of their affairs and other matters concerning the same Royal Society.

And further we will, and by these presents for us, our heirs, and successors do ordain, constitute, and appoint, that if any abuses or differences hereafter shall arise and happen concerning the government or other matters or affairs of the aforesaid Royal Society, whereby any injury or hindrance may be done to the constitution, stability, and progress of the studies, or to the matters and affairs, of the same; then that and so often, by these presents, for us, our heirs, and successors, we do authorize, nominate, assign, and appoint our aforesaid very well-beloved and very trusty Cousin and Councillor Edward, Earl of Clarendon, our Chancellor of our Realm of England, by himself during his life, and, after his death, then the Archbishop of Canterbury, the Chancellor or Keeper of the Great Seal of England, the Treasurer of England, the Keeper of the Privy Seal, the Bishop of London, and the two Principal Secretaries for the time being, or any four or more of them, to reconcile, compose, and adjust the same differences and abuses.

And further we will, and by these presents for us, our heirs, and successors, firmly enjoining, do order and command all and singular the Justices, Mayors, Aldermen, Sheriffs, Bailiffs, Constables, and other officers, ministers, and subjects whomsoever of us, our heirs, and successors, that they be from time to time aiding and assistant to the aforesaid President, Council, and Fellows of the Royal Society aforesaid, and to their successors for ever, in and by all things, according to the true intention of these our Letters Patent.

Although express mention of the true yearly value or of the certainty of the premises, or of any of them, or of other gifts or grants before these times made by us or by any of our progenitors or predecessors to the aforesaid President, Council, and Fellows of the Royal Society aforesaid, is not made in these presents; or any statute, act, ordinance, provision, proclamation, or restriction to the contrary thereof heretofore had, made, enacted, ordained, or provided, or any other thing, cause, or matter whatsoever, in any wise notwithstanding.

In witness whereof we have caused these our Letters to be made Patent. Witness Ourself, at Westminster, the twenty-second day of April, in the fifteenth year of our reign.

By writ of Privy Seal.

HOWARD.

APPENDIX II

STATISTICAL TABLES

A. COMPOSITION OF THE FELLOWSHIP

November of	Scientific Fellows		Non-scientific Fellows		Ratio of scientfiic to unscientific Fellows	Total of Ordinary Fellows	Peers, per-centage of the Ordin-ary Fellows	Foreign mem-bers
	Total	Per cent	Total	Per cent				
1663	44	32·2	93	67·8	1 : 2·1	137	10·8	—
1671	47	25·0	140	75·0	1 : 3·0	187	11·0	18
1698	36	30·6	83	69·4	1 : 2·3	119	8·4	28
1740	99	33·0	202	67·0	1 : 2·04	301	11·3	146
1770	113	29·0	271	71·0	1 : 2·4	384	10·4	153*
1800	149	28·6	376	71·4	1 : 2·52	525	11·2	77
1830	213	32·3	446	67·7	1 : 2·1	659	9·5	45
1860	330	52·6	300	47·4	1 : 0·9†	630	4·6	42

* 170 foreign members in 1766.   † 1663→1830 av. = 2·34.

B. PROFESSIONS OF SCIENTIFIC FELLOWS

Profession	1663		1671		1698	
	No.	Per cent	No.	Per cent	No.	Per cent
Mathematicians } Astronomers }	14	34·9	12	25·5	7	20·0
Physicists } Chemists } Engineers }	3	7·5	3	6·3	4	11·5
Physicians and Surgeons	26	55·1	31	66·0	19	54·3
Botanists } Zoologists } Geologists }	1	2·5	1	2·2	5	14·2
Total	44	100·0	47	100·0	35	100·0
Total fellowship	137	—	187	—	119	—

PROFESSIONS OF SCIENTIFIC FELLOWS (*cont.*)

Profession	1740 No.	1740 Per cent	1770 No.	1770 Per cent	1800 No.	1800 Per cent	1830 No.	1830 Per cent	1860 No.	1860 Per cent
Mathematicians	12	12·0	9	7·8	12	8·0	21	9·6	28	8·5
Astronomers	7	7·0	8	7·2	9	6·0	14	6·6	22	6·6
Chemists	2	2·0	4	3·5	14	9·3	13	6·2	27	8·2
Physicists	4	4·0	8	7·2	8	5·4	13	6·2	26	7·9
Engineers	—	—	2	1·8	7	4·5	7	3·3	30	9·1
Surveyors ⎫ Hydrographers ⎭	1	1·0	2	1·8	5	4·0	7	3·3	9	2·7
Instrument-makers ⎫ Opticians ⎭	2	2·0	—	—	1	0·6	1	0·5	1	0·3
Physicians ⎫ Surgeons ⎭	63	63·0	72	63·5	84	56·2	83	38·9	80	24·4
	—	—	—	—	—	—	23	10·4	37	11·2
Botanists	8	8·0	5	4·5	8	5·4	4	1·9	17	5·2
Zoologists ⎫ Naturalists ⎭	—	—	3	2·7	—	—	8	3·8	19	5·8
Geologists	1	1·0	—	—	1	0·6	19	9·3	34	10·1
Total of scientific Fellows	100	100·0	113	100·0	149	100·0	213	100·0	330	100·0
Total number of Fellows	301	—	384	—	525	—	659	—	630	—

C. COMPOSITION OF COUNCILS, NUMBER OF MEETINGS AND AVERAGE ATTENDANCE AT A MEETING

Period	Average number of members of Council Scientific	Average number of members of Council Non-scientific	Average number of meetings yearly	Average attendance at a meeting
1663–1680	8·6	12·4	12·5	9·0
1681–1700	7·8	13·2	7·6	8·5
1701–1720	8·1	12·9	7·4	9·0
1721–1740	11·0	10·0	6·6	10·1
1741–1760	8·5	12·5	7·5	10·2
1761–1780	9·4	11·6	17·1	11·8
1781–1800	6·9	14·1	11·5	9·2
1801–1820	9·4	11·6	10·0	10·2
1821–1840	17·0	4·0	*	*
1841–1860	18·9	2·1	16·4	13·2

* Numbers not accessible now.

APPENDIX III

A. AVERAGE NUMBER OF ORDINARY FELLOWS AND OF FOREIGN MEMBERS FOR EACH FIVE-YEAR PERIOD BETWEEN 1665 AND 1940

Five-year period	Ordinary Fellows average No.	Foreign members average No.	Five-year period	Ordinary Fellows average No.	Foreign members average No.
1663–1665	145	—	1801–1805	540	62
1666–1670	203	—	1806–1810	543	54
1671–1675	215	—	1811–1815	561	44
1676–1680	199	—	1816–1820	606	42
1681–1685	153	22	1821–1825	671	40
1686–1690	116	22	1826–1830	686	45
1691–1695	115	24	1831–1835	705	46
1696–1700	125	28	1836–1840	736	47
1701–1705	131	42	1841–1845	766	48
1706–1710	149	50	1846–1850	754	47
1711–1715	160	53	1851–1855	701	44
1716–1720	170	64	1856–1860	645	43
1721–1725	207	67	1861–1865	592	41
1726–1730	247	85	1866–1870	547	42
1731–1735	271	118	1871–1875	518	43
1736–1740	290	137	1876–1880	490	46
1741–1745	309	140	1881–1885	471	46
1746–1750	327	147	1886–1890	465	48
1751–1755	353	157	1891–1895	457	47
1756–1760	355	145	1896–1900	449	47
1761–1765	346	161	1901–1905	466	45
1766–1770	361	160	1906–1910	470	44
1771–1775	394	146	1911–1915	475	42
1776–1780	445	125	1916–1920	464	39
1781–1785	476	95	1921–1925	455	39
1786–1790	486	90	1926–1930	446	42
1791–1795	511	92	1931–1935	457	46
1796–1800	531	82	1936–1940	458	48

Note. The maximum and minimum numbers in each series are printed in italics.

B. AVERAGE NUMBER OF ORDINARY FELLOWS AND OF
FOREIGN MEMBERS FOR EACH FIVE-YEAR PERIOD
BETWEEN 1665 AND 1940

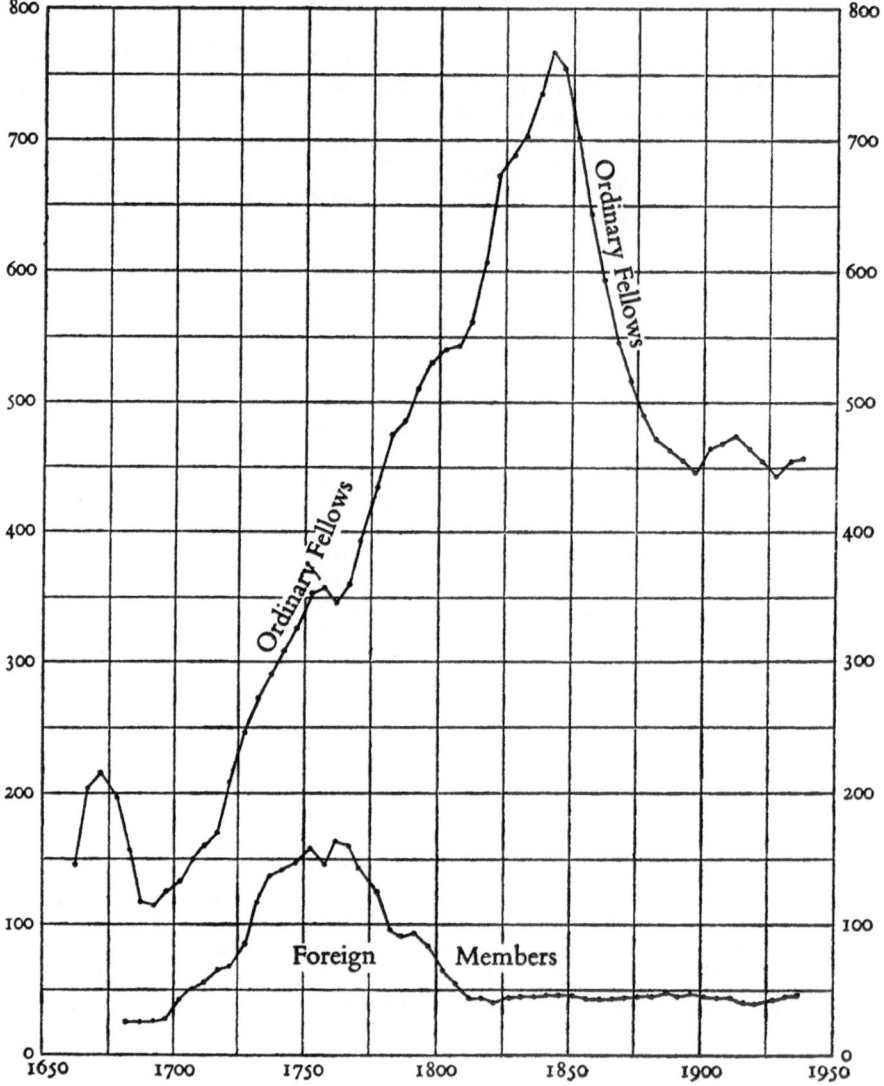

INDEX